OIL COMPANY SIGNS
A Collector's Guide

Scott Benjamin & Wayne Henderson

Motorbooks International
Publishers & Wholesalers

First published in 1995 by Motorbooks International Publishers & Wholesalers, 729 Prospect Avenue, PO Box 1, Osceola, WI 54020 USA

Motorbooks International books are also available at discounts in bulk quantity for industrial or sales-promotional use. For details write to Special Sales Manager at the Publisher's address

Library of Congress Cataloging-in-Publication Data Available

ISBN 0-7603-0073-9 (pbk.)

On the front cover: This quartet of highly collectible signs features a colorful Musgo porcelain dealer sign, a foreign Castrol Motor Oil porcelain pump sign, a Fort Pitt porcelain sign, and a Texaco lubester sign.

On the back cover: A small Sinclair Opaline porcelain sign carries the famous green dinosaur. It is displayed with a Mobil Gargoyle porcelain sign and a small Texaco truck door sign.

Printed and bound in the United States of America

Acknowledgments

Of all of the various types of items that fall within the realm of petroleum collectibles, signs, particularly porcelain signs, are by far, the most popular. Even so, the creation of this work would not have been possible without the assistance of the many people actively involved in the hobby today. So many of you believed in us enough to buy our previous books, and to take part in the creation of this work by your continued support in the form of information, photos, and comments. It would be impossible to try and name everyone that had a hand in making this book possible, but thanks to all who did, and to all of those who are interested in preserving the history of petroleum marketing through the collectibles hobby.

—Wayne Henderson and Scott Benjamin

Contents

Chapter 1

Collecting Signs

In the mainstream of oil company collectibles there are four distinct categories: signs, gas globes, pumps, and cans. There are several dozen other collectible categories but the above are the most popular.

Collecting signs can be very exciting and profitable. Signs also have definite advantages over other types of collectibles. It is probable that there are more signs collected than any other category of oil company collectibles. Even most of the collectors that we know that specialize in another category still have a few signs hanging around.

Signs are popular because they are easier to handle, easier to store and nearly unbreakable compared to gasoline globes and glass cylinder type gas pumps. With proper protection they can even be stacked atop one another or stood up several deep. They are simple to hang or display, except, of course, for the very large ones.

Many nice porcelain or tin signs can be purchased in the $75 to $300 range, although many go much higher than that.

Overall, signs are much more plentiful than a good pump or any gasoline globe. Oil cans share this abundance too. Some signs can be found at nearly any auto swap meet or antique market. This availability and popularity of signs keeps the sign market hopping.

Signs are a good investment. Like any collectible, the rarer or especially pretty signs appreciate in value the fastest. Over the last several years the better signs have increased dramatically in price. So this trend makes signs a very good investment. Signs are easy to sell too as many collectors are looking for them.

The only drawback in signs that we have noticed is the very large number of undated reproductions which are being mass produced. Unlike the other categories of oil company collectibles, very few are being dated so this becomes confusing to the novice collector. Some could say, "Well, that sign was never made in that size or it was never double sided, etc..." Yet, only an experienced collector or expert knows this and that doesn't help the majority of us. A date though, really helps.

Study the list of reproduction signs in this book, talk to people who know signs, and basically get involved in the hobby if you intend to collect good original signs. As stated above, signs have several advantages over other oil company collectibles.

Chapter 2

Glossary of Terms

Accessory Signs - signs used on battery racks, spark plug displays, or any service station accessories.

Architecture/building Signs - larger signs that hung directly on the building to identify a brand.

Can Type Sign - two single-sided signs held back to back by a wide metal band or ring.

Cookie-cutter Signs - signs that are raised outward and have a distinct edge around the border which can be a few inches high, like a large cookie-cutter.

Corner Sign - rarely seen in gas or oil signs, any signs that attach into a wall or building corner.

Crazing - fine cracks in the porcelain finish, caused by improper cooling in the manufacturing process.

Curb Signs - signs usually in the 18-30 inch range, often round which have a stand and base. These signs went out by the curb or pump island to attract customers.

Curved Pump Signs - earlier signs curved to fit the round clock-face or visible pumps.

Die Cut - any sign with a definite shape that was cut or stamped into such form.

Double-sided - any sign with two sides showing a logo on each side.

Embossed - any sign with a raised surface, often stamped from a mold.

Enamel - the type of paint that was applied to many non-porcelain signs.

Enameled Signs - any signs with enamel paint as opposed to porcelain or tin.

Flange Signs - signs that hang perpendicular to a wall. Their distinct feature is a right angle edge for mounting on one side only.

Frit - a combination of ground glass and other materials applied and then baked to form porcelain.

Grommets - brass/aluminum/zinc - small metal rings where a nail or screw would hold a sign and not damage the porcelain.

Hanging Signs - any free-swinging sign such as the larger dealer signs which hung on a pole.

Heavy Gauge Iron - heavy iron used in many early signs.

Hub Cap - any round convex sign.

Inside Bay Signs - signs that hang up inside garage bays such as "Lubrication."

Lubester Signs - small signs that were attached to lubrication carts often 6-9 inches in size.

Neon Signs - any sign with neon tubing.

One-sided - any single-sided sign.

Painted Signs - signs usually painted with enamels as opposed to porcelain signs.

Porcelain - any porcelain sign. Glass frit was applied to a metal screen which was placed over the sign. The screen was then removed and the sign was fired in a kiln.

Pressed Tin - a stamped/embossed metal sign.

Privilege Sign - any sign provided by a national or regional company with space to place a local retailer's name.

Pump Signs - smaller signs that attached directly to a gas pump identifying the brand.

Rack Sign - signs attached to oil racks identifying the brand of oil.

Repaired - any restored sign.

Rolled Iron - iron pulled through rollers to increase its strength. This gives some would-be flat signs a natural curve.

Self-framed Tins/Porcelains - any sign with a built-in frame made of tin or porcelain. The frame is actually part of the sign and cannot be removed.

Set-up for Neon - any sign with holes the size of a quarter to half-dollar for placement of neon tubes. Thus any sign set up for neon without the actual neon tubes.

Silk-screened - a process of creating a porcelain sign wherein a partially blocked fine mesh screen is used for applying the frit to the metal.

Single-sided - any sign with a logo on one side only. Same as "One Sided."

Stained Glass Signs - signs made of stained glass. The pieces are held together by lead.

Stenciled - the process where a metal sheet is used to create the lettering or design. Frit was then applied and later fired in a kiln.

Thin Gauge Steel - thinner metals used in later signs and in most reproduction signs.

Tin - some signs are made of this silver-white metal as different from iron.

Touched - any restored sign.

Truck Signs - signs that were attached to the doors or back of a delivery or tanker truck.

Wood Framed Porcelain/Tin - any porcelain or tin sign that originally came with a wooden frame around it.

Chapter 3

History of Oil Company Signs

They stand on every corner today. Huge internally illuminated "pylon" signs, supported by twin steel columns joined at the top. An upper section houses a translucent plastic square, complete with nationally recognized logos displayed on the color coordinated background. Below, smaller panels list site features...store name, diesel, auto repair, etc., while below illuminated panels display pricing levels for three, sometimes four fuels. Nothing is particularly distinctive about any one of them, as if intentionally designed to generically mean simply "gas" to the passing motorists. Very few examples of freeform design in the signmakers art remain as "approved" in oil company image manuals. Amoco torch & oval cutouts are still more common than those signs using only the Amoco logo on a blue or black background, Conoco "capsules" top their pylons, Hess still uses a few rotational signs atop their pylons, and Pure Oil stations in the south still use inserts in the round Union 76 frames, but aside from these, the only other gasoline signs still allowed under most image standards are those waiting to be replaced.

Whatever happened to the "banjo" sign? Those venerable steel and porcelain (or yes, sometimes plastic) structures that identified yesterday's filling station. Magnificent examples of "miracle mile" signmaker's art that today are, literally and figuratively, museum pieces. In all reality, modern image standards have mandated a "European" look to service station signage that has all but eliminated individuality, as well as the character of such signs as the Clark "Circle of Service" or the canted twin pole Kerr McGee shields. These signs of the immediate past have a relatively short history, although many stood long enough to become well established landmarks to roadside America.

The first signs advertising gasoline and motor oils were typical of those advertising all other products...painted "bulletin" style signs on building walls and fences near where gasoline was sold. The station itself was often identified by flanged signs projecting from the building entry doorpost, or from porch posts at rural locations where gasoline was sold. Curb pumps were topped by illuminated globes, common advertising devices of that era, to give a location nighttime identification as a filling station. These signs and globes were sized and placed for travel speeds of the day and designed mainly for curb pump locations. Texaco was the first to establish a national image standard that included a free standing identification sign. That design, introduced in 1915, consisted of a 42" double sided porcelain disc displaying the Texaco logo. Spotlights were provided for nighttime illumination. Most other oil companies quickly followed suit, including the various Standard companies, Gulf, Shell, and Sinclair. As drive-in filling stations, placed to the back of the lot, outnumbered curbside fueling sites, the free standing identification sign became an accepted method of station identification. Auxiliary products and services, motor oil brands or Ethyl grade fuels were often highlighted on smaller "curb" signs, signs mounted on a weighted stand that could be moved around the lot as necessary. Sometimes "advance" signs were placed on highways leading up to a station, encouraging the motorists to plan a stop ahead.

Prior to the mid 1930s, identification signs rarely exceeded 42" in diameter or 36x72 if rectangular. As speeds increased, oil companies introduced larger versions of earlier sign structures, often installing these newer signs along a main street and moving the smaller signs to side street corners of the lot. These signs, the Texaco banjo, Gulf sidemount, Pure banjo with fluted poles, and the mast arm signs from many other companies are the signs that most collectors identify with. Many were elaborate metal-work structures supporting the porcelain trademark inserts, and most featured spotlight illumination. Neon was used for gasoline signage at high volume stations, but these signs were short lived, as they were replaced with the first generation internally illuminated plastics after World War II.

Pump signs first came into common use in the 1930s as well, simply because most pumps prior to that time were round, making creation of porcelain signs a specialized process that involved rolling the sign. Building signage and other specialty signage really took hold in this era as well, as many gasoline marketers established "image standards" for the first time.

World War II interrupted gasoline marketing and restricted material availability immediately after the war, but many gasoline marketers began fully implementing image standards in this era, eliminating those older signs in favor of larger signs necessary for visibility at higher speeds. In 1947 Shell Oil, in conjunction with plastics manufacturer Rhom and Haas introduced the first plastic, internally illuminated station identification signs. Prior to this time, there had been several attempts at internal illumination, notable being Pure's domed glass signs and Gulf's opal glass units. The Shell sign was different, though, as it was lightweight, durable, with molded in colors that lit up evenly. Gas station imagining has never been the same. Within three years most oil companies had at least experimental plastic signage in place. Gone were spotlights on porcelain, as both sign and station illumination was improved. Discovering the design flexibility and relative inexpense of plastic over porcelain, the oil industry embraced the new material, made from their own petrochemicals, wholeheartedly. Pump signs, prone to damage from car doors, remained porcelain, although Shell, Pure, and Martin even experimented with plastic pump signs in the 1960s.

The necessary stopping distances for interstate highway travel speeds brought about the need for larger and larger high-rise signage. Most trademarks of the era did not reproduce well in plastic in small six foot sizes and certainly looked out of proportion in larger forms. Beginning with Mobil in 1955, virtually every major oil company replaced or seriously altered their existing logo to create designs more easily adapted to plastics. Gasoline images did not escape the extravagances in signmaker's art as was in evidenced in elaborate sign structures, some complete with rotators, rotating signs with stationary arrows, three dimensional signage, light animation and other characteristics typical of the day. High rise signs were often modular letter boxes, without logo of any form. As plastics technology improved, methods of forming large examples of specific sign faces led to larger and larger illuminated logos, some as high as one hundred feet in the air.

Today's pylon signage, and pumps with vinyl decal images will leave little of interest for tomorrow's collector. Probably most interesting are those smallest signs designed for canopy mounting, or used where zoning restrictions limit larger signs. These relative miniatures accurately reproduce in illuminated plastic those logos designed for this era of high speed visibility. Don't neglect these collectibles of tomorrow while searching for that rare porcelain beauty of yesterday.

Chapter 4

Types of Signs to Collect

Collecting signs offers one a variety of different types to collect. Porcelain signs are more popular overall than tin or painted signs. Yet, many tins offer several colors and more beauty. Then we have to consider the smaller pump signs, flange type signs, 30" diameter signs which are always popular, curb signs, no-smoking signs, accessory signs, among others. Some are single sided, some are double-sided, and some have wooden frames.

This gives you an idea of the choices available in collecting signs. Most collectors we know go after a variety of the above. And, some stick to one type or category. The more graphic or picture type signs are obviously the most popular. And, sign collecting has trends which we have noticed. Years ago the 24," 30," or 42" round porcelain signs were very popular. As more of these were found and space for collectors became scarcer, we have noticed a shift towards smaller signs, especially pump plate signs or signs 20" square/round and smaller. On a large wall one could display dozens of smaller signs. The same wall might hold only a few large signs.

There are a few people who collect the large 60" or 70" round porcelain dealer signs. We have seen them hanging on the sides of barns and such, but they take up too much room.

Neon signs look great anywhere, whether it be a den, garage, or display room.

Tin signs sometimes offer multiple colors, scenes, or interesting graphics, especially on the older ones. Self framed tins, tins with borders or frames built into the sign are very interesting to the collector.

Collections of a type of sign within a larger collection, we feel, are very interesting. Small signs, approximately 7" round, went on oil carts and are called lubsters; they are quite rare. A collection, say of even a dozen of these amongst a larger collection of signs looks great.

Flange signs, signs that hang perpendicular to a wall or post, have a neat appeal when displayed properly. They can be hung within other flat signs too, presenting an array of various signs.

Curb signs, medium size framed signs with a base on a short pedestal just sit on the floor. They take up some floor space but not a lot. Most curb signs are fairly rare excluding some of the major oil companies. A good size collection of these is rare, even among the better sign collections.

There are sign collectors, certain oil company collectors, porcelain collectors, each of which go after a certain type of sign. Obviously Standard Oil collectors would stick with Standard Oil signs, or a collector of Michigan oil companies would stick with any signs, whether they be porcelain, tin, small or whatever, pertaining to Michigan oil companies.

If space is limited one can stick with collecting the smaller signs. We feel the most interesting collections have a little bit of all the various types of signs.

Obviously the more graphic, colorful, rare, or picturesque a sign, the more unique the collection.

Sign collecting offers various avenues to pursue, and there is a little bit there for everyone interested to enjoy.

Chapter 5

Grading Signs

Collectors of signs, as well as dealers and investors, need to understand some type of a grading system. Most sign collectors we talk to usually use the simple "1 through 10" system, 10 being the best, in grading a sign. Numbers below 5 are rarely used and replaced by "very rough" or "one side washed out." So, the talk usually centers around numbers 6 through 10, including an exact description of the chips, color loss or damage. Mint signs are quite rare, especially in scarcer signs.

The biggest drawback in a "1 through 10" grading system is that it is all an opinion, varying from one person to another. So, always ask for an exact description of someone's 7, 8, 9 grading of a sign. Someone saying that "it is very good, a 9," varies from person to person. A 9 condition sign should have only a few very small chips, for instance, on the edge. One chip in the center of a detailed logo might ruin the overall appeal of a sign. So again, a 9 condition with only one chip might be nothing what you expect. Most people are honest and fair in grading signs, and most people have good judgment in grading a sign. But, get an accurate description or photo of both sides of a sign if applicable before any dealing is done. This is in the best interests of both parties involved. We could give you examples of signs grading from a 5 through 10 but this would also be an opinion and we'd probably never hear the end of it ad infinitum. Again, get that 8 grade sign described to you in detail covering color loss, rust, chips, creases, etc.

Chapter 6

Placing Values on Signs

Pricing a sign these days is a real talent to say the least. Sign prices can get very trendy, some prices sky-rocketing in a few months and some showing a typical, gradual increase. Certain types of signs become "popular" and these demand high prices. Currently, as of this writing, any smaller signs are very popular, especially pump plate signs. We have seen some of these categories of signs jump several hundred dollars in a matter of months. Auctions, hype and general sudden popularity will and do drive the prices up quickly in certain cases. Our point here is that it can be difficult to place specific values on certain signs that will be getting somewhat higher prices in a few months or a year. But, there are guidelines and rules to follow.

First of all, use this book to help you narrow down the prices of your signs. We tried to estimate as accurately as possible current values of many of the signs one might find. Note the patterns in value of the rarer, more colorful or graphic signs. These always command a higher price. The common mass-produced pump signs of Gulf, the later Sinclair Dino signs, Atlantic, Texaco Sky Chief and Fire Chief signs usually fall into the $35 to $75 range, some more or less. Rarer pump signs with colorful graphics often bring $500 to $1,500, or more. Many 24" to 42" diameter double-sided porcelain signs bring around $250 to $600 in good condition. Again, multi-colored, picturesque, and/or rarer versions of these bring several times the above amount. There are not infinite categories of signs. Pick a category, like "no smoking" signs, and notice that most fall in a certain range. Some lower in the range and some higher - say $250 to $500. Then, by asking a few questions and using this book you will have a guideline as to how to price your "no-smoking" sign. Some ranges of prices are not so extreme, yet some are more. Signs falling above this range are probably over-priced. Why be the first to set a new trend?!

Porcelain signs generally bring more than tin signs or painted metal sings, mainly because they are more popular. Most curb signs can be purchased for under $500. Yet, some can bring six times that amount. And the ones that bring $3,000 are rare, picturesque and in great condition. One of the best ways to learn sign prices is to attend the larger gas and oil shows such as Iowa Gas, the I.P.C.A. convention in Columbus, Ohio and other large meets. Many vendors carry a wide variety of signs and pricing them, asking question, being aware of signs, quality and trends will help you sort out signs and pricing. For those who cannot attend shows, study this book as thoroughly as possible.

Keep in mind a mint sign that brings $3,000 may only bring $1,200 as a condition 6 or 7. These condition/price ranges can vary greatly. We have noticed, especially lately, a large variation in prices that a sign will bring in mint condition versus a lower condition. Most collectors and dealers will gladly pay a premium for a sign in excellent condition, meaning 9, 9.5, or 10. Though a 7 condition sign might seem like a good deal because it is bringing half the going rate, ask around before buying, if possible. Only experience can help someone price most signs in various conditions. Use this book as a guide to narrow down the range, then go from there.

Good luck, but remember, if you like a sign and feel it is a fair value but not sure, buy it. Especially if you are at a swap meet, auction or such where the sign may be purchased quickly by someone else. Rarely do we hear complaints about someone paying way above retail for a sign they wanted. However, we often hear, "I wish I would've just bought that sign when I had the chance!"

Chapter 7

Investing in Signs

As we have stated, signs are the most popular of the four major categories of oil company collectibles. With that as a foundation, investing in signs has unlimited potential. As in any collectible, the prettier, rarer, more colorful signs always appreciate in value the fastest. So, if you are in this for an investment or want a collection that appreciates in value the quickest, then you would need to collect these types of signs. You need to speak to collectors, go to shows, or hire a consultant if you want to invest in signs. Again, follow a few simple rules to start:

Number one: Go for the best quality available. Retail value on a great sign will increase more, for example, in one year than retail value on a fair quality sign.

Number two: Go for the more graphic, more colorful or picturesque type signs. These increase faster in value too.

Number three: Note current trends in signs. If, for example, good pump plate signs are selling well and if they are "the item to buy right now in signs," they will become scarcer because everyone is looking for them. Say then that in three years you go and sell your collection of pump signs, you are likely to do well. See our point? It will take some phone calls and a few questions, but this is a good direction in which to go.

Often, you will get a situation where you have the chance to buy a whole collection of signs. Again, weigh the pluses and minuses. Don't buy a collection of junk just to have a lot of signs. Buy the collection if there are a few great signs and if the rest are pretty good. One can always sell off the lesser signs and upgrade later. Do not buy 100 signs to get one good one. Such decisions definitely need input of an experienced sign collector or expert in the field.

Use this book as a guide and buy below retail when possible. But, a great sign, a rare sign, or colorful sign, even at retail will go up in value much quicker and will be a good investment.

Chapter 8

Restoration of Signs

Any sign can be restored. We need to discuss which signs should be restored and which should be left alone. Some important questions are: Will a restored sign then be worth a lot more? Do collectors accept a "touched" sign? How good does a restoration have to be? How far can a restoration go?

Good sign restoration is not inexpensive. We are talking about the kind of repair where you honestly can't see what was restored. That is a good restoration. There are very few people that can restore a sign that good. If you have a sign that would be worth $150 in perfect condition, and yours is really chipped up, go find a better one. A cheap restoration doesn't help much in the way of looks, let alone value. A professional restoration would be cost prohibitive in this case. However, if one had a picturesque sign with some chips it might be a good idea to have it restored. A good example of this was a 30" diameter porcelain sign with a picture of a deer in the center of it. It had several chips that took away from the sign's beauty. This sign would look a lot better if restored professionally. Its value, in theory, should increase more than the cost of the repair. Yet, if this sign only had a few chips on the edge or in a less important area of the sign, most collectors would leave it alone. If you have a really great sign and it is in horrible shape, well, most people wouldn't bother with a restoration. A few collectors will, but perhaps it is too far gone. Restored with very little of it original doesn't appeal to most collectors.

As time passes, a trend is occurring wherein most collectors accept a restored sign, especially if it is well done. It is a judgment call on a case by case basis. Like a gasoline pump, you can replace missing parts and that is okay, but replacing *everything* would make it a reproduction. If your great sign is no more than a 5 out of 10, then I would think twice about restoring it.

In collecting signs, then, it is obvious to find the sign you want in the best condition possible. It is a better investment, it may need no restoration and if you decide it needs to be restored, the restoration will then cost less.

In speaking with collectors, some feel even a professionally restored sign is worth only 70 percent of the same sign in mint condition. Even though they will look nearly identical. Some people feel that that figure is 40 percent or 30 percent. As time passes, this figure is sure to increase as more collectors accept a "touched" sign. Few people enjoy looking at any collectible in rough condition. So, either find a better one, enjoy it as is, or consider a professional restoration. Again, this needs to be considered on a case by case basis. If it is a keeper, do what you want. If it is a trader or for resale, you must consider not everyone would accept a restored sign. Give it some careful thought before making that choice.

Chapter 9

Reproduction Signs

AUNTIE CHATTER MOTOR OIL FOR FORD CARS
porcelain
12" round

ARCHER MOTOR OIL
metal
10"x14"

ATLANTIC REFINING CO. AUTOMOBILE GASOLINE FOR SALE HERE
porcelain heavy gauge steel
9.5"x14.5"

AUBURN SALES & SERVICE
porcelain, single side, no flange
yellow, black, white
23.5"x15" (10.5")

BMW
porcelain on heavy gauge steel
12" diameter

BUICK Authorized Service
porcelain on heavy gauge steel
12" diameter

CADILLAC Authorized Service
porcelain on heavy gauge steel
12" diameter

CHAMPION SPARK PLUG SERVICE
porcelain on heavy gauge steel
8"x18"

CHAMPION SPARK PLUGS SUMMER AHEAD
metal
10"x12"

CHAMPION SPARK PLUGS WINTER AHEAD
metal
10"x12"

CHEVROLET CORVETTE
porcelain on heavy gauge steel
12" diameter

CHEVROLET GENUINE PARTS
porcelain, single/double sided die cut around "Chevrolet",
black and white
26"x18"

WE USE GENUINE CHEVROLET PARTS
porcelain on heavy gauge steel
14" diameter

CHEVROLET SUPER SERVICE
porcelain on heavy gauge steel
11" and 12" diameter

FOR ECONOMICAL TRANSPORTATION CHEVROLET QUALITY AT LOW COST
porcelain on heavy gauge steel
8.5"X22"

CHEVRON GASOLINE
metal
10" diameter

CHRYSLER PLYMOUTH APPROVED SERVICE
vinyl
12.5" diameter

CHRYSLER PLYMOUTH Factory Engineered Parts Accessories
porcelain on heavy gauge steel
12" diameter

CITIES SERVICE
Airplane Mechanic on Duty
porcelain on heavy gauge steel
13.5"x9.5"

CITIES SERVICE KOOLMOTOR
metal
10" round

CONOCO
porcelain on heavy gauge steel
Conoco minuteman
12" diameter

CONTAINS LEAD FOR USE as a Motor Fuel Only (Tetraethyl)
porcelain on heavy gauge steel
7"x5"

CONTAINS LEAD (Tetraethyl) and is to be used as motor fuel only, not for cleaning or any other use. Avoid spilling.
porcelain on heavy gauge steel

CONTINENTAL MOTORS America's Standard, Powerful as the Nation, picture of Capitol
porcelain on heavy gauge steel
12" diameter

DeSOTO PLYMOUTH
Approved Service
porcelain on heavy gauge steel
12" diameter

DIXIE - ETHYL LOGO
porcelain, yellow or ground rough back, aluminum grommets
12" diameter

DIXIE - REBEL FLAG LOGO
all are reprros, porcelain
12" diameter, tan, red, black

DODGE BROTHERS
Approved Service Stations
porcelain on heavy gauge steel
6"x15"

DODGE BROTHERS Sales Service
porcelain on heavy gauge steel
12" diameter

DODGE "Job Rated"
Trucks Sales and Service
porcelain on heavy gauge steel
9"x9"

DODGE PLYMOUTH
Approved Service
porcelain on heavy gauge steel
12" diameter

DODGE PLYMOUTH
Dependable Service Dodge Trucks
vinyl
14"x12.75"

ENARCO - WHITE ROSE GASOLINE - white, red, yellow porcelain,
4 holes-top/bottom/sides
12" diameter

ESSO
metal
10" diameter

ESSO YOU GET MORE
metal
12"x14"

ETHYL - With Ethyl logo sign
porcelain on heavy gauge steel
12" diameter

FERRARI
porcelain on heavy gauge steel
12"x8"

FISK Time to Re-tire
porcelain on heavy gauge steel
9.5"x13"

FLYING A GASOLINE
wing logo, 12" diameter,
red, white, black, background red,
border black.
never existed

FoMoCo GENUINE FORD PARTS
metal
18"14"

FORD Authorized Service
porcelain on heavy gauge steel
11"x18"

FORD Authorized Service Station
porcelain on heavy gauge steel
8.5"x18"

FORD THE UNIVERSAL CAR
porcelain on heavy gauge steel
7"x18"

FORD THE UNIVERSAL CAR
die cut pyramid with wings, porcelain,
single sided
30"x11.75" black, white

GENUINE FORD MECHANIC ON THE PREMISES
porcelain on heavy gauge steel
19"X11"

GENUINE FORD PARTS
porcelain on heavy gauge steel
2.5"x5. 2 sides

GENUINE FORD PARTS
porcelain on heavy gauge steel
10.25"X5.75" oval

GENUINE FORD PARTS
porcelain on heavy gauge steel
16"x11" oval

GENUINE FORD PARTS
porcelain, single/double sided
224"x16.5" oval

GENUINE FORD PARTS SOLD HERE
porcelain on heavy gauge steel
6"x22"

GENUINE FORD PARTS USED HERE
porcelain on heavy gauge steel
6"x22"

FORD MUSTANG
porcelain on heavy gauge steel
12" diameter

FORD ROADSTER PICKUP
metal
16"x10"

FORD THUNDERBIRD New for 1957
pressed tin
10"x13"

FORD TRACTOR
porcelain on heavy gauge steel
12"x19"

FORD TRUCKS V-8
Engine 75 Horsepower metal
16"x10"

FORD (crest) YOU CAN PAY MORE..
BUT YOU CAN'T BUY BETTER
metal
10"x12"

FRANKLIN MOTOR CARS
Service Station
porcelain, single sided,
24.5"x14"
black, white, yellow

FREEDOM MOTOR OILS & LUBRICANTS
metal
12"x15"

GAS TODAY DRINK COCA-COLA
WHILE YOU WAIT
metal
9"x12"

GAS OILS
metal
9"x24"

GENERAL MOTORS TRUCKS GMC
porcelain on heavy gauge steel
12" diameter

GENERAL PETROLEUM CORPORATION
Gasoline and Lubricants
porcelain on heavy gauge steel
12" diameter

GILMORE AIRCRAFT RED LION
porcelain on heavy gauge steel
12" diameter

GILMORE GASOLINE
lion logo, rounded top/bottom
porcelain, thin gauge metal
marked "Cameo"
24"x15" approx.
red, white, black, yellow

ROAR WITH GILMORE
porcelain on heavy gauge steel
12" diameter

MG
octagon
porcelain on heavy gauge steel
12"x12"

MACK SERVICE
porcelain on heavy gauge steel
12" diameter

MAGNOLIA GASOLINE
porcelain on heavy gauge steel
12" diameter

MAGNOLIA MOTOR OIL
porcelain , thin gauge steel
flower logo, double sided, 42"
diameter, red, white, green,
yellow, light blue

MAGNOLIA PETROLEUM COMPANY
Magnolia Gasoline Sold Here
porcelain on heavy gauge steel
14" diameter

MARATHON AIRCRAFT
porcelain on heavy gauge steel
12" diameter

MARATHON GASOLINE
metal
10" diameter

MARATHON PRODUCTS
running man, porcelain on heavy
gauge steel
12" diameter

McCORMICK DEERING SERVICE
porcelain on heavy gauge steel
12" diameter

MERCEDES BENZ
porcelain on heavy gauge steel
12" diameter

MOBILGAS
porcelain on heavy gauge steel
12" diameter

MOBILGAS and MOBILGAS SPECIAL
pump sign shields, porcelain
13"x11" and 12"x11,"
red, white, blue

MOBILGAS
Gargoyle, Aero W, single sided
12" diameter, white, black, red
never existed

MOBILOIL Aircraft
Gargoyle, Aero W, porcelain on heavy
gauge steel
12" diameter

MOBILOIL
rack sign w/gargoyle, single sided,
porcelain
white, red, black 24"x19.5"

MOBILOIL Gargoyle - Stop Here
Authorized Service
pressed tin
7"x9"

MOBIL PEGASUS
die cut, flat, 3.5" and screw hole on
nose of horse, porcelain, screw holes on
edge
4.5" approx., red, white

MOPAR CHRYSLER ENGINEERED PARTS
vinyl
17.75"x18.5" oval

NASH Service Parts
porcelain on heavy gauge steel
9"x9"

NO AUTOS FILLED WHILE ENGINES ARE RUNNING OIL OR GAS LIGHTS BURNING OR OCCUPANTS SMOKING
porcelain on heavy gauge steel
9.5"x15.5"

OLDSMOBILE SERVICE
porcelain on heavy gauge steel
12" diameter

OILZUM
orange, black, white porcelain, no known originals in this size, brass grommets
11" and 12" diameter

OILZUM NON-CHATTER FOR FORD CARS
metal
16'x6"

PACKARD Approved Service
porcelain on heavy gauge steel
12" diameter

PACKARD PARTS
die cut radiator, porcelain, same size as original,
29.75"x28.5"
opaque blue, back is greenish

PACKARD SERVICE
die cut radiator, black, white porcelain, opaque blue
much smaller than original

PHILLIPS 66
die cut shield, orange, black porcelain, 3 screw holes each side of top and on bottom, thin gauge metal
12" across

PHILLIPS 66 Airplane Pilot onDuty
porcelain on heavy gauge steel
13.5"x9.5"

PHILLIPS 66 MOTOR OIL
rack sign, green, orange black porcelain, single sided
23.75"x11"

PIERCE ARROW
service depot
porcelain, star on lower right
36"x18"
double sided, white, red, black

1932 PLYMOUTH Convertible Coupe
pressed tin

PONTIAC SERVICE
porcelain on heavy gauge steel
12" diameter

POP'S GARAGE AUTOMOBILES
metal
15"x11"

PORSCHE
porcelain on heavy gauge steel
12"x12"

POWERLUBE
motor oil, double sided, porcelain
28"x20"
black, yellow, white

POWER-LUBE Motor Oil
pressed tin
13.5"x10"

QUAKER CITY MOTOR OIL
porcelain on heavy gauge steel
18.5"x13"

RAMBLER Parts Service
porcelain on heavy gauge steel
8.5" diameter

RED CROWN STANDARD OF INDIANA
porcelain
30," 42" and possibly 48"

RED CROWN STANDARD OF OHIO
porcelain
30," 42" and possibly 48"

RED CROWN GASOLINE
porcelain on heavy gauge steel
12" diameter

RED CROWN GASOLINE
porcelain on heavy gauge steel
11" diameter, 2 sides, flange

RED CROWN GASOLINE AND POLARINE
porcelain on heavy gauge steel
14"x20"

RED INDIAN MOTOR OILS metal
13"x14"

ROCK ISLAND GASOLINE
porcelain
23.5" diameter, double sided,
red, white, blue

SAVIN' SAMS GASOLINE
porcelain
10" diameter approx., red, white, black

SHELL
Golden shell Motor Oil
die cut Shell, porcelain,
2 screw holes on top
10," yellow, red, black

SHELL U.S. GOVERNMENT LICENSED FLIGHT INSTRUCTOR ON DUTY
porcelain on heavy gauge steel
14"x10"

SHELL GASOLINE
porcelain on heavy gauge steel
12"x12" die-cut

SIGNAL GASOLINE
porcelain on heavy gauge steel
12" diameter

SIGNAL GASOLINE
porcelain
11" and 12" diameter, black, red, yellow, white

SINCLAIR AIRCRAFT
porcelain, thin gauge metal
12" and 42" diameters

SINCLAIR DINO GASOLINE
metal
10" diameter

SINCLAIR DINO GASOLINE
porcelain on heavy gauge steel
12" diameter

SINCLAIR GASOLINE
pressed tin
11"x9.5"

SINCLAIR GASOLINE
porcelain pump sign
14"x10," red, white, green

SINCLAIR H-C GASOLINE
metal
10" diameter

SINCLAIR H-C GASOLINE
porcelain on heavy gauge steel
12" diameter

SINCLAIR MOTOR OILS

Best for Motor Cars
pressed tin
5"x16"

SINCLAIR NO SMOKING AT PUMPS
metal
5"x13"

SINCLAIR OPALINE MOTOR OIL
metal
9"x14"

SINCLAIR
Opaline and Pennsylvania Motor Oils with Dinosaurs, porcelain
12" diameter, red, black, green, white

NEW ADVANCED FORMULA SINCLAIR MOTOR OILS
The Sinclair Law of Lubrication

pressed tin
16.25"x8.5"

die-cut Dinosaur
porcelain on heavy gauge steel
17.5"x10"

SKELLY
No Smoking in Hangar or Ramp Area, porcelain on heavy gauge steel
12"x14"

SOHIO
metal
14"x9"

SOCONY AIRCRAFT
plane logo, red, white, blue porcelain
12"x20" or 10"x14"

SOCONY MOTOR GASOLINE
porcelain on heavy gauge steel
12" diameter

STANOCOLA POLARINE
flange, porcelain
22"x20," green, orange, white, black, red

STUDEBAKER Authorized Service
porcelain on heavy gauge steel
12" diameter

STUDEBAKER service station
porcelain, single sided
30"x13.75" black, white

STUDEBAKER TRUCKS
porcelain on heavy gauge steel
12"x7"

STUTZ
porcelain

SUNRAY DX
porcelain
10" approx., octagon.white, gray, yellow, orange, black, red

TEXACO AVIATION Gasoline
porcelain on heavy gauge steel
12" diameter

TEXACO NO SMOKING
porcelain
23"x4" strip, Texaco logos on each end, white, green, black, red

TEXACO FIRE CHIEF
metal
10" diameter

TEXACO FIRE CHIEF
porcelain on heavy gauge steel
11.5"x16"

TEXACO FIRE CHIEF, SKY CHIEF and DIESEL CHIEF
porcelan pump plates, smaller than the 12"x18" originals
dated
white, green, black, red

TEXACO
black "T" logo white, green, black, red
porcelain
12" and 8" diameters

TEXACO CERTIFIED RESTROOM
porcelain
white, green, black
9"x12"

TEXACO GASOLINE FILLING STATION
metal
10" diameter

TEXACO FILLING STATION
porcelain, white, green, black, red
36" diameter with 1 1/2" lip.

TEXACO GLIDE-AWAY STARTS
metal
11"x14"

TEXACO RED HOT GETAWAY
metal
11"x14"

WE SELL TEXACO
Texaco man
metal
14"x18"

TEXACO STAYS FULL LONGER
The "FURFURAL'D FILM" does it!
porcelain
16"x9.5"

TRI-STAR RED HAT GASOLINE
porcelain on heavy gauge steel
12" diameter

UNION 76 GASOLINE
porcelain on heavy gauge steel
12" diameter

UNITED SERVICE MOTORS
porcelain on heavy gauge steel
11"x16" oval

USED OK CARS
porcelain on heavy gauge steel
12" diameter

WHITE EAGLE
metal
10" diameter

WHITE EAGLE
porcelain
30" diameter, red, white, blue

WHITE ROSE GASOLINE
porcelain on heavy gauge steel
12" diameter

WHITE STAR GASOLINE
porcelain on heavy gauge steel
12" diameter

WHITE TRUCKS
Tomorrow's Truck Today,
porcelain on heavy gauge steel
9"x12"

Chapter 10

How to Use the Sign Listings in this Book

The sign listings in the following pages are arranged into three groups:

1) Major Oil Companies and their affiliates
2) Independent Oil Companies that marketed gasoline & oil.
3) Independent Oil Companies that marketed only lubricants.

In order to find a particular sign listing, you must first determine which classification it will appear in. A table of contents follows and an alphabetical index of brandnames listed in the major oil company section follows that section. If the brandname as it appears on the sign does not appear in the index, then proceed to the independent section. Signs are listed alphabetically by brandname in the independent sections, not by company name. In some instances some signs must be listed out of alphabetical order in order to be with other signs from the same company. In that case the signs from that particular company appear under the company's primary brandname. An example is Midwest Oil Company's "Ace High" brand. The company's primary brandname was "Midwest," but for a 20 year period the Ace High name alone was used. In that case all signs from Midwest, including those that say only "Ace High," as well as those from an affiliate that say "Trophy" appear under "Midwest." Both Ace High and Trophy are listed in their respective sections with the note to "See Midwest."

If your sign does not appear in either the major or independent section, it could be one of the few signs that were unknown to us at the time of this writing, or it falls into one of the classifications which have been specifically omitted. Sign classifications specifically omitted include:

1. Cardboard or Paper signs
2. Banners
3. Local Made Signs - signs a local dealer contracted with a local sign painter to create, as opposed to mass produced signs. Even the smallest companies used signs that were "mass produced."
4. Signs too large to be considered collectible. There are some spectacular wall murals, but you'd have to buy the building to get the sign, and Boston's landmark Citgo sign is just a little large for the garage wall.
5. Plastic signs & letter sets after about 1960, except pump signs. Pure, Shell, and Martin all used plastic pump signs after 1960 that are considered collectible.
6. Production, transportation, or refining signs. Pipeline markers, property lease markers, signs from ocean going tanker are not included. Signs in this book are from marketing divisions only.
7. Signs that could not be established to any particular brand. If ABC Oil Company order stock "MEN" and "WOMEN" signs from a manufacturer for their stations, then they are not brand specific and are not included. If they ordered them in specific company colors, then they have been included where identifiable.

If your sign is not listed and you feel that it falls within the scope of this book, please contact:

Scott Benjamin	Wayne Henderson
411 Forest St.	20 Worley Road
LaGrange, OH 440 50	Marshall, NC 28753

We will see that your sign is listed in all future editions of this reference work. The major oil company section is broken into 47 headings and within each of those classifications signs are listed in something of a chronological order in conjunction with the order in which affiliates were purchased by the major company listed. It is broken down in an order that collectors use, that is, all brands related to, say the Pure Oil Company are listed with Pure: Hi-Speed, Woco-Pep, etc., while Union 76 globes are listed under Union. Although Union Oil purchased Pure in 1965 and the Union 76 brandname replaced Pure in 1970, collectors tend to look at Pure items as separate from Union items. The same principle is used in Amoco/Standard. Amoco and Pan-Am signs are listed in the "AMOCO" section while Standard, Utoco, and related brands are listed under "STANDARD." The reason for this is, again, most collectors who specialize in Amoco have only a passing interest in Standard, and most Standard (IN) collectors have little interest in Amoco. The table of contents breaks down the company headings and the brands that appear therein.

Once you find the listing for the particular sign you are looking for, a great deal of information is given about that sign. A typical listing looks like this:

COMPANY NAME and LOCATION: Then a brief history and explanation as to why it appears with this particular major heading.

SIGN NAME - SIZE - RARITY GUIDE NUMBER - YEARS USED - PRICING

Description text follows the listing.

The SIGN NAME is either the actual wording that appears on the sign, as in "SINCLAIR HC GASOLINE" or it is the term by which collectors refer to the globe as in "SINCLAIR TRIPLE CHECK ROUND" (Round Sinclair Triple-Check experimental sign).

The rarity guide is a series of numbers, one through five, that specifies a level of rarity. It is interpreted as follows:

(1) Very common sign - readily available in collectors' circles.

(2) Common sign - fairly available among most collectors.

(3) Available - these signs, though available, are scarcer and harder to find.

(4) Rare or very rare - very few known to exist, seldom offered for sale.

(5) Extremely rare or unique - less than 10 known, many are unique and a few signs listed as such are known only in old photographs.

The "years used" section is a listing of the time period that this sign could have appeared in official use. Many logos introduced in the 1930s were carried over into the 1950s and later, even though stations of more recent construction carried the newer signage. It was not at all unusual to find a country store selling Pure gasoline at the time Pure rebranded to Union 76 in 1970, still displaying the 42" round porcelain "The Pure Oil Company, USA" sign first used in 1930. Indeed, specific fiberglass Union 76 inserts were made to fit these signs, just like the larger inserts made for the 72" signs of later vintage.

Likewise, Cities Service, Shell, and others did little to standardize their trademarks until, in the case of Cities Service they converted to Citgo, or with Shell, a late 1960s imaging program was introduced, eliminating many still functional 1947 plastic cut out Shell signs.

In the pricing section, the prices listed are flat retail. The sign listed found priced to the lower side would be considered a "good deal," while those priced at the listed amount would simply be "at full retail." Market factors have determined a much wider range between wholesale and retail on the higher priced signs than on signs of lesser value. The pricing is relative to the design and desirability of the signs and not necessarily the rarity. The numbers are current as of spring 1995 and the relative values will remain the same for years to come. The pricing combines many years of experience in collecting petroleum collectibles and is thought to be realistic throughout the listings.

Within each brand classification, types of signs are broken down as follows:

IDENTIFICATION SIGNS: These are the pole mounted signs used to identify a retail station.

BUILDING SIGNS AND LETTERS: These are the removable signs and letters that were physically attached to the station structure. Not included are those signs that were part of the porcelain paneling on the exterior of the station.

PUMP SIGNS: Simply, signs designed to attach to gas pumps.

CURB SIGNS: Those signs that are less than 42" that were installed or set out at curbside to identify a gas station or particular product available there.

In the description text, a simple written description of the sign is given. Special marks within the text include:

Quotation Marks - "Flite Fuel" - This is the actual wording in the design.

All upper case or lower case in quotes - "regular GASOLINE" - lettered as it appears in the design.

Slash Mark - "REGULAR/GASOLINE" - This wording appears as one word over the other.

Ring or outline ring: any color markings circular on the sign face.

Band or line: any color markings horizontal on the sign face.

Ethyl (EGC) and (EC) logo: The (EGC) refers to an Ethyl logo with the words "ETHYL GASOLINE CORPORATION" in the banner at the bottom. Used from 1926 on, although officially in April 1942 by "ETHYL CORPORATION" listed (EC) in the text. While a sign with the (EGC) logo could have been used any time after 1926, a sign with the (EC) logo would have to date from 1942 and later.

PGCOA seal: In 1926 manufacturers of lubricants made from superior grade Pennsylvania lubricants formed an association called the Pennsylvania Grade Crude Oil Association, and licensed the use of "Pennsylvania," "Penn," or related brandnames so as to protect the integrity of the products. Their seal was adopted from a logo first used by Sterling Oil Company in 1924 and licensed manufacturers displayed the seal and permit number on signage and product containers. Hence, any item with the PGCOA seal and permit number cannot pre-date 1926.

REFERENCE LIST

AMOCO: Amoco and Pan-Am
ASHLAND: Pepper, Ashland, Tri-State, National/White Rose, Aetna, Frontier, Freedom and Valvoline, IQS, Payless and Dance, Superamerica, Oskey Bros., North Star, Red Head, Rotary, Hoods, Tresler Comet and Pyroil.
ATLANTIC: Atlantic
CHAMPLIN: Champlin, Harbor and Eason
CHEVRON: Standard (CAL), Chevron, and Calso
CITIES SERVICE: Cities Service, Lincoln Oils, Red Parrot, Loreco, Crew-Levick, Warner-Quinlan, Shoemaker, Jenney
CLARK: Clark, Target, and Owens
COLONIAL: Colonial and American
CONOCO: Conoco, Marland, Western (Husky, Mileage, DS), Malco Valley and Numex), Douglas, Kriegler, Kayo
CO-OP: Farmland (Co-op and Falcon), Cenex, Midland, MFA, Missouri Farm Bureau, Southern States, Tennessee Co-Op, Growmark (Aladdin and FS), Fargo, Central Co-Op, Statex, GLF, Landmark, Farmoil, Iowa Farm Service
CROWN: Crown, Thrift, and Peoples
DERBY: Derby and Sovereign
DIXIE DISTRIBUTORS: Dixie
DX: DX, Diamond, Sunray, Premier, Barnsdall, Monamotor, Harkeye, Iowa Gas
EXXON: Standard(NJ), Esso, Exxon, Stanocola, Humble, TP, Colonial-Beacon, Powerine, Grizzly, Litening, Oval-E, Carter, Enco, Pate, Oklahoma, Gaseteria-Bonded, Johnson, Hoosier-Pete, Ryans, Senco, Deem, Sun-Flash, Alert
FINA: Fina, Panhandle, Amlico, Elreco, Pemco, Golden Rule, Cosden, Col-Tex, Onyx, Wides
FLYING A: Tydol, Flying A, Associated, Seaside
GULF: Gulf, Wilshire, Superior 400, Coryell 70, Bulko, Traffic
HESS: Hess, Olixir, Southern, Delhi-Taylor/Billups Eastern
HUDSON: Hudson, Silver King, Hornet, majors, Highway, Fisca
HUSKY: Husky, Hecco, Husky Hi-Power, Frontier, Beeline
IOMA: Independent Oils, Red Hat, High Hat, Royal 400
KERR McGEE: Deep Rock, Kerr McGee, Kermac, Triangle, Mileage Mart, Trackside, Peoples, Mutual, Sparky
KENDALL: Kendall

LION: Lion
MARATHON: Linco, Marathon, Tower, Speedway 79, Old Dutch, Republic, Hi-Way, Bonded, Consolidated, Cheker, Sun-Glo, Imperial, Imperial of Michigan, Rock Island, Tulsa, Colonial-Progressive, Duro, United, Zephyr (Naph-Sol), Osceola
MARTIN: Martin, Zephyr (J.D. Street), Martin Oil Service
MOBIL: Socony, Magnolia, General petroleum, White Eagle, Vacuum-Mobilgas, Wadhams, Bartles-McGuire, White Star, Independent of PA, Metro, Gilmore, Mobilgas, Mobil
PENNZOIL: Pennzoil and Pennzip, Wolf's Head, Keystone, United, Fleet-Wing, Penn-Drake
PHILLIPS: Phillips 66, Benzo-Gas and Unique, Independent (Tulsa), Manhattan, Paraland, Wasatch, Red X
PURE: Pure, Woco-Pep, Hi-Speed, W.H.Barber (Fyre Drop and Meteeor), Save More, Super Par, Pacer, Dickey
QUAKER STATE: Quaker State and Sterling
RICHFIELD: Richfield, Richfield of New York, Richfield of California, Walburn, Betholine, Lamson, Rio Grande, Rocket
SHAMROCK: Shamrock and Cliff Brice
SHELL: Shell
SIGNAL: Signal, Hancock, Watson, Regal, Billups, Spar, Rose, Supertest, SOC, Dixie Vim-Charter
SINCLAIR: Sinclair, Pierce-Pennant, Parco, Larco and Covey, Simpson, Stoll, Okaw
SKELLY: Skelly and Surfco
SOHIO: Sohio and Boron, Caldwell and Taylor, Fleet-Wing, Refiners, Canfield
SPUR: River States and RS Royal, Webb, Spur, Wood River, Ingram, Hosco
STANDARD: Standard(IN), Red Crown, Standard of Nebraska, Pep 88 and Utoco, Arro, Stanavo, Rainbow
STANDARD OIL: Kyso, Standard Oil
SUNOCO: Sunoco
TENNECO: Bay, Citizens, Speed, Gulf Coast, Tenneco, Direct, Red Diamond, Dixie, Sunset
TEXACO: Texaco and Indian
TOTAL: McClanahan, Leonard, Total, Roosevelt, Michigan Maid, Mid-West and Mid-West/White Rose, Cavalier, Col-Tex, Anderson-Prichard and Apco, Kanotex, Lesh, Vickers, Bell, Knight, Ben Franklin
UNION: Union, Union 76
ALPHABETICAL INDEX TO BRANDNAMES LISTED WITH THE MAJORS
INDEPENDENTS: Listed Alphabetically

Chapter 11

Directory of Sign Listings

Major Oil Companies

AMOCO and PAN AM

AMERICAN OIL COMPANY - BALTIMORE, MD/PAN-AM SOUTHERN CORP - NEW ORLEANS, LA/STANDARD OIL COMPANY OF INDIANA - CHICAGO, IL

The American Oil Company was founded in Baltimore in 1910, refining and marketing kerosene. In 1915 they introduced a gasoline/benzol blend they called "AMOCO-GAS." Without crude oil production of their own and forced to buy on the open market, they operated with the possibility of shortage at any time. This was remedied in 1923 when a supply agreement was signed with Pan American Petroleum, based in New Orleans, giving 50 percent interest in American Oil to Pan-Am in return for the assurance of crude oil supply. Pan American, primarily a production company with trace marketing in several southern states, began its operations in 1916 and by the time of the 1923 merger was already an international company. In 1925 Standard Oil of Indiana purchased a controlling interest in Pan-Am, in the wake of Pan-Am's involvement in the "Teapot Dome" Scandal. In 1933 Pan-Am, now a subsidiary of Standard of Indiana, purchased the remaining 50 percent of American Oil and converted the Pan-Am marketing facilities in Georgia and Florida to the Amoco brand, retaining the Pan-Am name in Alabama, Louisiana, Mississippi, and Tennessee. Arkansas and Kentucky were added at a later date, while the Amoco brandname was used in all states from Maine to Florida as well as West Virginia and Ohio. Standard bought the remaining interest in Pan-Am in 1954 and in 1956 the Pan-Am name was replaced by Amoco in the six southern states. About the same time the Amoco brand had been introduced in Texas. While the Eastern seaboard states and Texas retained the old red and black Amoco oval, marketing in the former Pan-Am states was under an Amoco torch and oval logo similar to the ones used by Standard and affiliates since 1946. In 1961 the entire Amoco marketing area was converted to the "American" brandname and stations were re-imaged. American signs replaced Amoco in the east, Amoco in the former Pan-Am territory in the south and Utoco in the west. Standard stations remained branded Standard but were reimaged as well. In 1971 the brandname Amoco was introduced for gasolines and motor oils sold by both American and Standard stations and in 1974 the Amoco brand replaced the American brand as primary station identification. Since that time Amoco has also replaced the Standard name in many midwestern locations. See Standard and Utoco section for more corporate history.

AMERICAN OIL COMPANY - BALTIMORE, MD

(see Amoco introduction for American Oil history.)

IDENTIFICATION SIGNS

AMOCO oval w/yellow border 48x96
(5) 1923-1932 $425
Steel framed porcelain sign. Red, white, black & yellow. 1923 logo.

AMOCO oval no border 48x96
(5) 1923-1932 $350
Steel framed porcelain sign. Red, white & black. 1923 logo.
AMERICAN AMOCO GAS oval 60x100
(2) 1932-1961 $150
Steel framed porcelain sign. Red, white, black & grey. 1932 logo.
AMERICAN AMOCO GAS oval 40x80
(4) 1932-1961 $200
Steel framed porcelain sign. Red, white, black & grey. 1932 logo.
AMERICAN torch & oval 60x100
(1) 1961-1974 $100
Steel framed porcelain sign with formed porcelain two-piece flame. Red, white & blue. 1961 logo.
AMERICAN torch & oval 30x54
(3) 1961-1974 $100
Tin litho diecut sign. Red, white & blue. 1961 logo.
AMERICAN torch & oval 60x100
(5) 1971-1974 $100
Steel framed porcelain sign with formed porcelain two-piece flame. Red, white & blue. 1971 logo.
AMOCO torch & oval 60x100
(2) 1974-PRES $100
Steel framed porcelain sign with formed two-piece flame. Red, white & blue. 1974 logo.

SIGN POLES

AMOCO LARGE MAST-ARM POLE
(1) 1932-1974 $200
Cast octagonal pole with pipe mast arm and two rolled steel support brackets.
AMOCO SMALL MAST-ARM POLE
(2) 1932-1974 $200
Cast round pole with pipe mast arm and two rolled steel support brackets.

ILLUMINATED SIGNS

AMOCO GAS W/ GLOBE INSERT
(5) 1925-1935 $600
Opal glass "AMOCO-GAS" lettering with die cut surrounding panel. Frame at top for installation of globe lens. Two sided can-type assembly.
AMERICAN AMOCO GAS oval 60x100
(5) 1954-1961 $150
Formed plastic face for internally lit sign. 1932 logo.

BUILDING LETTERING & SIGNAGE

"AMOCO SERVICE"
(4) 1946-1961 $250
Green porcelain lettering. Various sizes available.

NEON SIGNS

AMERICAN AMOCO GAS oval 60x100
(5) 1935-1961 $500
Flat 1932 oval signs listed above assembled back-to-back on a can type frame. Red neon "AMOCO" and green neon "AMERICAN" and "GAS."
AMOCO neon lettering 24" and 30"
(4) 1954-1961 $200
Assembled galvanized channel letters, painted red, with red neon fill line tubing. Several sizes available, mounted on rooftop channel and with and without red plastic covers.

CURB SIGNS

AMERICAN ETHYL 30"RD
(5) 1935-1938 $550
Round porcelain sign with stand. Red, yellow, black & white with large Ethyl
(EGC) logo.
AMERICAN ETHYL 30"RD
(5) 1935-1938 $350
Round porcelain sign with stand. Large "AMERICAN" arched over Ethyl
(EGC) logo.
STOP FOR ORANGE AMERICAN GAS
(5) 1926-1932 $250
Pipe framed tin litho sign. White upper area with orange "STOP/for." Orange lower area with black "ORANGE/AMERICAN/GAS."
AMOCO AT YOUR SERVICE CHALKBOARD
(5) 1935-1940 $250
Framed chalkboard with small Amoco oval at top. White band above oval with blue "At Your Service." Sanitary rest rooms logo at bottom.
AMOCO MOTOR OIL
(5) 1926-1932 $350
Pipe framed green tin litho sign. Upper yellow area with white "AMOCO/MOTOR OIL/as good as the gas" logo. White "100% PURE/PARAFFIN/BASE" below.
PENN-AMO MOTOR OIL
(5) 1926-1932 $350
Pipe framed blue tin litho sign. Large yellow area covers most of sign with red "PENN-AMO" above PGCOA seal.

LUBRICANTS SIGNAGE

AMOCO MOTOR OIL DIE CUT
(5) 1926-1932 $250
Die cut green porcelain sign with red panel covering most of sign. White "AMOCO/MOTOR OIL" on red panel. White "100% PURE/PARAFFIN/BASE" at top, "AMERICAN OIL COMPANY" across bottom.
PERMALUBE AMOCO SERVICE BACKBAR SIGN
(5) 1939-1950 $150
Wood framed tin backbar sign. Large Amoco oval to left with white "Permalube" in upper oval area, "Service" below. "APPROVED/LUBRICATION" on ribbon to right.
PERMALUBE AMOCO SERVICE BACKBAR SIGN
(5) 1950-1961 $500
Die-cut porcelain backbar sign with large Amoco oval at top. White "Permalube" in upper oval area, "Service" below. Blue lower area with white "APPROVED/LUBRICATION" below logo.

OIL RACK SIGNS

QUALITY-ECONOMY AMOCO MOTOR OIL
(5) 1932-1935 $75
Rectangular light green tin litho sign. Large Amoco oval in center. Dark green semi-circles top & bottom with white "QUALITY-ECONOMY" at top, "MOTOR OIL" at bottom.
SEALED IN QUALITY AMOCO OIL
(5) 1935-1946 $75
Rectangular white tin litho sign. Large Amoco oval in center with blue bands top & bottom. White "SEALED IN QUALITY" across top, "100% PARAFFIN BASE DEWAXED" across bottom.

PENN-AMO
(5) 1930s $75
Rectangular white tin litho sign. Red/white ribbon at upper left with black "PENN," blue/white ribbon at lower right with black "AMO." Other black lettering on sign.

PERMALUBE MOTOR OIL
(3) 1940s $75
Rectangular white tin litho sign. Small Amoco oval in center with red and blue bands below. Blue "Permalube/MOTOR OIL across top.

SECONDARY SIGNS

MEN
(2) 1930's-1961 $20
Rectangular green porcelain sign with rounded ends and white lettering.

LADIES
(2) 1930's-1961 $20
Rectangular green porcelain sign with rounded ends and white lettering.

SANITARY REST ROOMS 10"
(2) 1934-1961 $75
Blue outlined white porcelain round sign with blue center circle and red & blue lettering.

AUTHORIZED AMERICAN OIL COMPANY STATION 20x34
(4)1920-1926 $500
Green porcelain hanging sign with white lettering & small Amoco round sign in center.

AUTHORIZED DEALER AMERICAN OIL COMPANY 12x30
(4)1926-1932 $650
White porcelain two-sided hanging sign with green lettering & green band across bottom. Small Amoco oval in center of sign.

CHARGE ACCOUNTS OF AMERICAN OIL COMPANY HONORED HERE 12x24
(4) 1932-1939 $200
White porcelain hanging sign with all green lettering.

AMERICAN AMOCO COURTESY CARDS HONORED HERE 12x24
(2) 1939-1956 $150
White porcelain hanging sign with American Amoco oval at top above black "COURTESY CARDS/HONORED HERE."

AMOCO COURTESY CARDS HONORED HERE 12x24
(4) 1956-1961 $150
White porcelain hanging sign with Amoco oval at top above black "COURTESY CARDS/HONORED HERE."

AMERICAN CREDIT CARDS HONORED
(4) 1961-1965 $75
White outlined red vertical rectangular porcelain sign with American logo at top over white "CREDIT/CARDS/HONORED."

BULK PLANT SIGNAGE

AMOCO loading rack marker 6x12
(5) 1950's $50
Red & black porcelain with white "AMOCO" lettering similar to gas pump ad panel.

AMERICAN loading rack marker 6x12
(5) 1950's $50
Red & blue porcelain with white "AMERICAN" lettering similar to gas pump ad panel.

TRUCK SIGNS

AMOCO truck door sign 6x12
(5) 1930's $200
Small red & black porcelain oval with white "AMOCO" lettering.

PAN-AM SOUTHERN CORPORATION - NEW ORLEANS, LA

(see Amoco introduction for Pan-Am history).

IDENTIFICATION SIGNS

PAN AM no burst 42"RD
(5) 1925-1930 $300
Green outlined yellow porcelain sign with red "PAN-AM" across center, "GASOLINE" arched around top, "MOTOR OILS" around bottom.

PAN AM GASOLINE/MOTOR OILS 30"RD
(5) 1925-1930 $300
Red outlined beige porcelain sign with red sunburst in center and green band across burst. Green "GASOLINE" arched around top, "MOTOR OILS" around bottom, beige "PAN-AM" on band.

PAN AM GASOLINE/MOTOR OILS 48"
(3) 1930-1938 $250
Red outlined white porcelain sign with red sunburst in center and blue band across burst. Blue "GASOLINE" arched around top, "MOTOR OILS" around bottom, white "PAN-AM" on band.

PAN AM red & blue 48"
(4) 1938-1947 $250

PAN-AM red & blue 66"
(5) 1938-1947 $200
Same as above, with revised burst and without "GASOLINE" or "OILS."

PAN-AM torch & oval white sm
(4) 1947-1950 $300
Die-cut porcelain PAN-AM torch and oval with white flame with red details.

PAN-AM torch & ova white med
(4) 1947-1950 $350
Steel framed oval porcelain PAN-AM torch and oval with detachable one piece porcelain white flame bolted to frame.

PAN-AM torch & oval red sm
(4) 1950-1957 $300
Die-cut porcelain PAN-AM torch and oval with red flame with white details.

PAN-AM torch & oval red med
(4) 1950-1957 $350
Steel framed oval porcelain PAN-AM torch and oval with detachable two piece red porcelain flame bolted to sign.

PAN-AM torch & oval red lg
(4) 1947-1957 $300
Pole framed oval porcelain PAN-AM torch an oval with bolt on torch bowl and internally lit plastic or glass flame.

AMOCO torch & oval sm
(4) 1957-1961 $300
Die-cut porcelain AMOCO torch and oval with red flame with white details.

AMOCO torch & oval lg
(5) 1957-1961 $300
Pole framed oval porcelain AMOCO torch an oval with bolt on torch bowl and internally lit plastic or glass flame.
AMERICAN torch & oval lg
(5) 1961-1962 $300
Pole framed oval porcelain AMERICAN torch an oval with bolt on torch bowl and internally lit plastic or glass flame.

BUILDING LETTERS & SIGNS

PAN-AM
(5) 1946-1957 $150
Blue porcelain lettering. Various sizes available.
PERMALUBRICATION
(5) 1946-1957 $100
Blue porcelain lettering. Various sizes available.
TORCH assembly
(5) 1954-1961 $250
Formed porcelain "torch" with internally lit plastic or glass flame. Mounted on building pylon.

PUMP SIGNS

PAN-AM GAS & OIL CURVED 12"RD
(5) 1920's $225
Curved white porcelain sign with green and red outer rings. White "GASOLINE" arched around top of green ring, "MOTOR OILS" around bottom with black "PAN-AM" across center white area.
PAN-AM GAS & OIL FLAT 12"RD
(5) 1920's $225
White porcelain sign with green and red outer rings. White "GASOLINE" arched around top of green ring, "MOTOR OILS" around bottom with black "PAN-AM" across center white area.
PAN AM QUALITY 13x17
(2) 1947-1955 $125
White porcelain vertical rectangle with PAN-AM torch and oval with white flame at top. Blue "QUALITY" over red "Gasoline" below.
PAN-AM ETHYL 13x17
(3) 1947-1950 $150
Same as above with PAN-AM logo offset to left and Ethyl logo offset to right at top. Blue "PREMIUM" replaces "QUALITY."
PAN-AM PREMIUM 13x17
(4) 1950-1954 $125
Same as above with blue "PREMIUM" replacing "QUALITY."
PAN AM 1954 PREMIUM 13x17
$100
Yellow tin litho sign with large white burst in center. Pan-Am torch and oval logo at top above blue "1954 PREMIUM." Smaller red "Balanced Energy/Gasoline" across bottom.
PAN-AM QUALITY 13x17
(2) 1954-1957 $100
Same as above, with red flame logo.
PAN-AM PREMIUM 13x17
(2) 1954-1957 $100
Same as above, with red flame logo.
PAN-AM DIESEL 13x17
(5) 1954-1957 $125
Same as above, with red flame logo and blue "DIESEL" replacing "QUALITY."
PAN-AM COVER PLATE 5x6
(5) 1947-1954 $150
Porcelain die cut torch and oval with white flame designed as a cover plate for sight glass openings in Martin & Schwartz pumps when sight glasses were eliminated.

NEON

PAN AM NEON LETTERS
(5) 1950'S $100
Assembled galvanized channel letters, painted red, with red neon fill line tubing. Several sizes available, mounted on rooftop channel and with and without red plastic covers.

CURB SIGNS

PAN AM PURPLE 30"
(5) 1920'S $750
White porcelain sign with purple "lightning bolt" areas at upper left & lower right. Large purple "PURPLE" across center with smaller "Fill Up With" above and "PAN-AM/Gasoline" below.
GASOLINE/PAN-AM/MOTOR OIL 24"
(5) 1920'S $400
White porcelain sign with green and red outer rings. White "GASOLINE" arched around top of green ring, "MOTOR OILS" around bottom with black "PAN-AM" across center white area.
PAN AM ETHYL 30"
(5) 1930-1938 $550
Blue outlined white porcelain sign with blue outlined white center circle. Red "PAN-AM" arched around top, "GASOLINE" around bottom with Ethyl
(EGC) log with yellow rays in center.

LUBRICATION/MOTOR OIL SIGNS

PAN AM MOTOR OIL red/green 15"RD
(5) 1920's $300
White porcelain sign with green and red outer rings. White "PAN-AM" arched around top of green ring, "MOTOR OILS" around bottom with black "PAN-AM" across center white area.
PAN-AM MOTOR OIL SOLDIER 30"RD
(5) 1930-1934 $2000
Red outlined white porcelain sign with blue outlined red center circle. Blue "PAN-AM" arched around top, "MOTOR OIL" around bottom with detailed blue and white soldier and white "STANDS/UP/UNDER/FIRE" in center circle.
PANOCO MOTOR OILS TIN RACK SIGN
(5) 1930's $125
Rectangular yellow tin litho sign with blue "PANOCO" over "MOTOR OILS."

SECONDARY SIGNS

PAN-AM THERMOMETER 16x72
(5) 1925-1930 $450
Cream embossed porcelain vertical strip with green border area. Red/cream/green Pan-Am burst logos top and bottom with thermometer with green markings in center.
WHITE TORCH REST ROOMS
(4) 1950'S $250
Blue & white die cut torch with red flame. White square interrupts torch shaft with red "CLEAN/REST/ROOMS" on square & red "AWARD" on torch bowl.
WHITE TORCH REST ROOMS

(4) 1950'S $250

Blue & white die cut torch with red flame. Red "CLEAN REST ROOMS" vertical on torch shaft & red "AWARD" on torch bowl.

APPROVED CREDIT CARDS HONORED

(4) 1950'S $75

Black tin litho sign with gold "APPROVED/CREDIT CARDS/ HONORED."

DISTRIBUTOR OF AMOCO PRODUCTS

(4) 1954-1957 $150

White porcelain sign with small red, black and white logo in center. Black "Distributor of" above logo, "Products" below. Used in transition between Pan-Am & Amoco.

MEN

(2) 1946-1961 $40

White rectangular sign with blue lettering.

WOMEN

(2) 1946-1961 $40

White rectangular sign with blue lettering.

COLORED

(5) 1946-1961 $100

White rectangular sign with blue lettering.

CAPITOL OIL COMPANY - RICHMOND, VA

Richmond, VA based independent purchased by and rebranded Amoco in 1926.

No collectible size Capitol signs known at this time.

SING PETROLEUM - THOMASVILLE, GA

Founded in 1935 as another southern discount service station chain, most of Sing's branded marketing was purchased by Amoco in 1990. Most locations have remained with the Sing brand.

IDENTIFICATION SIGNS

SING GASOLINE 60"RD

(4) 1960-1990 $200

Blue outlined white porcelain sign with red "SING" across upper sign face, over blue stylized falcon design. Small blue "PRODUCTS" below falcon.

PUMP SIGNS

SING 10x10

(4) 1965-1980 $45

Blue tin litho diamond shaped sign with blue outlined white center circle with red "SING" across upper sign face, over blue stylized falcon design.

ASHLAND OIL and REFINING - ASHLAND, KY

Ashland Oil was founded in 1924 as a refining division of the Swiss Oil Company of Lexington, KY, which had numerous ties to National Refining. Ashland began marketing gasoline under the brandname "Pepper," with the gasolines dyed red and green as well as clear to market the famous Red Pepper, Green Pepper, and White Pepper products. Stations under the Pepper brandname were operated in an area of about 250 miles around the Ashland, KY, headquarters. Branded gasoline marketing was all but discontinued during WWII, and after the war when retail gasoline sales resumed, the brandname Ashland was chosen, Ashland Oil having replaced the corporate name Swiss Oil in about 1935. Ashland has grown through the many mergers listed below, with branded Ashland outlets found within about a 200 mile radius of Ashland, KY, and with company owned and operated stations, mostly under the SuperAmerica brand, throughout the central and eastern United States.

PEPPER

Refined products manufactured by Ashland Oil were originally sold under the brandname "Pepper." The trademark was introduced in 1924 and continued in use until 1946. The "Ashland" name was also used, in limited ways, beginning about 1940.

IDENTIFICATION SIGNS

PEPPER GASOLINES-THEY'RE PEPPIER 47x60

(5) 1935-1946 $375

White porcelain elongated octagon signs assembled can-type back to back with stainless steel band. Large "PEPPER" over smaller "GASOLINES" on upper sign area. Script "They're Peppier" below.

CURB SIGNS

RED PEPPER ETHYL 30"RD

(5) 1930-1946 $400

Green outlined white tin sign with green outlined red "RED/PEPPER" across upper sign area. Large Ethyl (EGC) logo with green rays on lower sign face.

SECONDARY SIGNS

PEPPER GASOLINES-THEY ARE PEPPIER 36x72

(5) 1925-1930 $450

Wood framed tin rectangular sign. "ASHLAND /REFINING/ COMPANY" on large pepper to left. Large "PEPPER/GASOLINES" to right over smaller "THEY ARE PEPPIER."

ASHLAND

In 1946 the "PEPPER" brandname at service stations was replaced by the corporate "ASHLAND" brandname. The Ashland logotype, "Ashland" in stylized lettering arched with descending letter height from left to right, was introduced at this time.

IDENTIFICATION SIGNS

ASHLAND PRODUCTS 46x70

(4) 1946-1954 $300

White porcelain elongated octagon signs assembled can-type back to back with stainless steel band. Red/black/white "Ashland" logotype over black outlined green "PRODUCTS" on sign face.

ASHLAND OIL PRODUCTS 72x96

(4) 1954-1960 $200

Flat two sided pole framed porcelain sign same as above with "OIL" added above "PRODUCTS."

ASHLAND OIL PRODUCTS 72x96

(4) 1960-1968 $200

Flat two sided pole framed porcelain sign same as above with blue "OIL/PRODUCTS."

ASHLAND OIL PRODUCTS OVAL 42x84

(5) 1960-1968 $350

Porcelain oval shaped sign with logotype similar to above.

Used as a replacement face in existing Aetna sign frames when the Ashland brand replaced Aetna.
ASHLAND 47x60
(5) 1968-1995 $150
Porcelain elongated octagon with red/white/blue/black 1968 Ashland logo.
ASHLAND 72x96
(5) 1968-1995 $100
Porcelain elongated octagon pole framed porcelain sign with red/white/blue/black 1968 Ashland logo.

BUILDING SIGNS

ASHLAND PRODUCTS
(5) 1950-1968 $350
Cast aluminum "Ashland/PRODUCTS" logotype designed to attached to station pylon.

PUMP SIGNS

RED/BLUE STRIPES 6x18
(2) 1968-1985 $25
Porcelain vertical twin stripes. Left stripe red/right blue.

SECONDARY SIGNS

ASHLAND AETNA FRONTIER CREDIT CHECKS 12x18
(5) 1950-1960 $400
White porcelain sign with red "WE HONOR" across top, "CREDIT CHECKS" across bottom and full color Ashland, Aetna and Frontier logos in center.

BULK PLANT SIGNAGE

ASHLAND OIL & REFINING COMPANY 46x70
(4) 1946-1954 $350
White porcelain elongated octagon signs assembled can-type back to back with stainless steel band. Red/black/white "Ashland" logotype over black outlined green "OIL AND REFINING COMPANY" on sign face.

TRI-STATE - KENOVA, WV

Tri-State, an independent refiner serving much the same territory as Ashland did in its earliest days, merged with Ashland in 1930. The Tri-State gasoline brands were retained until the early 1960s as an Ashland secondary brand.

No collectible size Tri-State signage is known at this time.

NATIONAL REFINING - CLEVELAND, OH

The National Refining Company was a large refiner/marketer based in Cleveland, Ohio and founded in 1882. Marketing was concentrated throughout the Eastern and Central states with a subsidiary in Canada that was spun off as a separate company in the 1930s and survived until it's 1964 merger with Shell. National marketed the famous White Rose Gasoline and for many years their boy and slate logo was one of the best recognized trademarks in the country. After WWII, National was greatly weakened in the marketplace and the major portion of the company merged with Ashland in 1950. National White Rose stations were rebranded Ashland about 1957. Unique in the gasoline business, the White Rose brandname was franchised to Mid-West Refineries of Muskegon, Michigan, about 1946 and although Mid-West merged with Leonard in 1955, the White Rose name remained in use in Michigan through the early 1970s.

IDENTIFICATION SIGNS

WHITE ROSE BOY & SLATE ON CIRCLE 42"RD
(5) 1926-1937 $2000
Porcelain black outlined white circle with multi-color "boy & slate" logo. White "WHITE/ROSE" on slate.
ENARCO MOTOR OIL BOY & SLATE ON CIRCLE 42"RD
(5) 1926-1937 $2000
Porcelain black outlined white circle with multi-color "boy & slate" logo. White "EN-AR-CO/MOTOR OIL" on slate.
WHITE ROSE GASOLINE BOY & SLATE ON CIRCLE 42"RD
(5) 1926-1937 $2000
Porcelain black outlined white circle with multi-color "boy & slate" logo. White "WHITE ROSE/GASOLINE" on slate.
ENARCO MOTOR OIL BOY & SLATE ON CIRCLE 42"RD
(5) 1926-1937 $750
Tin litho black outlined white circle with multi-color "boy & slate" logo. White "EN-AR-CO/MOTOR OIL" on slate.
WHITE ROSE GASOLINE BOY & SLATE ON CIRCLE 42"RD
(5) 1926-1937 $750
Tin litho black outlined white circle with multi-color "boy & slate" logo. White "WHITE ROSE/GASOLINE" on slate.
ENARCO WHITE ROSE DEALER
(5) 1937-1946 $125
Green porcelain diamond shaped with cream colored band across center. Cream "EN-AR-CO" in upper green area, "DEALER" below with green "WHITE/ROSE" in center.
NATIONAL WHITE ROSE DEALER
(5) 1946-1957 $150
Green porcelain diamond shaped with cream colored band across center. Cream "NATIONAL" in upper green area, "DEALER" below with red "WHITE/ROSE" in center.
NATIONAL WHITE ROSE DEALER
(5) 1946-1957 $200
Same as above, embossed form and lettering.

BUILDING SIGNS AND LETTERS

GASOLINE PANEL
(5) 1930-1937 $150
Black/yellow outlined self framed porcelain panel with red "GASOLINE" across center.
MOTOR OIL PANEL
(5) 1930-1937 $150
Black/yellow outlined self framed porcelain panel with red "MOTOR OIL" across center.

CURB SIGNS

DIE CUT BOY & SLATE
(4) 1925-1957 $750
Die cut framed multicolor "boy and slate" logo with real slate for changeable messages.
DIE CUT SCRIPT ENARCO "WHAT A DIFFERENCE"
(5) 1925-1937 $1250
Same as above,, with white script "EN-AR-CO" across slate. Black "WHAT A DIFFERENCE" on underline.
DIE CUT WHITE ROSE ETHYL
(5) 1925-1937 $1250
Same as above,, with white "WHITE ROSE/ETHYL" across slate.
DIE CUT ROYAL GASOLINE
(5) 1925-1937 $1250

Same as above,, with white "ROYAL/GASOLINE" across slate.
ENARCO PRODUCTS 24"RD
(5) 1930-1937 $350
Round white porcelain sign with "boy & slate" logo. White "EN-AR-CO/PRODUCTS" on slate. Black "WHITE ROSE/ETHYL/GASOLINE" to lower left below slate.
ENARCO PENN MOTOR OIL W/PGCOA SEAL 24"RD
(5) 1930-1937 $350
Round white porcelain sign with "boy & slate" logo. White "EN-AR-CO/PENN MOTOR/OIL" on slate. PGCOA seal to lower right.
EN-AR-CO MOTOR OIL 30"RD
(5) 1946-1957 $200
Round cream colored tin sign with red outline and lower area. Red "EN-AR-CO" over green "MOTOR OIL" across upper face. Red "FOR PEAK PERFORMANCE" above small boy & slate logo below.
EN-AR-CO C1 MOTOR OIL 30"RD
(5) 1954-1957 $200
Similar to above, with large "C1" in center.

LUBRICATION/MOTOR OIL SIGNS

EN-AR-CO MOTOR OIL 6x7.5
(5) 1930-1937 $175
Square sign with "boy & slate." White "EN-AR-CO/MOTOR OIL" on slate. Clamp on lubester sign.
ENARCO MOTOR OIL 8"
(5) 1930-1937 $350
Black outlined white porcelain sign with "boy & slate" logo. White script "EN-AR-CO" over "MOTOR OIL" on slate.

SECONDARY SIGNS

WHITE ROSE MOTOR GASOLINE/NATIONAL CARBONLESS OIL 9.5x14
(4) 1910-1917 $425
Tin sign with upper half black, lower half light green split diagonal. White "WHITE ROSE/MOTOR/GASOLINE" with rose in upper area black "AND/NATIONAL/CARBONLESS/MOTOR OIL" in lower area.
WHITE ROSE DIE CUT BOY & REAL SLATE 58x77
(2) 1917-1957 $1200
Three piece porcelain. Upper piece is black & white boy's head. center section is boy's hands holing real slate with red/yellow/black border. Black "EN-AR-CO MOTOR OIL" in upper slate border "WHITE ROSE GASOLINE" below. Lower section is legs & feet in black/white/red.
WHITE ROSE DIE CUT BOY & REAL SLATE 58x77
(2) 1917-1957 $750
Three piece tin litho. Upper piece is black & white boy's head. center section is boy's hands holing real slate with red/yellow/black border. Black "EN-AR-CO MOTOR OIL" in upper slate border "WHITE ROSE GASOLINE" below. Lower section is legs & feet in black/white/red.
WHITE ROSE DIE CUT BOY & EN-AR-CO SLATE 58x77
(5) 1917-1937 $1200
Three piece porcelain. Upper piece is black & white boy's head. center section is boy's hands holing real slate with red/yellow/black border. White script "EN-AR-CO" over "MOTOR OIL" on slate. Lower section is legs & feet in black/white/red.
BOY & SLATE "X MILES TO WHITE ROSE SERVICE STATION" 39x59
(5) 1917-1937 $850
One piece tin. Die cut boy & slate with white "MILES TO/WHITE ROSE/SERVICE STATION" on slate. Area to fill in mileage.
NATIONAL CREDIT CARDS HONORED 14x14
(5) 1937-1950 $250
Green tin litho sign with cream band across center. Green "CREDIT/CARDS" in center band with cream "NATIONAL/REFINING CO" in green area above and "HONORED/HERE" below.
ENARCO ROAD MAPS 12x16
(5) 1937-1940 $200
Red background tin litho sign with boy holding US map. Black "FREE" over red ""EN-AR-CO" over black "ROAD MAPS" on map.
ROYAL WHITE ROSE ETHYL 18x96
(5) 1930-1935 $350
Yellow porcelain panel with white arrow at left, red at right. Blue "WHITE ROSE" on white arrow, white "ETHYL" on red arrow with blue "ROYAL" in center above arrow points. Exact use of this sign unknown.
USE THE BEST-WHITE ROSE GASOLINE 24x40
(5) 1950-1957 $275
Self framed cream tin litho sign with red border. Large red "WHITE ROSE/GASOLINE" across center with smaller green "Use The Best" above and "To Get The Best" below. Small boy & slate logo in lower left.
NATIONAL ALERT SERVICE 30x40
(5) 1950-1957 $175
Die cut irregular shape sign. Green upper area with cream "NATIONAL" across top. Red "ALERT/SERVICE" below with green ribbon arched around bottom. Cream "AUTHORIZED DEALER" on ribbon.

WHITE ROSE/MID WEST REFINING - GRAND RAPIDS, MI

No collectible size signs identifiable as specifically Mid-West White Rose are known at this time. Signage basically the same as National Refining Company White Rose signage.

AETNA OIL - LOUISVILLE, KY

Aetna Oil, dating from the 1910s, marketed through service stations in central and western Kentucky and southern Indiana. Aetna merged with Ashland in 1950. The Aetna brand was replaced by Ashland in the early 1960s.

IDENTIFICATION SIGNS

AETNA OILS 42x60
(5) 1925-1935 $400
White outlined red die-cut oval with circle at top. White "AETNA/OILS" across center of sign with white "mountain" logo at top.
AETNA PRODUCTS 42x60
(5) 1935-1940 $400
White outlined red die-cut oval with circle at top. White "AETNA/PRODUCTS" across center of sign with white "mountain" logo at top.
AETNA 30x60

(4) 1946-1955 $250
Dark green outlined tan tin litho sign. Red "AETNA" with light green/dark green dropshading across center.

AETNA 42x84
(4) 1946-1955 $175
Dark green outlined tan porcelain oval sign. Red "AETNA" with light green/dark green dropshading across center.

AETNA 30x60
(5) 1955-1960 $250
Dark green outlined white tin litho oval sign. Red "AETNA" with light green/dark green dropshading across center.

AETNA 42x84
(3) 1955-1960 $175
Dark green outlined white porcelain oval sign. Red "AETNA" with light green/dark green dropshading across center.

AETNA 42x84
(5) 1955-1960 $200
Dark green outlined white porcelain oval sign. Red "AETNA" with light green/dark green dropshading across center. Second "A" is reversed.

CURB SIGNS

AETNA ETHYL 30"RD
(5) 1926-1935 $375
Red outlined white porcelain sign with red "AETNA" arched across top over large Ethyl
(EGC) logo.

FRONTIER REFINING - BUFFALO, NY

Founded in Buffalo in 1931, Frontier marketed gasoline in Western New York and Pennsylvania. Frontier merged with Ashland in April, 1950, and Frontier stations were rebranded Ashland about 1960.

No Frontier signage is known at this time.

FREEDOM OIL WORKS - FREEDOM, PA
FREEDOM-VALVOLINE - OIL CITY, PA

Freedom was an early Pennsylvania lubricants manufacturer who had also marketed gasoline since the 1920s. They sold gasoline under the "Freedom" brand and lubricants under the Freedom, Beaver Penn, Penn Trump and Fort Pitt brands until after the 1944 merger with Valvoline Oil Co. of Oil City, PA. At that time the few remaining Valvoline stations were rebranded Freedom, while most lubricants were branded Valvoline. Freedom was in the process of discontinuing gasoline marketing, selling it's stations to Esso, when the company merged with Ashland Oil in early 1950. Since that time the primary product has been Valvoline Oils, marketed by all the Ashland retail network and throughout the world.

CURB SIGNS

FREEDOM ETHYL 30"RD
(5) 1926-1935 $575
Outlined white sign with "FREEDOM" arched around top, "GASOLINE" around bottom and large Ethyl
(EGC) logo in center.

LUBRICATION/MOTOR OIL SIGNS

FREEDOM PERFECT MOTOR OIL 24"RD
(4) 1926-1940 $450
Yellow porcelain sign with large black "FREEDOM" over smaller "PERFECT/MOTOR OIL." PGCOA seal at top.

FORT PITT MOTOR OIL 18x24
(4) 1930s $750
Blue porcelain sign with large yellow keystone covering most of sign face. Blue "FORT PITT/MOTOR OIL" across center, with drawing of Ft Pitt at top and PGCOA seal at bottom.

FREEDOM MOTOR OILS 20x30
(5) 1920s $275
Blue porcelain sign with yellow band across bottom. Blue band on lower yellow area with white "FREEDOM" over "100% PURE/PENNSYLVANIA/MOTOR OILS." Red Freedom bulldog logo in yellow area above band. White "IN ALL THE WORLD NO FINER OIL" across top above dog.

PENN TRUMP MOTOR OILS 24x30
(5) 1936-1940 $450
Black trapezoidal sign with large white playing card covering most of sign. Large spade in center of card, with white "PENN/TRUMP" in upper area, with red banner arched below. White "MOTOR OIL" in banner. PGCOA seal & other lettering at bottom.

BEAVER PENN 18x22
(5) 1936-1940 $600
Yellow porcelain sign with black/yellow border around black center area. White oval at top with black lined drawing of beaver. Yellow "BEAVER-PENN/100% PURE PENNSYLVANIA/MOTOR OIL/THE FREEDOM OIL WORKS CO./FREEDOM, PENNSYLVANIA" across black area. Yellow/black Freedom bulldog on white square at lower left, PGCOA seal at lower right.

VALVOLINE OIL COMPANY - OIL CITY, PA

Pioneer lubricants manufacturer and owner of the oldest trademark in the petroleum industry today. Valvoline was founded as the Continuous Oil Refining Company in Binghampton, NY in 1866 and introduced the famous trademark "Valvoline" in 1873. So well known were the company's Valvoline lubricants that the company name changed to Valvoline Oil Co. in 1902. In the early 1920s Valvoline began marketing gasoline in New York and Pennsylvania and continued to do so until the 1944 merger with Freedom Oil Co. of Freedom, PA. Since that time the primary product has been Valvoline Oils, and after the 1950 merger with Ashland, Valvoline products were marketed by all the Ashland retail network and throughout the world.

IDENTIFICATION SIGNS

VALVOLINE GASOLINE 30x60
(5) 1920S $450
Green self framed tin litho sign with white bowtie "VALVOLINE" and valve logo across center above white "GASOLINE."

CURB SIGNS

VALVOLINE ETHYL 30"RD
(5) 1926-1935 $1300
White outlined green porcelain sign with large white bowtie "VALVOLINE" across white "valve" logo in center. Red outlined white "GASOLINE" arched around top, Large Ethyl
(EGC) logo with yellow rays below.

PENNSYLVANIA VALVOLINE MOTOR OILS 30"RD
(3) 1925-1943 $700

White outlined green porcelain sign with large white bowtie "VALVOLINE" across white "valve" logo in center. Red outlined white "PENNSYLVANIA" arched around top, "MOTOR OILS" around bottom.

100% PENNSYLVANIA VALVOLINE MOTOR OIL 30"RD
(4) 1949-1955 $300
Green/yellow outlined green tin litho sign. Yellow "VALVOLINE" across center with yellow script "100%/Pennsylvania" diagonal at top, yellow "MOTOR OIL" at bottom.

ASK FOR VALVOLINE GREEN/YL 30"RD
(4) 1955-1960 $300
Green/yellow outlined green tin litho sign. Yellow "VALVOLINE" across center with yellow script "Ask For" diagonal at top, yellow "MOTOR OIL" at bottom.

ASK FOR VALVOLINE R/W/B 30"RD
(1) 1960-1976 $100
Blue outlined white tin litho sign with large red/blue "V" in upper area. Red bowtie "VALVOLINE" above blue "MOTOR OIL"" below "V." Red "Ask For" at top.

VALVOLINE ON RED SQUARE 30"SQ
(1) 1970s $75
Red tin litho sign with blue outlined white circle in center. Large red/blue "V" in upper area. Red bowtie "VALVOLINE" above
blue "MOTOR OIL" below "V." Red "Ask For" at top.

LUBRICATION/MOTOR OIL SIGNS

100% PURE PENNSYLVANIA VALVOLINE 24x48
(5) 1930's $250
Wood framed black tin sign. Large orange "VALVOLINE" and valve across center. Green "100% PURE PENNSYLVANIA" across top, "MOTOR OILS" across bottom.

VALVOLINE LUBRICANTS CUSTOM BUILT 24x48
(5) 1930's $300
Black sign with green band across top and orange green stripes below. Black outlined orange "VALVOLINE" on upper green band with green script "lubricants" below. Orange "CUSTOM BUILT" across bottom below stripes.

COSTS LESS TO USE VALVOLINE 18x30
(4) 1930's $300
Self framed green porcelain with black border. Black outlined white "VALVOLINE" across center. White outlined red "COSTS LESS TO USE" across top, "THE ORIGINAL/PENNSYLVANIA OIL" across bottom.

VALVOLINE MOTOR OIL 12x60
(5) 1930's $200
Self framed vertical strip. Dark green with red/white outline and red outlined white "VALVOLINE" vertical on sign over small white "MOTOR/OIL" at bottom.

VALVOLINE PENNSYLVANIA MOTOR OIL 10x48
(5) 1930's $200
Self framed horizontal strip. Dark green with yellow outline. Yellow "VALVOLINE" to left with smaller yellow "PENNSYLVANIA/MOTOR OIL" to right.

VALVOLINE MOTOR OIL 12x60
(5) 1940's $200
Self framed vertical strip. Dark green with yellow outline. Yellow "VALVOLINE" vertical on sign over small yellow "MOTOR/OIL" at bottom.

BEST THING GOING FOR ANY CAR VALVOLINE 36x60
$125
Self framed tin sign. Blue outline around white signface, with large red bowtie "VALVOLINE" across center. Blue "Best Thing Going For Any Car" across top, "MOTOR OIL" across bottom.

WORLDS FIRST-WORLDS FINEST VALVOLINE 36x60
(2) 1960's $125
Self framed tin sign. Blue outline around white signface, with large red bowtie "VALVOLINE" across center. Blue "World's First - World's Finest" across top, "MOTOR OIL" across bottom.

VALVOLINE MARINE LUBRICANTS 24"RD
(4) 1960'S $125
Embossed white circular bubble sign. Large red/blue/black "V" at top, with red bowtie "VALVOLINE" below. Blue "MARINE LUBRICANTS" across bottom.

FLANGE SIGNS

VALVOLINE MOTOR OILS 8x18
(5) 1910s $250
White outlined blue pennant shaped flange sign with white "VALVOLINE" across valve logo in center. White "MOTOR" across top, "OILS" across bottom.

VALVOLINE PENNSYLVANIA 18"RD
(5) 1920s $250
Green/white outlined tin litho sign with green center and white band. White "Pennsylvania" arched around top, "Motor Oil" around bottom and "VALVOLINE" bowtie lettering with valve logo in center band.

LUBESTER SIGNS

VALVOLINE LUBESTER SIGN 7"RD
(2) 1920s $250
White outlined green tin litho sign with large white VALVOLINE" bowtie lettering across center. "PENNSYLVANIA" arched around top, "MOTOR OIL" around bottom. White area near bottom for grade identification LIGHT MEDIUM.

INDEPENDENT QUALITY SERVICE

Independent Quality Service was an Ashland secondary brand operated for independent, unbranded dealers from the mid 1940s through the 1960s.

IDENTIFICATION SIGNS

INDEPENDENT QUALITY SERVICE 46x70
(4) 1946-1954 $300
White porcelain elongated octagon signs assembled can-type back to back with stainless steel band. Red band across top, blue across bottom with red "INDEPENDENT" at top, blue "QUALITY SERVICE" at bottom. Blue bell with red stars in center.

PAYLESS - NEW ALBANY, IN

Payless is the postwar creation of James Thornton, operating discount service stations "from Canada to the Gulf Coast" as their slogan claimed. In the middle 1950s Payless acquired the marketing of Louisville's Dance Oil Company, Mr. Thornton having married Mr. Dance's daughter, Bonnie. Ashland Oil, long a supplier to both Dance and Payless, bought the operation in about 1970. A few Payless outlets remain today, although most have been converted to SuperAmerica. James Thornton formed Louisville based Thornton Oil Corporation after selling Payless, and continues to operate Thornton stations in six states today.

PUMP SIGNS

RED/BLUE STRIPES 6x18
(2) 1968-1985 $25
Porcelain vertical twin stripes. Left stripe red/right blue.

SECONDARY SIGNS

PAYLESS fence sign
(3) 1975-PRESENT $50
Tin sign with red stripe at top, blue at bottom and white across center. Black "Payless" across center.

SUPERAMERICA - MINNEAPOLIS, MN

Northwestern Refining Co., the St. Paul Park refining unit of the Erickson family, began operating the SuperAmerica gasoline/c-store units about 1962. In 1970, SuperAmerica, a marginal c-store operation, was sold to Ashland Oil, and by the middle 1970s began to expand out of it's north central base. Many of the Ashland affiliate brand locations, closed during the gasoline shortages of 1973 and 1978, were rebuilt as SuperAmerica convenience stores and the operation today is considered the model of efficient and customer service oriented gasoline c-store chains.

SECONDARY SIGNS

SA AIR 12x12
(3) 1970-1985 $25
Red tin litho embossed sign die cut in the shape of the "SA" shield. White "SA" on red over white "AIR." Used on air stands.
SA DOOR PUSH 9x12
(4) 1985-PRESENT $50
Cast aluminum door push panel with cast and painted "SA SUPERAMERICA" logo and "PUSH" below.

OSKEY BROS. PETROLEUM - ST. PAUL, MN

Oskey Bros. has operated the North Star brand stations in the north central states since the 1920s. Ashland acquired the chain about 1980 and has since closed many locations or converted them to SuperAmerica. The company has no connection with Canada's North Star Oil Company.

IDENTIFICATION SIGNS

NORTH STAR GASOLINE 42x84
(4) 1946-1960 $200
Blue outlined white porcelain oval with red "NORTH STAR" arched around top, "GASOLINE" around bottom and blue star in center.
NORTH STAR 42x84
(3) 1960-1985 $200
White outlined oval porcelain sign with blue field to left and white to right split by sweeping curve. White "N" on blue area, red star on white area.

RED HEAD OIL CO. - WOOSTER, OH

Red Head, operating discount gasoline stations with service bays, unique in the industry, was founded in Wooster, Ohio, in the mid-1930s and operated stations in Ohio, Pennsylvania, and West Virginia when they were bought out by Ashland in about 1970. Many have since been converted to SuperAmerica, but some Red Head stations remain today.

PUMP SIGNS

RED/YELLOW STRIPES 6x18
(2) 1968-1985 $25
Porcelain vertical twin stripes. Left stripe red/right yellow.

NEON SIGNS

RED HEAD
(5) 1936-1975 $750
White outlined black keyhole shaped porcelain can type sign. White outlined red "RED/HEAD" on upper round area with change board area below.

SECONDARY SIGNS

RED HEAD FENCE SIGN
(3) 1975-PRESENT $50
Red tin rectangular sign with black outlined yellow "RED/HEAD" logotype lettering across center.

SOUTHERN OIL OF NY - HORSEHEADS, NY

Southern Oil operated the Rotary brand stations in New York and Pennsylvania from the mid-1920s until the 1970s when the operation was acquired by Ashland Oil.

IDENTIFICATION SIGNS

ROTARY ROTA-ZOL BENZOL
(5) 1930's $400
White outlined red porcelain oval with white outlined blue band across center. White "ROTARY" across top with white "ROTA-ZOL" across center blue band. White "BENZOL" in lower red area.

HOOD OIL CO. - NORCROSS, GA

Hood's was a postwar southern operation, rebranded Fleet about 1960 and acquired by Ashland about 1967. In the 1980s, Fleet was sold to the son of the founder and continues in operation today.

SECONDARY SIGNS

FLEET fence sign
(3) 1975-PRESENT $50
Tin sign with yellow stripe at top, black at bottom, and white across center. Black "FLEET" across center.

TRESLER OIL - CINCINNATTI, OH

Tresler was founded in Cincinnati in 1926 and operated retail service stations in Ohio, Indiana and Kentucky until 1985 when they were purchased by Ashland. All have since been converted to SuperAmerica.

IDENTIFICATION SIGNS

TRESLER'S COMET GASOLINE
(5) 1930-1940 $375
Rectangular sign with white oval diagonal across center. "Comet Gasoline" uphill in oval. Outlined diamond at upper left with "TRESLER".

NEON SIGNS

COMET
(5) 1939-1950 $600
Blue porcelain round end rectangle with yellow/white/blue

comet across sign. Blue outlined yellow "COMET" on comet tail. Punched for neon.

ILLUMINATED SIGNS

TRESLER GASOLINE

(5) 1930-1935 $450

Large die-cut can type porcelain sign with opal glass lettering "TRESLER/GASOLINE" across center. Cutout for globe lens insert at either end.

CURB SIGNS

TRESLER ETHYL GASOLINE 30"RD

(5) 1930's $375

"TRESLER" arched around top, "GASOLINE" around bottom with large Ethyl (EGC) logo in center.

CINCINNATI OIL WORKS - CINCINNATI, OH

Early refiner marketer that operated under the White Rose brand in Cincinnati and surrounding area. No known ties to En-Ar-Co White Rose. Purchased by Tresler Oil Company in 1956 & rebranded Tresler Comet.

NEON SIGNS

WHITE ROSE 48x96

(5) 1940's $500

Blue outlined yellow porcelain can type sign with white/red outlined red circle with yellow center in center. Blue "WHITE ROSE" in bowtie lettering across center with blue outlined white "DEPENDABLE" arched around top, "GASOLINE" around bottom.

SECONDARY SIGNS

WE SELL PREMIUM HIGH TEST KEROSENE 24x30

(5) 1910's $200

Yellow embossed tin litho sign with red and black lettering.

PYROIL CHEMICAL CO. - LA CROSSE, WI

Pyroil marketed a chemical additive for gasoline, and stations adding this lubricant to their products sold it under the Pyroil Lubrication Gasoline trademark. Pyroil was bought out by Ashland in 1988 and continues to market automotive chemicals today.

No collectible size Pyroil signs are known at this time.

HI-FY GASOLINE STATIONS - COLUMBUS, OH

Columbus based independent purchased by Ashland in the early 1970's. Most stations converted to SuperAmerica in the 1980's.

SECONDARY SIGNS

HI-FY fence sign

(3) 1975-PRESENT $50

Tin sign with red stripe at top blue at bottom and white across center. Black "HI-FY" across center.

PUMP SIGNS

RED/BLUE STRIPES 6x18

(2) 1968-1985 $25

Porcelain vertical twin stripes. Left stripe red/right blue.

BI-LO OIL COMPANY - LAWRENCEVILLE, IN

Discounter purchased by Ashland in the early 1970's. Most stations converted to SuperAmerica in the 1980's.

PUMP SIGNS

RED/GREEN STRIPES 6x18

(2) 1968-1985 $25

Porcelain vertical twin stripes. Left stripe red/right g

SECONDARY SIGNS

BI-LO FENCE SIGN

(3) 1975-PRESENT $50

Tin sign with green stripe at top, black at bottom and white across center. Red "BI-LO" across center.

SAVE MORE and SAVE MART

Ashland secondary operations developed in the 1970's for convenience store marketing before the SuperAmerica brand came into use in the Kentucky/West Virginia marketing area. Successful locations with sufficient space converted to SuperAmerica in the 1990's.

SECONDARY SIGNS

SAVE MORE FENCE SIGN

(3) 1975-PRESENT $50

Tin sign with red upper "teardrop" area, blue lower. White "Save" above "More."

SAVE MART FENCE SIGN

(3) 1975-PRESENT $50

Tin sign with red upper "teardrop" area, blue lower. White "Save" above "Mart."

THE ATLANTIC REFINING CO. - PHILADELPHIA, PA

The Atlantic Refining Company was founded in Philadelphia in 1866 from earlier interests of the various founders and brought together as a production and refining company. In 1874 Atlantic sold out to Standard Oil and until the breakup of Standard in 1911, functioned as a refining division. Atlantic began marketing in 1915, after the breakup of Standard, and throughout most of its history was in constant search for crude oil to keep its refineries and service stations operating. In 1965, Atlantic, then operating service stations all along the Eastern Seaboard, began merger talks with Richfield Oil Corporation of California, a west coast refiner/marketer that traces its roots back to the turn of the century. Atlantic merged with Richfield in 1966, and from 1966 until 1970 marketed under the Atlantic name in the east and Richfield in the west. With the 1969 merger with Sinclair, a new identity was needed, resulting in the introduction of the Arco brandname. Arco replaced both the Atlantic and Richfield brands as well as some of the Sinclair brand. By the 1980s, Sinclair had been sold off and was operating as a successful independent and Arco had abandoned marketing in the east, where the name Atlantic was reintroduced by the independent marketing company that had purchased the former Arco marketing. In 1988 the new Atlantic was sold to Sunoco, and stations operate under the Atlantic name in the northeast today. See separate Richfield and Sinclair listings for more information on these affiliated companies.

IDENTIFICATION SIGNS

ATLANTIC REFINING COMPANY/GASOLINE/POLARINE/OILS AND GREASES/FOR SALE HERE 20x28

(5) 1915-1918 $375
White lettering on blue porcelain rectangle, as described.
THE ATLANTIC REFINING COMPANY/AUTOMOBILE/GASOLINE/FOR SALE HERE 20x28
(4) 1918-1924 $375
White lettering on blue porcelain rectangle, as described.
ATLANTIC GASOLINE W/ CROSSED ARROWS 36x52
(2) 1924-1932 $350
Blue outlined white vertical porcelain rectangle. Large red and white Atlantic "crossed arrows" logo in center, with blue "ATLANTIC" across top, "GASOLINE" across bottom.
ATLANTIC MOTOR OIL W/ CROSSED ARROWS 36x52
(5) 1924-1932 $350
Same as above, "MOTOR OIL" replacing "GASOLINE."
ATLANTIC WHITE FLASH 36x52
(3) 1926-1932 $300
Blue outlined white vertical porcelain rectangle. Red "ATLANTIC" above red line at top, with large blue "WHITE/FLASH" on lower sign face.
ATLANTIC GASOLINE/MOTOR OIL 43x73
(5) 1932-1935 $300
White porcelain horizontal rectangle with blue outline area. Large red "ATLANTIC" above smaller blue "GASOLINE/MOTOR OIL."
ATLANTIC WHITE FLASH 43x73
(4) 1932-1935 $300
White porcelain horizontal with blue outline area. Red "ATLANTIC" above red line at top, blue "WHITE/FLASH" on lower sign face.
ATLANTIC
(1) 1935-1966 $100
ATLANTIC 28x48
(3) 1935-1966 $175
Red porcelain horizontal rectangle with white lines above and below white "ATLANTIC" with blue dropshading.
ATLANTIC
(2) 1966-1970 $75
ATLANTIC
(3) 1966-1970 $75
ATLANTIC 26x48
(3) 1966-1970 $100
Red porcelain elongated hexagon with white band across center. Blue italic "ATLANTIC" across white band.

SIGN POLES

ATLANTIC MAST-ARM POLE
(3) 1932-1956 $500
Round cast pole with pass-through mast arm. Detailed wrought iron work at intersection of pole and arms, with wrought iron brackets from pole to arm on either side. Stepped & fluted cast base. Probably the most ornate identification sign pole used in the industry.

ILLUMINATED SIGNS

ATLANTIC GASOLINE W/ LENS INSERT
(5) 1920s $600
Red die cut porcelain faces with ornate white porcelain frame. White opal glass lettering with cut out for globe lens at top.

BUILDING SIGNS & LETTERS

"ATLANTIC" Red porcelain lettering 12"
(4) 1940-1960 $175
Red 12" formed porcelain "ATLANTIC" lettering on white rails.
"WASHING" Red porcelain lettering 12"
(5) 1940-1960 $100
Red 12" formed porcelain "ATLANTIC" lettering on white rails.
"LUBRICATION" Red porcelain lettering 12"
(5) 1940-1960 $100
Red 12" formed porcelain "ATLANTIC" lettering on white rails.

PUMP SIGNS

THE ATLANTIC REFINING COMPANY CURVED 13x13
(4) 1915-1925 $250
Curved white porcelain sign with red/white/black Atlantic crossed arrows logo in center. Narrow irregular red margin and thick black lettering on logo.
THE ATLANTIC REFINING COMPANY CURVED 13x13
(4) 1920-1925 $225
Same as above, with wider irregular border and thinner black lettering on logo.
ATLANTIC DIE CUT CROSSED ARROWS 9.25"
(3) 1925-1932 $400
Red, white, black porcelain diecut Atlantic crossed arrows logo with irregular red border.
ATLANTIC DIE CUT CROSSED ARROWS WHITE OUTLINE 9.25"(4) 1925-1932 $400
Same as above, with white outline around logo. Production variation.
ATLANTIC PANEL 15x41.5
(3) 1932-1953 $100
Atlantic "panels" are combined signs and porcelain lower doors designed to fit Wayne Model #60 pumps. Red porcelain panel with white strip vertically down center & thin white line to either side. Red "ATLANTIC" in deco lettering vertically down strip. Limited use for this panel after 1936.
ETHYL PANEL 15x41.5
(3) 1932-1936 $150
Red porcelain panel with Atlantic logo with lines at top above large Ethyl.
(EGC) logo on white outlined yellow circle. White stripe down from logo.
CAPITOL PANEL 15x41.5
(4) 1932-1935 $150
Red porcelain panel with Atlantic logo with lines at top above black outlined white circle. Red/blue logo in circle with black "CAPITOL" across center. White stripe down from circle.
WHITE FLASH PANEL 15x41.5
(1) 1936-1953 $100
Red porcelain panel with Atlantic logo with lines at top above white outlined White Flash logo. Red/blue logo in circle with white "WHITE/FLASH" across center. White stripe down from circle.
HI-ARC PANEL 15x41.5
(1) 1936-1953 $100
White porcelain panel with Atlantic logo with lines at top. Arched blue "HI-ARC" banner with white lettering below logo. Double red lines above & below banner. Vertical red stripe below.
ATLANTIC WHITE FLASH 13x17
(1) 1939-1953 $85
Small sign identical to White Flash panel, above. No stripe.
ATLANTIC HI ARC 13x17

(1) 1939-1953 $85
Small sign identical to Hi-Arc panel, above. No stripe.
ATLANTIC DIESEL FUEL 13x17
(2) 1939-1966 $75
White porcelain sign with red Atlantic logo with lines at top. Blue "DIESEL/FUEL" below.
ATLANTIC KEROSENE 13x17
(3) 1953-1966 $75
White porcelain sign with red Atlantic logo with lines at top. Blue "KEROSENE" below.
ATLANTIC KEROSENE 13x17
(5) 1939-1953 $100
White porcelain sign with red Atlantic logo with lines at top. Blue "KEROSENE" below with red lines arched above & below.
ATLANTIC 9x13
(1) 1953-1966 $50
Red porcelain pump sign in form of Atlantic logo with lines. Red margin area above & below logo.
ATLANTIC NO MARGINS
(2) 1953-1966 $45
Red porcelain pump sign in form of Atlantic logo with lines. No margin area.
ATLANTIC SPLIT PUMP
(4) 1956-1966 $75
Red porcelain pump sign in form of Atlantic logo with lines. For use on two-product twin pumps.
ATLANTIC PREMIUM 11x13
(1) 1953-1957 $125
White porcelain sign with Atlantic logo hanging from mastarm pole diagonal on sign. Blue "PREMIUM" across bottom.
ATLANTIC IMPERIAL SHIELD 13x17
(1) 1957-1966 $75
Embossed brass die-cut shield with litho Atlantic logo at upper left above blue script "Imperial."
ATLANTIC IMPERIAL SHIELD SPLIT PUMP 9x11.75
(4) 1957-1966 $100
Same as above, for use on two-product twin pumps.
ATLANTIC FLAT 9x13
(1) 1966-1970 $40
Red porcelain sign with red/white/blue Atlantic elongated hexagon logo across center. Margin area above and below logo.
ATLANTIC FLAT NO MARGINS 7.5x13
(1) 1966-1970 $50
Same as above, no margin area.
ATLANTIC FLAT SPLIT PUMP 5.25x9
(5) 1966-1970 $75
Same as above, for use on two-product twin pumps.
ATLANTIC STUDDED 7x13
(1) 1966-1970 $50
Die cut red/white/blue porcelain Atlantic elongated hexagon logo. Studs welded to back for attachment to pump.
ATLANTIC STUDDED SPLIT PUMP 5x9
(4) 1966-1970 $75
Die cut red/white/blue porcelain Atlantic elongated hexagon logo. Studs welded to back for attachment to pump. For use on two-product twin pumps.
ATLANTIC IMPERIAL 13x17
(1) 1966-1970 $75
Same as Imperial, above, with decaled hexagon logo.
ATLANTIC IMPERIAL SPLIT PUMP 9x11.75
(4) 1966-1970 $100
Same as Imperial, above, with decaled hexagon logo.
ATLANTIC IMPERIAL 13x17
(1) 1966-1970 $75
Same as Imperial, above, with embossed litho hexagon logo.
ATLANTIC IMPERIAL SPLIT PUMP 9x11.75
(4) 1966-1970 $100
Same as Imperial, above, with embossed litho hexagon logo.
ATLANTIC DIESEL 10x13
(2) 1966-1970 $50
White porcelain die-cut sign with studs on reverse. Red/white /blue hexagon logo at top, above blue "DIESEL."
ATLANTIC KEROSENE 10x13
(3) 1966-1970 $50
Same as above, "KEROSENE" replacing "DIESEL."

NEON SIGNS

ATLANTIC STRIPED NEON
(5) 1933-1940 $500
Elaborate red porcelain can type sign with multiple white stripes and white "ATLANTIC" across center.

CURB SIGNS

ATLANTIC #1 CURB SIGN 12x19
(5) 1935-1940 $325
Red rectangular porcelain sign in form of Atlantic logo with stripes.
ATLANTIC WHITE FLASH CURB SIGN 30"RD
(5) 1926-1940 $250
Tin white outlined red/blue White Flash logo with white "WHITE/FLASH" across center. White "ATLANTIC" arched to upper left.
ATLANTIC ETHYL CURB SIGN 30"RD
(3) 1926-1936 $375
Blue outlined white porcelain sign. Blue "ATLANTIC" arched around top, "GASOLINE" around bottom with large Ethyl (EGC) logo with yellow rays in center.
ATLANTIC WHITE FLASH PLUS CURB SIGN 30"RD
(4) 1934-1940 $250
Same as White Flash, above, with white "PLUS" added at bottom.

LUBRICATION/MOTOR OIL SIGNS

ASK FOR ATLANTIC MOTOR OIL 14x18
(4) 1926-1930 $100
White/blue outlined white tin litho sign. Red "ATLANTIC" across center above blue "PARAFFIN BASE" and red "MOTOR OIL." Small blue "ASK FOR" at top.
ATLANTIC NC MOTOR OIL STOPS FORD CHATTER INSTANTLY W/PARROT $1800
Self framed black porcelain sign with yellow border. Large yellow/orange/white parrot on white perch at left. White band around parrot's mouth with tag "N.C." off band. Red "ATLANTIC" across top above white "NC." Red "MOTOR OIL" below, above white "STOPS FORD CHATTER INSTANTLY" Beautiful and detailed sign.
ATLANTIC LUBRICATION SERVICE 28x62
(4) 1926-1935 $200
Red outlined white tin litho sign. Blue band across lower sign face with Atlantic crossed arrows logo in center. Red "AT-

Jenney Aero 12"x9" porcelain pump sign. *Dunbar's Gallery.*

Husky 3'x4' porcelain sign. *Dennis Leith.*

Mobilgas 4'x4" porcelain hanging sign. *Jim Hollabaugh.*

Gulf 6" porcelain sign. *Dunbar's Gallery.*

Atlantic Gasoline 48"x36" porcelain sign.

Pennzoil Outboard Motor Fuel 36" porcelain sign. *Jim Hollabaugh.*

Mobilgas 42" very rare porcelain sign with late horse. *Jim Hollabaugh.*

Texaco Kerosene heating fuels metal sign. *Jim Hollabaugh.*

Wasatch 60" porcelain sign. *Jeff Decker.*

Husky 5'x5' die-cut porcelain sign. *Dick Theurer.*

RPM 28" porcelain sign. *Bruce Bednar.*

Oilzum porcelain sign. *Bruce Bednar.*

Miscellaneous pump plate porcelain sign. *Bruce Bednar.*

Phillips 66 - 60" embossed porcelain sign. *Jim Hollabaugh.*

Conoco 48" die-cut porcelain sign. *Jim Hollabaugh.*

LANTIC" above blue "LUBRICATION/SERVICE" across upper sign face.

WE SELL ATLANTIC PARAFFIN BASE MOTOR OIL/AVIATION A PENNSYLVANIA OIL/PHILADELPHIA/PITTSBURGH/FRANKLIN/ PENNSYLVANIA 14x44
(3) 1932-1939 $450
Wood framed tin sign in red, white, & blue with wording as described. Early airplane in center.

WE SELL/ATLANTIC/MOTOR/OIL/AVIATION/HIGH FILM STRENGTH/A PENNSYLVANIA OIL/THE ATLANTIC/REFINING COMPANY/PHILADELPHIA USA 12x38
(3) DATED 11/39 $400
Wood framed tin sign in red, white, and blue with wording as described. Blue wing design in center.

ATLANTIC PARAFFIN BASE MOTOR w/ HEART 26x29
(3) 1926-1934 $250
White tin litho sign with blue band across lower face. Red, white & black Atlantic crossed arrows logo on band with white lettering to right. Large light blue heart above band, with red "ATLANTIC" above blue "PARAFFIN BASE" above red "MOTOR OIL" across heart. Blue "the Life of the Crude" below heart.

ATLANTIC QUALITY LUBRICANTS 26x29
(3) 1934-1946 $175
White tin litho sign with large red "a" in center. Red "Atlantic" across top, with blue "QUALITY/Lubricants" across "a." Blue "KEEPS UPKEEP DOWN" across bottom.

ATLANTIC QUALITY A MOTOR OIL CAN 13x17
(4) 1946-1950 $200
Porcelain die-cut sign in shape of oil can. White "ATLANTIC" in red upper area, with large "a" in center with white "MOTOR OIL" in blue band around bottom.

LUBESTER SIGNS

ATLANTIC PARAFFIN BASE MOTOR OIL EXTRA HEAVY 6x8
(5) 1920s $200
Blue outlined white tin litho sign with red and white Atlantic seal logo on blue band across bottom. Red "ATLANTIC" across top above blue "PARAFFIN BASE" and red "MOTOR OIL" Blue "EXTRA HEAVY" below, just above band. Small white lettering on band.

HEAVY ATLANTIC MOTOR OIL 6x8
(5) 1920s $200
Blue outlined white tin litho sign with red and white Atlantic seal logo on blue band across bottom. Red "ATLANTIC" across top above blue "PARAFFIN BASE" and red "MOTOR OIL" Blue "EXTRA HEAVY" below, just above band. Small white lettering on band.

AVIATION ATLANTIC MOTOR OIL 6x8
(5) 1920s $200
Blue outlined white tin litho sign with red and white Atlantic seal logo on blue band across bottom, with small white lettering on band. Red "ATLANTIC" across top above blue "PARAFFIN BASE" and red "MOTOR OIL" Blue "AVIATION/(MEDIUM)" above ATLANTIC" at top.

OIL RACK SIGNS

ATLANTIC QUALITY LUBRICANTS 12x22.5
(5) 1930s $135
White tin litho sign with large red "a" in center. Red "Atlantic" across top with blue "Quality/Lubricants" across center. Blue "KEEP UPKEEP DOWN" across bottom.

ATLANTIC PARAFFIN BASE AVIATION 12x22.5
(5) 1930s $275
White tin litho sign with "ATLANTIC PARAFFIN/BASE MOTOR OIL" across top with blue biplane circling just above blue "AVIATION" across center. "A PENNSYLVANIA OIL" across bottom.

ATLANTIC/ATLANTIC AVIATION MOTOR OIL 10.5x18.5
(3) 1950-1957 $50
Reverse is blue outlined white tin litho with blue "wing" in center above blue "MOTOR OIL." Red "ATLANTIC" in upper left corner, with white "AVIATION" across wing and red "KEEPS/ENGINES/CLEAN" to right of wing. Dated 6/50.

ATLANTIC PREMIUM MOTOR OIL 10x17.5
(3) 1957-1963 $50
Blue tin litho sign with red/gold/blue Atlantic Premium Motor Oil can at left. White "OUR/FINEST/MOTOR/OIL" to right of can.

ATLANTIC IMPERIAL MOTOR OIL 10x17.5
(3) 1963-1966 $35
White tin litho sign with red vertical band on left side. Blue/white/gold Atlantic Imperial Motor Oil can with old logo diagonal on red area. Blue "IN A/CLASS/BY/ITSELF" on white area to right. Dated 1963.

ATLANTIC IMPERIAL MOTOR OIL 10x17.5
(2) 1966-1970 $35
White tin litho sign with red vertical band on left side. Blue/white/gold Atlantic Imperial Motor Oil can with old logo diagonal on red area. Blue "IN A/CLASS/BY/ITSELF" on white area to right. Dated 3/66.

FLANGE SIGNS

ATLANTIC GASOLINE/PUTS PEP IN YOUR MOTOR 17.5x21.5
(5) 1915-1920 $225
White outlined blue tin sign with red "ATLANTIC/GASOLINE" across upper area. White "PUTS PEP IN YOUR MOTOR/SOLD HERE" below.

ATLANTIC MOTOR OILS/KEEP UPKEEP DOWN 17.5x21.5
(5) 1915-1920 $225
White outlined blue tin sign with red "ATLANTIC/MOTOR OILS" across upper area. White "KEEP UPKEEP DOWN/SOLD HERE" below.

SECONDARY SIGNS

ATLANTIC WHITE FLASH 32x76
(5) 1926-1930 $275
Horizontal rectangle wood framed white tin sign. Blue border with blue outlined red "ATLANTIC" above blue "WHITE FLASH" across sign face.

ATLANTIC HI-ARC 36x60
(5) DATED 5/53 $250
Horizontal rectangle wood framed white tin sign. Blue banner across center with large white "HI-ARC" on banner. Red banner arched above at left with white "ATLANTIC." Smaller red banner below.

ATLANTIC CERTIFIED LUBRICATION 20x20
(2) 1953-1960 $125
Square tin litho sign. Red outline with blue pentagon in center. Red Atlantic logo and star in pentagon. Blue "Certified" at top, "Lubrication" below.

ATLANTIC CREDIT CARDS HONORED 10x15
(3) 1950-1960 $100

Tin litho sign with red upper band and blue lower band. White "ATLANTIC" in red area, white "CREDIT CARDS/HONORED HERE" in blue area.

BULK PLANT SIGNAGE

ATLANTIC HEATING OILS 36x36
(4) 1946-1966 $300
Wood framed tin litho sign. Red Atlantic logo at top, with blue area below. White "HEATING OILS" above white line on lower blue area.

ATLANTIC HEATING OILS 36x52
(4) 1946-1966 $300
Same as above, porcelain.

ATLANTIC OIL HEAT
(4) 1966-1970 $100
White porcelain sign with red/white/blue elongated hexagon logo at top above blue "OIL HEAT."

ATLANTIC DIESEL
(5) 1966-1970 $100
Same as above, "DIESEL" replacing "OIL HEAT."

CHAMPLIN OIL AND REFINING - ENID, OK

Champlin Oil and Refining was an Enid Oklahoma based refiner/marketer founded in 1920. Marketing was concentrated throughout the mid-continent, eventually expanding to 15 states from Tennessee to Colorado. Champlin had long been involved in unbranded marketing and with the purchase of Champlin by the government owned oil company from Venezuela, the decision was made to withdraw from branded marketing. Retail marketing was sold off to American Petrofina and the Fina brand replaced Champlin in 1984. The production and refining divisions survive as an Irving, Texas based subsidary of Citgo Petroleum which is also owned by the Venezuelan government.

IDENTIFICATION SIGNS

CHAMPLIN GASOLINE USE CHAMPLIN OILS 36x36
(4) 1930-1940 $350
White porcelain sign with Champlin cross logo in center. White "CHAMPLIN" on blue notched band across center, with white "USE" above and "OILS" below. Red outlined blue "CHAMPLIN" above logo, "GASOLINE" below.

CHAMPLIN "USE CHAMPLIN OILS" 48x48
(3) 1935-1955 $350
Steel framed white porcelain sign with large Champlin cross logo covering most of sign. White "CHAMPLIN" in notched blue band across center, with white "USE" above and "OILS" below.

CHAMPLIN CROSS & OVAL 72"RD
(3) 1955-1962 $100
Pole framed white porcelain sign with large Champlin logo covering most of sign. White "CHAMPLIN" in rounded blue band across center.

PUMP SIGNS:

CHAMPLIN REGULAR 6x10
(3) 1962-1965 $75
Blue outlined white porcelain die-cut rounded rectangle. Thin blue vertical stripes on white area, with blue "CHAMPLIN" over red "REGULAR."

REGULAR SHIELD 6x10
(2) 1965-1970 $50
Same as above, with red "REGULAR" across center.

DeLUXE SHIELD 6x10
(2) 1965-1970 $50
Same as above, with red outline and vertical lines and blue "DeLuxe" across center.

CURB SIGNS:

CHAMPLIN GASOLINE 30"RD
(3) 1930s $350
White porcelain sign with small Champlin cross logo covering most of sign. White "CHAMPLIN" in notched blue band across center, with white "USE" above and "OILS" below. Red outlined black "CHAMPLIN" arched around top, "GASOLINE" around bottom.

CHAMPLIN MOTOR OILS 30"RD
(3) 1930s $350
White porcelain sign with small Champlin cross logo covering most of sign. White "CHAMPLIN" in notched blue band across center, with white "USE" above and "OILS" below. Red outlined black "CHAMPLIN" arched around top, "MOTOR OILS" around bottom.

CHAMPLIN ETHYL GASOLINE 30"RD
(4) 1926-1940 $375
White porcelain sign with red outline ring and red ring around center circle. Blue "CHAMPLIN" arched around top, "GASOLINE" around bottom, with large Ethyl (EGC) logo with yellow rays in center.

CHAMPLIN HI-V.I. 20x32
(3) 1930's $250
Yellow porcelain rectangular sign with rounded corners. Large red "HI-V.I." across center, with blue "CHAMPLIN" across top and blue script "Motor Oil" below. Red "ON THE GROUND OR IN THE SKY" across bottom.

SECONDARY SIGNS:

USE CHAMPLIN OILS SMOOTH DEPENDABLE POWER 12x18
(5) 1920-1930 $175
Yellow embossed tin with Champlin cross logo in center. Red outlined blue "CHAMPLIN" arched across top, "GASOLINE" across bottom, with red "SMOOTH/DEPENDABLE/POWER" to left and right of logo.

ALL CREDIT CARDS HONORED 18x30
(4) 1962-1970 $100
Red outlined white porcelain die cut rounded rectangle. Blue "CREDIT CARDS" across center, with blue "ALL APPROVED" across top, "HONORED" across bottom.

CLEAN REST ROOMS 18x30
(4) 1962-1970 $100
Blue outlined white porcelain die-cut rounded rectangle. Blue "RESTROOMS" across center, with red "CLEAN" at top, "HOT WATER" below.

MEN 6x10
(3) 1962-1970 $40
Flanged blue outlined white porcelain die-cut rounded rectangle. Thin blue vertical stripes on white area, with blue "MEN" across center.

WOMEN 6x10
(3) 1962-1970 $40
Flanged blue outlined white porcelain die-cut rounded rec-

tangle. Thin blue vertical stripes on white area, with blue "WOMEN" across center.

HARBOR - TULSA, OK

Harbor is a Tulsa based Champlin secondary brand from the 1960s.
No collectible size Harbor signs known to exist.

EASON OIL CO. - ENID, OK

Eason Oil Co was an Oklahoma based refiner marketer that was one of the early participants in the Sovereign Service marketing program. By the early 1950s they were based in Enid, OK and eventually merged with Champlin in the mid 1950s.

IDENTIFICATION SIGNS

EASON GASOLINE 30"RD
(5) 1930s $400
Green/white outlined green porcelain sign with white circle in center. White "EASON" arched around top, "GASOLINE" around bottom, with large green outlined red "E" in center.

CURB SIGNS

EASON ETHYL 30"RD
(5) 1930s $425
Red/white outlined red porcelain sign with white circle in center. White "EASON" arched around top, "GASOLINE" around bottom, with large Ethyl (EGC) logo with yellow rays in center.

CHEVRON AND AFFILIATES STANDARD OIL CO. OF CALIFORNIA

Standard Oil of California, known in the industry as "Socal," is the west's largest gasoline marketer. Tracing it's orgins to the 1879 founding of the Pacific Coast Oil Company. Standard Oil, which had maintained a San Francisco marketing office since 1879 purchased Pacific Coast Oil Co in 1900 and reorganized it as Standard Oil of California.
Standard of California is credited with one of the first retail gasoline stations, opened at the company's bulk plant in Seattle, Washington in 1907. Always an innovator, Standard was soon operating hundreds of retail stations throughout the west. By the mid 1930s company stations were under the brand "Standard Stations Inc." and dealer operated locations were "Standard Oil Dealers." In 1945 the Chevron name was introduced for Standard gasolines and independent dealers in the Standard western territory, Standard Texas southern territory and Calso territory became "Chevron Gas Stations," while company owned and operated stations remained "Standard" stations, selling Chevron and Chevron Supreme gasolines. In 1961 Standard of California bought Standard of Kentucky and from 1961 on Chevron gasolines were sold at Standard Oil stations in the southeast. About 1977 the Standard name came down from all but a very small number of service stations and was replaced by the Chevron name. Standard of Kentucky globes can be found under "Standard Oil" in this book.

STANDARD STATIONS, INC. AND STANDARD OIL/CHEVRON DEALERS

IDENTIFICATION SIGNS:

STANDARD OIL DEALER "CLOUD" 30x60
(4) 1930-1939 $800
White outlined blue cloud shaped sign with white black "STANDARD OIL" above white "Dealer" across sign face.
STANDARD OIL PRODUCTS "CLOUD" 30x60
(4) 1930-1939 $800
White outlined blue cloud shaped sign with white black "STANDARD OIL" above white "Products" across sign face.
STANDARD STATIONS INC HALLMARK 30x60
(4) 1939-1957 $350
Formed porcelain faces assembled back to back over frame can-style. Each face is die-cut hallmark with blue upper area and blue/white/red chevron stripes below. White "STANDARD/STATIONS/INC." in upper blue field.
CHEVRON GAS STATIONS CAN 30x72
(4) 1946-1955 $300
Green/white outlined maroon porcelain sign face with green outlined "CHEVRON"
(Winged "V") above green outlined white "GAS STATION."
CHEVRON GAS STATIONS CAN TYPE 60x108
(4) 1946-1955 $300
Formed porcelain faces assembled back to back over frame can-style, with separate end caps. Green/white outlined maroon porcelain sign face with green outlined "CHEVRON" (Winged "V") above green outlined white "GAS STATION."
CHEVRON MAROON HALLMARK FLAT 48x72
(5) 1955-1970 $350
Die-cut hallmark with maroon upper area and blue/white/red chevron stripes below. White "CHEVRON/DEALER" in upper maroon field.
CHEVRON MAROON HALLMARK CAN TYPE 48x72
(5) 1955-1970 $550
Sectional formed can-type porcelain sign. Lower section is porcelain chevron stripes blue/white/red. Upper half is maroon field with white "CHEVRON/DEALER" across face and white chevron line across bottom. Two halves join in center to create Chevron hallmark logo.
CHEVRON BLUE HALLMARK 48x72
(5) 1955-1970 $350
Die-cut porcelain hallmark with blue upper area and blue/white/red chevron stripes below. White "CHEVRON/DEALER" in upper blue field.
CHEVRON BLUE HALLMARK CAN TYPE 48x72
(5) 1955-1970 $550
Sectional formed can-type porcelain sign. Lower section is porcelain chevron stripes blue/white/red. Upper half is blue field with white "CHEVRON/DEALER" across face and white chevron line across bottom. Two halves join in center to create Chevron hallmark logo.
CHEVRON 24X26
(5) 1970-PRESENT $175
Extremely late porcelain sign, used in parking garage gas station settings in the 1970's. White porcelain die-cut hallmark with blue and red chevron stripes. Black "Chevron" across top of sign.

BUILDING SIGNS & LETTERS

STANDARD OIL PRODUCTS 12x288

(4) 1946-1957 $200
Formed maroon porcelain panel sign. Green outlined cream "STANDARD OIL PRODUCTS" across sign.
CHEVRON DEALER 12x144
(4) 1946-1957 $200
Formed maroon porcelain panel sign. Green outlined cream "CHEVRON DEALER" across sign.
ATLAS TIRES 12x144
(4) 1946-1957 $75
Formed maroon porcelain panel sign. Green outlined cream "ATLAS TIRES" across sign.
TIRES & BATTERIES 12x144
(4) 1946-1957 $75
Formed maroon porcelain panel sign. Green outlined cream "TIRES & BATTERIES" across sign.
LUBRICATION 12x144
(4) 1946-1957 $75
Formed maroon porcelain panel sign. Green outlined cream "LUBRICATION" across sign.
PRIVILEDGE PANELS 12x144
(4) 1946-1957 $100
Formed maroon porcelain panel sign. Green outlined cream custom dealer name or station name across sign.

PUMP SIGNS:
RED CROWN ATTACHMENT SIGN 14"
(3)1915-1930 $300
Blue outlined white porcelain sign with blue outlined white burse forming center circle. Red and blue crown with white lines in center. Blue outlined red "RED CROWN" arched around top, "GASOLINE" around bottom. White mounting tab for attachment to gas pump cylinder rods.
RED CROWN ATTACHMENT SIGN 10"
(3)1915-1930 $300
Blue outlined white porcelain sign with blue outlined white burst forming center circle. Red and blue crown with white lines in center. Blue outlined red "RED CROWN" arched around top, "GASOLINE" around bottom. White mounting tab for attachment to gas pump cylinder rods.
CHEVRON GASOLINE 11"RD
(4) 1946-1950 $165
Blue porcelain sign with white band across center. Blue outlined red "CHEVRON"
(winged "V") on white band over red "GASOLINE" on white center band.
CHEVRON SUPREME GASOLINE 11"RD
(4) 1946-1950 $165
Blue outlined white porcelain sign with blue outlined red "CHEVRON"
(winged "V") over over blue "SUPREME" over red "GASOLINE" across sign face.
CHEVRON GASOLINE 11"RD
(4) 1950-1955 $165
Blue porcelain sign with white band across center. Red "CHEVRON"
(winged "V") on white band over red "GASOLINE" on white center band.
CHEVRON SUPREME GASOLINE 11"RD
(4) 1950-1955 $165
Blue outlined white porcelain sign with red "CHEVRON" (winged "V") over over blue "SUPREME" over red "GASOLINE" across sign face.
CHEVRON W/HALLMARK BLUE 11x13
(4) 1955-1962 $165
Blue/white outlined blue porcelain sign with white band across center and blue/white/red hallmark logo projecting from bottom of sign. Red "CHEVRON"
(winged "V") over blue "GASOLINE" on white center band.
CHEVRON W/HALLMARK WHITE 11x13
(5) 1955-1962 $150
Blue outlined white porcelain sign with blue/white/red hallmark logo projecting from bottom of sign. Red "CHEVRON" (winged "V") over blue "GASOLINE" across sign face.
CHEVRON SUPREME W/HALLMARK 11x13
(4) 1955-1962 $165
Blue outlined white porcelain sign with blue/white/red hallmark logo projecting from bottom of sign. Red "CHEVRON" (winged "V") over blue "SUPREME" over red "GASOLINE" across sign face.
CHEVRON DIESEL FUEL 11x13
(5) 1955-1962 $165
Blue porcelain sign with white band across center and blue/white/red hallmark logo projecting from bottom of sign. Red "CHEVRON" (winged "V") over blue "DIESEL FUEL" on white center band.
CHEVRON AVIATION 11x13
(5) 1955-1962 $350
Blue outlined white porcelain sign with blue/white/red hallmark logo projecting from bottom of sign. Red "CHEVRON" (winged "V") over blue "AVIATION" over red "GASOLINE" across sign face.
CHEVRON W/HALLMARK WHITE 11x13
(4) 1962-1970 $100
Blue outlined white tin litho sign with blue/white/red hallmark logo projecting from bottom of sign. Red "CHEVRON" (winged "V") over blue "GASOLINE" across sign face.
CHEVRON SUPREME W/HALLMARK 11x13
(4) 1962-1970 $100
Blue outlined white tin litho sign with blue/white/red hallmark logo projecting from bottom of sign. Red "CHEVRON" (winged "V") over blue "SUPREME" over red "GASOLINE" across sign face.

NEON
STANDARD OIL PRODUCTS 30x60
(5) 1930-1939 $1000
Formed porcelain faces assembled back to back over frame can-style. White outlined blue cloud shaped sign with white black "STANDARD OIL" above white "Products" across sign face.
STANDARD STATIONS, INC. HALLMARK 30x60
(4) 1939-1955 $500
Formed porcelain faces assembled back to back over frame can-style. Each face is die-cut hallmark with blue upper area and blue/white/red chevron stripes below. White "STANDARD/STATIONS/INC." in upper blue field. Neon stripes.
CHEVRON GAS STATIONS 60x108
(4) 1946-1955 $500
Formed porcelain faces assembled back to back over frame

can-style, with separate end caps. Green/white outlined maroon porcelain sign face with green outlined "CHEVRON" (Winged "V") above green outlined white "GAS STATION." Neon "CHEVRON" and "GAS STATION"
CHEVRON MAROON HALLMARK CAN TYPE 48x72
(5) 1955-1970 $300
Sectional formed can-type porcelain sign. Lower section is porcelain chevron stripes blue/white/red. Upper half is maroon field with white "CHEVRON/DEALER" across face and white chevron line across bottom. Two halves join in center to create Chevron hallmark logo. Neon chevron stripes.
CHEVRON BLUE HALLMARK CAN TYPE 48x72
(5) 1955-1970 $500
Sectional formed can-type porcelain sign. Lower section is porcelain chevron stripes blue/white/red. Upper half is blue field with white "CHEVRON/DEALER" across face and white chevron line across bottom. Two halves join in center to create Chevron hallmark logo. Neon Chevron stripes.

CURB SIGNS:

RED CROWN GASOLINE 30"RD
(3) 1915-1930 $850
Blue outlined white porcelain sign with blue outlined white burst forming center circle. Red and blue crown with white lines in center. Blue outlined red "RED CROWN" arched around top, "GASOLINE" around bottom. Small white "PATENDED OCT 12 1915" at bottom of blue area.
ZEROLENE 24"RD
(5) 1920-1935 $1400
Red/blue outlined white tin litho sign with blue center circle and blue band across center. White polar bear with blue details in center circle below band. White "ZEROLENE" across band with blue "The Standard Oil" arched around top, "For Motor Cars" around bottom.
ZEROLENE 30"RD
(5) 1920-1935 $1400
Red/blue outlined white porcelain sign with blue center circle and blue band across center. White polar bear with blue details in center circle below band. White "ZEROLENE" across band with blue "The Standard Oil" arched around top, "For Motor Cars" around bottom.
RED CROWN GASOLINE ZEROLENE 30"RD
(5) 1929-1935 $1000
Blue outlined white porcelain sign with blue outlined band across center over blue center circle. Red and blue crown with white lines above band, blue and white polar bear below. Blue outlined red "RED CROWN" arched around top, "GASOLINE" around bottom. White "POLARINE" across center band..
STANDARD ETHYL GASOLINE 21x28
(4) 1930-1939 $600
Blue porcelain die-cut hallmark shaped sign with red band over white band across top. White "STANDARD/ ETHYL/ GASOLINE" across upper blue area with Ethyl (EGC) logo in white circle at bottom
STANDARD UNSURPASSED MICKEY MOUSE 24"RD
(4) 1940-1942 $500
Blue tin litho sign with multi-color Mickey Mouse holding sign with blue "For Extra Service" across top, red "STANDARD/ GASOLINE" across center and blue "Unsurpassed" across bottom.
RPM MOTOR OIL DONALD DUCK 24"RD
(4) 1940-1942 $500
Blue tin litho sign with multi-color Donald Duck and snowman scene. Blue "A KNOCKOUT/FOR WINTER" across top, RPM spinner with blue outlined white "RPM/MOTOR OIL" at bottom.
CHEVRON GASOLINE 30"RD
(4) 1946-1950 $500
Blue outlined white porcelain sign with blue outlined red "CHEVRON" (winged "V") over over blue outlined over red "GASOLINE" across sign face.
CHEVRON SUPREME GASOLINE 30"RD
(4) 1946-1950 $500
Blue outlined white porcelain sign with blue outlined red "CHEVRON" (winged "V") over over blue "SUPREME" over red "GASOLINE" across sign face.

LUBRICATION/MOTOR OIL SIGNS:

USE ZEROLENE FOR CORRECT LUBRICATION 14x28
(4) 1920s $175
White porcelain sign with red/blue outlined white circular sign with blue center circle and blue band across center. White polar bear with blue details in center circle below band. White "ZEROLENE" across band with blue "The Standard Oil" arched around top, "For Motor Cars" around bottom, at right. Blue "Use" at upper left and blue "the CORRECT LUBRICATION."
THE NEW ZEROLENE DIAMOND 27x27
(4) 1920-1925 $150
White rounded corner diamond shaped wgn with blue and white polarine
(with bear) logo at bottom of sign. Blue "THE NEW/AEROLENE" across center with blue "Money/can't buy a/better oil than" at top.
STANDARD OIL CO OF CAL-ZEROLENE 12"RD
(5) 1920s $225
Blue/red outlined white porcelain sign with blue "STANDARD OIL COMPANY" arched aorund top, "OF CALIFORNIA" around bottom with Red/blue outlined white circle with blue center circle and blue band across center. White polar bear with blue details in center circle below band. White "ZEROLENE" across band with blue "The Standard Oil" arched around top, "For Motor Cars" around bottom.
ZEROLENE 10"
(4) 1920s $185
Red/blue outlined white porcelain sign with blue center circle and blue band across center. White polar bear with blue details in center circle below band. White "ZEROLENE" across band with blue "The Standard Oil" arched around top, "For Motor Cars" around bottom.
MODERN CRANKCASE CLEANING SERVICE 14x22
(5) $350
Die-cut shield shaped porcelain sign with light blue outline around dark blue center. Red "MODERN/CRANKCASE/CLEANING/SERVICE" on upper sign above Calol and Zerolene logos. Smaller red lettering below.

RPM THERMO-CHARGED MOTOR OIL 24"RD
(3) 1936-1945 $250
Blue background sign with red "rotation" logo in center. Large

white "RPM" with blue black dropshading diagonal across "rotation" above "MOTOR OIL." White script "Thermo-Charged" across top, "100% PURE PARAFFIN BASE" across bottom.

RPM MOTOR OIL 34"RD
(4) 1945-1947 $200
Thick blue outline around white porcelain sign with red "rotation" logo in center. White "RPM" across center with blue dropshading. Blue "MOTOR OIL" arched around bottom.

RPM HEAVY DUTY MOTOR OIL 34"RD
(5) 1947-1952 $200
Thick blue outline around white porcelain sign with red "rotation" logo in center. White "RPM" across center with blue dropshading. Blue/white outlined blue square at upper left with white "HEAVY/DUTY" in square. Blue "MOTOR OIL" arched around bottom.

RPM MOTOR OIL 28"RD
(2) 1952-1955 $175
Blue outlined white porcelain sign with red "rotation" logo in center. White "RPM" across center with blue dropshading. Blue "100% PURE PARAFFIN BASE" arched around top, "MOTOR OIL" around bottom.

RPM MOTOR OILS/AND LUBRICANTS 28"RD
(2) 1952-1955 $175
Blue outlined white porcelain sign with red "rotation" logo in center. White "RPM" across center with blue dropshading. Blue "MOTOR OILS" arched around top, "AND LUBRICANTS" around bottom.

MOTOR OIL RPM W/ HALLMARK 30"
(3) 1955-1970 $100
Blue outlined white tin litho sign with red "rotation" logo in center. White "RPM" across center with blue dropshading. Blue "MOTOR OILS & LUBRICANTS" arched around top, with small blue/white/red hallmark logo projecting from center bottom.

FLANGE SIGNS

RED CROWN GASOLINE FLANGED 24"RD
(4) 1915-1925 $850
Blue outlined white porcelain sign with blue outlined white burse forming center circle. Red and blue crown with white lines in center. Blue outlined red "RED CROWN" arched around top, "GASOLINE" around bottom.

PERFECTION KEROSENE FLANGED 12x18
(5) 1915-1925 $500
White/red outlined white tin litho flange sign. Red "PERFECTION/KEROSENE" across upper sign face with blue oil lamp in center. Red flame and rays out from lamp. Blue "MADE" to left of lamp, "BY THE" to right with "STANDARD OIL COMPANY" across bottom.

PEARL OIL FLANGE 12x19
(5) 1915-1920 $550
Die cut gray porcelain in the shape of a kerosene carry can with bail handle. Red area across most of can with with white "PEARL/OIL/(KEROSENE)" across red area. Blue "for Oil Burners/Cookstoves/and Lamps" below.

PEARL OIL FLANGE 12x19
(4) 1920-1925 $550
Die cut gray porcelain in the shape of a kerosene carry can with bail handle. Blue area across most of can with red oval at top. Blue "WE SELL" in oval with white "PEARL/OIL" across upper sign above white/red/white lines. White "KEROSENE" above grey "for HEATING/LIGHTING/COOKING/STANDARD OIL COMPANY OF CALIFORNIA" below white lines.

MODERN CRANKCASE CLEANING SERVICE 14x22
(5) 1920s $450
Die-cut shield shaped porcelain sign with light blue outline around dark blue center. Red "MODERN/CRANKCASE/CLEANING/SERVICE" on upper sign above Calol and Zerolene logos. Smaller red lettering below. White flange area.

ZEROLENE OILS & GREASES 20x23
(5) 1920S $1400
Irregular shaped die-cut white outlined blue porcelain flange sign. White die-cut polar bear on top edge of sign with blue details. White "ZEROLENE/Oils & Greases" across sign face.

ZEROLENE FOR MOTOR CARS 22"RD
(5) 1920S $750
Red/blue outlined white porcelain sign with blue center circle and blue band across center. White polar bear with blue details in center circle below band. White "ZEROLENE" across band with blue "The Standard Oil" arched around top, "For Motor Cars" around bottom. White flange area.

ZEROLENE FOR MOTOR BOATS 22"RD
(5) 1920S $750
Red/blue outlined white porcelain sign with blue center circle and blue band across center. White polar bear with blue details in center circle below band. White "ZEROLENE" across band with blue "The Standard Oil" arched around top, "For Motor Boats" around bottom. White flange area.

RPM THERMO-CHARGED MOTOR OIL FLANGE 22"RD
(3) 1936-1945 $275
Blue background sign with red "rotation" logo in center. Large white "RPM" with blue black dropshading diagonal across "rotation" above "MOTOR OIL." White script "Thermo-Charged" across top, "100% PURE PARAFFIN BASE" across bottom.

SECONDARY SIGNS:

RED CROWN GASOLINE 10x32
(5) 1915-1920 $300
Black porcelain sign with white outlined red "RED CROWN" over white "THE GASOLINE OF QUALITY" across sign.

ZEROLENE 10x32
(5) 1915-1920 $300
Black porcelain sign with white outlined red "ZEROLENE" over white "THE STANDARD OIL FOR MOTOR CARS" across sign.

STANDARD OIL COMPANY 12x18
(5) 1915-1920 $300
Black porcelain sign with white outlined red "STANDARD/OIL/COMPANY" across sign.

RED CROWN GASOLINE 18x120
(5) 1915-1920 $300
Black porcelain sign with white outlined red "RED CROWN" over white "THE GASOLINE OF QUALITY" across sign.

RED CROWN GASOLINE SCRIPT ACCEPTED 14x14
(4) 1925-1930 $300
Keyhole shaped porcelain sign with red/white/blue Red Crown logo at top over blue outlined white panel with blue band across center. Red "SCRIPT" above band, "ACCEPTED" below with white "STANDARD OIL COMPANY/(CALIFOR-

NIA)" on center blue band. White tab for attachment to gas pump cylinder rods.

STANDARD OIL CREDIT CARDS 24x24
(4) 1935-1946 $175
Red/blue outlined white diagonal oval with blue "CREDIT CARDS" across center. Red script "Standard Oil" across top, "Honored here" across bottom

CHEVRON NATIONAL CREDIT CARDS ACCEPTED 18x24
(4) 1946-1950 $250
Cream porcelain sign with maroon outline across top and bottom. Green outlined maroon "CHEVRON" (winged "V") across top above maroon "NATIONAL/CREDIT CARDS/HONORED"

CHEVRON CREDIT CARDS ACCEPTED 30"RD
(3) 1950-1962 $200
Blue porcelain sign with red bands across top and bottom. White "CHEVRON" (winged "V") across top with white "CREDIT CARDS/HONORED" below.

BULK PLANT SIGNAGE

CHEVRON GASOLINES 48x58
(4) 1955-1965 $500
Blue outlined white embossed porcelain sign with blue/white/red hallmark logo extending from bottom center. Red "CHEVRON" with winged "V" across top over blue "GASOLINES."

RPM MOTOR OILS 30x72
(2) 1960-1970 $100
Aluminum framed blue/red outlined horizontal tin litho sign with blue/white outlined red RPM "rotation" logo at left. Blue "RPM/MOTOR/OILS" to right.

RPM MOTOR OILS 30x72
(2) 1960-1970 $100
Aluminum framed blue/red outlined vertical tin litho sign with blue/white outlined red RPM "rotation" logo at top over blue "RPM/MOTOR/OILS"

CHEVRON AND RPM MOTOR OILS 30x72
(2) 1970-1978 $75
Self framed tin litho sign with white upper half and light blue lower half. Red/blue/black Chevron hallmark in upper white area with white "CHEVRON/and/RPM/Motor Oils" in lower blue area.

TRUCK DOOR SIGNS:

STANDARD OIL PRODUCTS 4x8
(4) 1930-1946 $350
Red/white outlined blue cloud shaped sign with white black "STANDARD OIL" above white "Products" across sign face.

STANDARD OIL PRODUCTS 5x10
(4) 1930-1946 $350
Red/white outlined blue cloud shaped sign with white black "STANDARD OIL" above white "Products" across sign face.

STANDARD OIL PRODUCTS 6x12
(4) 1930-1946 $350
Red/white outlined blue cloud shaped sign with white black "STANDARD OIL" above white "Products" across sign face.

CHEVRON 6"RD
(4) 1946-1960 $100
White outlined blue porcelain sign with blue circle in center with white band across. Blue outlined red "CHEVRON" (winged "V") on white band over hallmark logo.

CHEVRON 6"RD
(4) 1960-1970 $75
Blue outlined white porcelain sign with blue "CHEVRON" across top of sign above blue/white/red hallmark logo.

STANDARD BURNER OILS
(4) 1930S $250
Irregular shaped die-cut porcelain sign. Red/white outlined notched end blue band across center with red triangle top and bottom. White "STANDARD/BURNER OILS" across band.

STANDARD HEATING OILS
(4) 1940S $250
Irregular shaped die-cut porcelain sign. Red/white outlined notched end blue band across center with red triangle top and bottom. White "STANDARD/HEATING OILS" across band.

STANDARD HEATING OILS 11x13
(2) 1946-1970
Blue/white outlined red porcelain sign with white band across center and blue/white/red hallmark logo projecting from bottom of sign. Blue "STANDARD" over blue "HEATING OILS" on white center band.

TRUCK TANK SIGNS

STANDARD OIL PRODUCTS 12x18
(4) 1930-1946 $300
Red/white outlined blue cloud shaped sign with white "STANDARD OIL" above white "Products" across sign face.

STANDARD HEATING OILS 15x24
(4) 1946-1950 $200
Irregular shaped die-cut porcelain sign. Red/white outlined notched end blue band across center with red triangle top and bottom. White "STANDARD/HEATING OILS" across band.

STANDARD HEATING OILS
(3) 1950S $200
Blue/white outlined red porcelain sign with white band across center and blue/white/red hallmark logo projecting from bottom of sign. Blue "STANDARD" over blue "HEATING OILS" on white center band.

CHEVRON AVIATION GASOLINE
(4) 1950S $450
Blue outlined white porcelain sign with blue/white/red hallmark logo projecting from bottom of sign. Red "CHEVRON" (winged "V") over blue "AVIATION" over red "GASOLINE" across sign face.

THE CALIFORNIA COMPANY - ROCKY MTN. MARKETING

In the late 1930s Standard of California expanded it's marketing into the Rocky Mountain area. Not being able to use the Standard name in this area due to its use by Standard of Indiana, gasoline was marketed under the "Calso" (California Standard Oil) name. In 1946 the Calso stations, all dealer/jobber operated, were rebranded with the new "Chevron Gas Station" logo. Most of the Chevron signage that is listed above was used at all Standard Stations in the west and Calso stations in the Rocky Mountain area. Dealer operated stations in the western region also used the "Chevron Gas Station" signage.

IDENTIFICATION SIGNS

CHEVRON GAS STATIONS CAN TYPE 60x108

(4) 1946-1955 $300

Formed porcelain faces assembled back to back over frame can-style, with separate end caps. Green/white outlined maroon porcelain sign face with green outlined "CHEVRON" (Winged "V") above green outlined white "GAS STATION."

BUILDING SIGNS AND LETTERS

CHEVRON DEALER 12x144

(4) 1946-1955 $200

Formed maroon porcelain panel sign. Green outlined cream "CHEVRON DEALER" across sign.

ATLAS TIRES 12x144

(4) 1946-1955 $75

Formed maroon porcelain panel sign. Green outlined cream "ATLAS TIRES" across sign.

TIRES & BATTERIES 12x144

(4) 1946-1955 $75

Formed maroon porcelain panel sign. Green outlined cream "TIRES & BATTERIES" across sign.

LUBRICATION 12x144

(4) 1946-1955 $75

Formed maroon porcelain panel sign. Green outlined cream "LUBRICATION" across sign.

PRIVILEDGE PANELS 12x144

(4) 1946-1955 $100

Formed maroon porcelain panel sign. Green outlined cream custom dealer name or station name across sign.

PUMP SIGNS

CHEVRON NATIONAL CREDIT CARDS ACCEPTED 18x24

(4) 1946-1950 $250

Cream porcelain sign with maroon outline across top and bottom. Green outlined maroon "CHEVRON" (winged "V") across top above maroon "NATIONAL/CREDIT CARDS/HONORED"

CALIFORNIA OIL COMPANY - NORTHEAST MARKETING

In 1945-46 Standard of California purchased New York Lubricating Oil Co. and set up "The California Oil Company" in Perth Amboy, NJ as an East Coast marketing subsidary. Stations were branded Calso, same as in the Rockies, and were found from Maine to Virginia. In 1957 Calso Stations were rebranded Chevron.

IDENTIFICATION SIGNS

CALSO SUPREME BUTTON 60"RD

(4) 1949-1957 $325

Formed domed red porcelain sign halves assembled back-to-back. Large blue outlined white "CALSO" over smaller white "GASOLINE" across sign face.

CHEVRON BLUE HALLMARK 48x72

(5) 1957-1970 $300

Formed porcelain faces assembled back to back over frame can-style. Each face is die-cut porcelain hallmark with blue upper area and blue/white/red chevron stripes below. White "CHEVRON" in upper blue field.

PUMP SIGNS

CALSO GASOLINE 11"RD

(3) 1949-1955 $185

White outlined red porcelain sign with large blue outlined white "CALSO" above white "GASOLINE" across sign face.

CALSO SUPREME GASOLINE 11"RD

(4) 1949-1955 $185

White outlined red porcelain sign with large blue outlined white "CALSO" above white "SUPREME/GASOLINE" across sign face.

CALSO GASOLINE W/ HALLMARK 11x13

(3) 1955-1957 $200

White outlined red porcelain sign with blue/white/red hallmark logo projecting from bottom of sign. Large blue outlined white "CALSO" above white "GASOLINE" across sign face.

CALSO SUPREME GASOLINE W/ HALLMARK 11x13

(4) 1955-1957 $200

White outlined red porcelain sign with blue/white/red hallmark logo projecting from bottom of sign. Large blue outlined white "CALSO" above white "SUPREME/GASOLINE" across sign face.

STANDARD OF TEXAS - EL PASO, TX: Subsidary assigned to Standard of California with the breakup of Standard in 1911. Standard of Texas marketing remained disticntly separate from other Calso & Standard operations until the 1960's. Originally branded "Pasotex," the stations had branded "Chevron Gas Station" in 1946 and became simply Chevron about 1957.

SECONDARY SIGNS

STANDARD OIL PRODUCTS 30"RD

(4) 1950-1960 $225

Blue outlined white porcelain sign with red "STANDARD/OIL/PRODUCTS" across sign.

STANDARD OIL OF TEXAS PRODUCTS 30"RD

(4) 1950-1960 $275

Blue outlined white porcelain sign with red "STANDARD" across top over blue "OIL/OF TEXAS"over red "PRODUCTS."

NEW YORK LUBRICATION OIL CO-PERTH AMBOY, NJ: Lubricants manufacturer that marketed the Monogram Motor Oils all over the country. In 1945 Standard of California purchased their refining and storage operation to establish their "California Oil Company" subsidary.

GREASES MONOGRAM OILS 15x24

(3) 1930s $300

Red porcelain flange sign with black outlined white band diagonal across center. Black outlined red "MONOGRAM" on band, with black outlined white "GREASE" above and "OILS" below.

MONOGRAM OILS AND GREASES

(4) 1940s $250

Cream color flange sign with red "MONOGRAM" across from left. Small red "OILS AND/GREASES" at right.

CITIES SERVICE OIL COMPANY - TULSA, OK

Cities Service was founded in 1910 as a public utility, operating municipal natural gas, lighting, ice, and other utility type services, hence the name "Cities Service." Venturing into petroleum marketing about 1914, Cities Service was a scattered collection of refining and marketing companies operating gas stations throughout the east and midwest. By the mid-1930s the various concerns were all marketing under the Cities Service brandname and black and white colors which gave way to green and white stations in 1946. In the 1950s many elaborate service stations were built but Cities Service continued to have an image problem. In 1965 the time-tested Cities Service brandname was replaced by Citgo and the entire marketing program was revitalized. Many new stations were added, only to close in record numbers during the 1970s gas shortages. In 1981 Southland Corporation, operator of the 7-Eleven convenience stores, and the government owned oil company in Venezuela purchased the marketing division, Citgo Petroleum, while Occidental Petroleum purchased the remaining divisions of Cities Service Oil Company. In the late 1980s, Southland, in bankruptcy, sold their share of Citgo to Venezuela and the company, with the assured supply of a government owned oil company, again began agressively marketing. Citgo now supplies thousands of stations in 46 states and new locations are added every day.

IDENTIFICATION SIGNS

CITIES SERVICE GASOLENE AND MOTOR OILS 30x50
(5) 1915-1920 $325
Black outlined white tin litho sign with black band across center. Cities Service Oils Once Always logo at upper left with black "CITIES/SERVICE" on upper white band. White "GASOLENE" on center band, with black "AND MOTOR OILS" on lower white area.

CITIES SERVICE OILS 30"RD
(3) 1920-1932 $350
CITIES SERVICE OILS 32"RD
(3) 1920-1932 $350
CITIES SERVICE OILS 36"RD
(3) 1920-1932 $350
CITIES SERVICE OILS 42"RD
(3) 1920-1932 $325
CITIES SERVICE OILS 48"RD
(4) 1920-1932 $325
Black outlined white porcelain sign with black clover and triangle logo in center. Black "CITIES SERVICE OILS" arched around top. Black outlined white pointed end rectangle below clover logo with black "ONCE-ALWAYS" on rectangle

CITIES SERVICE CLOVER 42"
(2) 1932-1946 $375
Steel framed black outlined white porcelain die-cut clover shaped sign with black "CITIES/SERVICE" across center. Small black clover and triangle logo at top of sign.

CITIES SERVICE CLOVER 42"
(3) 1946-1965 $300
Steel framed green outlined white porcelain die-cut clover shaped sign with green "CITIES/SERVICE" across center.

CITIES SERVICE CLOVER 72"
(2) 1946-1965 $200
Pole framed green outlined white porcelain die-cut clover shaped sign with green "CITIES/SERVICE" across center.

CITIES SERVICE TRIANGLE CLOVER 72"
(4) 1955-1957 $250
Pole framed green outlined white porcelain die-cut clover shaped sign with large green triangle, point up, in center. Small green "CITIES SERVICE" across center just below triangle.

CITIES SERVICE CLOVER & BAR 96"
(4) 1957-1965 $150
Pole framed green outlined white porcelain die-cut clover shaped sign with large green triangle, point up, in center. Bolt on green porcelain panel below clover with white "CITIES SERVICE" across panel

BUILDING SIGNS AND LETTERS

CAST ALUMINUM CITIES SERVICE LOGO & LETTERS
(4) 1946-1965 $175
Cast aluminum "CITIES/SERVICE" with large clover & triangle logo to right. Assembled on frame for rooftop mounting.

CITIES/LOGO/SERVICE BUILDING PANEL 12x96
(4) 1946-1965 $175
White porcelain panel sign with green "CITIES SERVICE" across panel. Green outline clover and triangle logo between "CITIES" and "SERVICE."

PUMP SIGNS

CITIES SERVICE OILS ONCE ALWAYS 10.5"RD
(5) 1920-1932 $350
Black outlined white porcelain sign with black clover and triangle logo in center. Black "CITIES SERVICE OILS" arched around top. Black outlined white pointed end rectangle below clover logo with black "ONCE-ALWAYS" on rectangle

CITIES SERVICE OILS ONCE ALWAYS CURVED
(5) 1920-1932 $300
Black outlined white porcelain sign with black clover and triangle logo in center. Black "CITIES SERVICE OILS" arched around top. Black outlined white pointed end rectangle below clover logo with black "ONCE-ALWAYS" on rectangle

CITIES SERVICE OILS ONCE ALWAYS DIE CUT
(5) 1926-1932 $475
Black outlined white porcelain die-cut clover sign with black clover and triangle logo in center. Black "CITIES SERVICE OILS" arched around top. Black outlined white pointed end rectangle below clover logo with black "ONCE-ALWAYS" on rectangle

NEON SIGNS

Neon outlined rings were added to standard indentification signs.

CURB SIGNS

CITIES SERVICE OILS 16"RD
(3) 1920-1932 $350
Black outlined white porcelain sign with black clover and triangle logo in center. Black "CITIES SERVICE OILS" arched around top. Black outlined white pointed end rectangle below clover logo with black "ONCE-ALWAYS" on rectangle

CITIES SERVICE OILS 24"RD
(3) 1920-1932 $350
Black outlined white porcelain sign with black clover and triangle logo in center. Black "CITIES SERVICE OILS" arched around top. Black outlined white pointed end rectangle below clover logo with black "ONCE-ALWAYS" on rectangle

CITIES SERVICE OILS ONCE ALWAYS CLOVER 24x24
(4) 1925-1932 $425
Black outlined white porcelain die-cut clover sign with black clover and triangle logo in center. Black "CITIES SERVICE OILS" arched around top. Black outlined white pointed end rectangle below clover logo with black "ONCE-ALWAYS" on rectangle

HIGH TEST KOOLMOTOR ANTI-KNOCK 24"RD
(4) 1926-1930 $500
Black outlined green porcelain sign with black outlined white bar & circle band across center. White "HIGH-TEST" arched around top, "ANTI-KNOCK" around bottom with black "KOOLMOTOR" over red "A CITIES SERVICE PRODUCT" on center band
CITIES SERVICE OILS CLOVER 24x24
(5) 1936-1940 $475
Steel framed black outlined white porcelain die-cut clover shaped sign with black "CITIES/SERVICE/OILS" across center. Small black clover and triangle logo at top of sign.
KOOLMOTOR THE PERFECT PENNA. OIL 24x24
(4) 1920-1925 $500
Black porcelain die-cut sign with elaborate green framing around white center and around red arched band at top. White "KOOLMOTOR" on red band with with "OIL" in black area below. Black "THE PERFECT/PENNA.OIL" on white center area with black "CITIES SERVICE OILS ONCE ALWAYS" logo in white area at bottom.
KOOLMOTOR OIL THE PERFECT PENNA. OIL 24"RD
(4) 1925-1930 $450
Black outlined green porcleain sign with black outlined white bar and circle design across center. Black outlined white "THE PERFECT" arched aorund top, "PENNA OIL" around bottom with black "KOOLMOTOR" above red "A CITIES SERVICE PRODUCT" across white band in center.
KOOLMOTOR A CITIES SERVICE PRODUCT 24"RD
(3) 1930-1932 $475
Black outlined green porcleain sign with black outlined white bar and circle design across center. Black outlined white "THE PERFECT" arched aorund top, "PENNSYLVANIA OIL" around bottom with black "KOOLMOTOR" above red "A CITIES SERVICE PRODUCT" across white band in center.
KOOLMOTOR A CITIES SERVICE PRODUCT 24x24
(2) 1932-1946 $500
Die cut clover shaped black outlined green porcleain sign with black outlined white bar and circle design across center. Black outlined white "THE PERFECT" arched aorund top, "PENNSYLVANIA OIL" around bottom with black "KOOLMOTOR" above red "A CITIES SERVICE PRODUCT" across white band in center.

LUBESTER SIGNS

CITIES SERVICE OILS ONCE ALWAYS 6"RD
(5) 1920s $450
Black outline white porcelain sign with black clover and triangle logo in center. Black "CITIES SERVICE OILS" arched around top. Black outlined white pointed end rectangle below clover logo with black "ONCE-ALWAYS" on rectangle

LUBRICATION/MOTOR OIL SIGNS

CS OILS ONCE ALWAYS KOOLMOTOR THE PERFECT PENNSYLVANIA MOTOR OIL 10x24
(5) 1915-1920 $325
White tin litho sign with black square at right side. White "CS Oils Once Always" logotype on black area. Red outlined black "Koolmotor" above red line on top of sign face with black "THE PERFECT/PENNSYLVANIA" and red outlined black "MOTOR OIL" below.

TROJAN MOTOR OIL BACKBAR SIGN 18x36
(5) 1935-1940 $175
Red porcelain sign with black band across center. White "TROJAN/MOTOR OIL" across band with white trojan helmet above and circle below.
TROJAN PREMIUM OIL BACKBAR SIGN 30x60
(4) 1946-1950 $175
White self-framed tin litho sign with large red rectangle in center, interrupting three red horizontal stipes across top. White "TROJAN/PREMIUM/Motor Oil" on red area with red "CITIES/SERVICE" & clover logo to left and red trojan helmet logo to right of rectangle.
PREMIUM KOOLMOTOR OIL BACKBAR SIGN 30x60
(4) 1946-1950 $200
White self-framed tin litho sign with red band across top. Red striped white "PREMIUM/KOOLMOTOR" oil can on left with red and white angled view of 1940s car and truck to right of can below red "PREMIUM/KOOLMOTOR OIL". White "FOR LONGER ENGINE LIFE" on upper red band.

OIL RACK SIGNS

REFILL WITH KOOLMOTOR 12x24
(3) 1937-1940 $275
Black outlined green tin litho sign with black band across bottom. White script "Refill with" at upper left over white "KOOLMOTOR" with red dropshading. Small black and white Cities Service clover logo at left on black band with white "The Perfect Pennsylvania Oil" to right.
REFILL WITH CITIES SERVICE OILS 12x24
(4) 1937-1940 $300
Black outlined green tin litho sign with white "REFILL WITH" with black dropshading at upper left beside large black and white Cities Service clover logo at left. Black "CITIES/SERVICE/OILS" on clover logo.

FLANGE SIGNS

CITIES SERVICE MOTOR OIL SOLD HERE 12x22
(3) 1935-1938 $275
Green tin litho flange sign with white line above black band across bottom. Black outlined white "CITIES SERVICE/MOTOR OIL/SOLD HERE" to left on sign face with black, white and green Cities Service quart can pictured at right. White "IF IT'S CITIES SERVICE - IT HAS TO BE GOOD" on black band at bottom.
CITIES SERVICE KEROSENE SOLD HERE 12x22
(3) 1935-1938 $275
Green tin litho flange sign with white line above black band across bottom. Black outlined white "CITIES SERVICE" clover logo to left on sign face with black outlined white "KEROSENE/SOLD HERE" at right. White "IF IT'S CITIES SERVICE - IT HAS TO BE GOOD" on black band at bottom.

SECONDARY SIGNS

CITIES SERVICE NATIONAL CREDIT CARDS 12x24
(5) 1936-1946 $150
White porcelain sign with black line at top and bottom. Large black "CITIES SERVICE" above smaller italic "NATIONAL" across upper sign face over "CHARGE CARDS." Small

Richfield painted metal sign. *Jim Hollabaugh.*

Richfield 48" round porcelain sign. *Jim Hollabaugh.*

Colonial pump plate sign. *Mark McKeown.*

Sinclair Opaline porcelain truck door sign. *Tom Kleeberg.*

Miscellaneous 12" pump plate signs. *Bruce Bednar.*

Miscellaneous porcelain pump plate signs.

Texaco 12"x14" small truck door sign. *Dunbar's Gallery.*

Collection of Texaco pump plate signs. *Tom Martin.*

Sky Chief scarce round porcelain sign. *Neil Stockwell.*

Texaco Motor Oil 30" porcelain sign.

Texaco Marine Lubricants porcelain sign. *Leo Mathieu.*

Texaco porcelain flange sign.

Texaco Diesel Chief 12"x18" porcelain pump sign.

"Accepted Here" across bottom of sign.

CITIES SERVICE NATIONAL CHARGE CARDS 12x24

(4) 1946-1957 $175

White tin litho sign with three green horizontal stripes forming top and bottom borders. Green "CITIES SERVICE" (with clover triangle between) across top above red "NATIONAL/CHARGE CARDS" in center. Smaller green "ACCEPTED HERE" across bottom.

C.S. ACME TIRES VERTICAL STRIP 24x72

(5) 1930s $225

Black outlined green self-framed vertical tin litho sign with black and white Cities Service clover logo at top. Black outlined white "ACME" vertical on sign with "TIRES" across bottom.

ACME CITIES SERVICE TIRES 24x72

(5) 1930s $225

Black outlined green self-framed tin litho sign with black and white Cities Service clover logo in center. Black outlined white "ACME" to left of logo, "TIRES" to right, with white "THE ACME BOND IS YOUR PROTECTION" across bottom.

CITIES SERVICE AHEAD 36x60

(5) 1950s $200

Green outlined white tin litho die-cut clover over bar shaped sign with green "CITIES/SERVICE" across center. Green outline across bottom on lower bar with green "AHEAD" in white band between.

CS CLOVER & BAR TIN 30x42

(5) 1957-1964 $100

Green outlined white tin litho die-cut clover over bar shaped sign with large green triangle, point up, in center. Lower white outlined green panel below clover with white "CITIES SERVICE" across panel.

TRUCK SIGNS

CITIES SERVICE 12"

(5) 1934-1946 $550

Black outlined white porcelain die-cut clover shaped sign with black "CITIES/SERVICE" across center. Small black clover and triangle logo at top of sign.

CITGO

In 1965 Cities Service began a rebranding and reimaging program that introduced the brandname "Citgo" to replace "Cities Service." Image colors were changed from green, white, and red to the typical oil industry red, white, and blue. Stations were painted in grays and earth tones.

IDENTIFICATION SIGNS

CITGO 42"

(4) 1965-1980 $200

Steel-framed white porcelain sign with large three-shade red triangle on upper sign face. Blue "CITGO" below trinagle across bottom of sign.

CITGO 72"

(4) 1965-1980 $150

White porcelain can type sign with large three-shade red triangle on upper sign face. Blue "CITGO" below trinagle across bottom of sign.

BULK PLANT SIGNS

CITIES SERVICE COMPANY/CITGO STRIP 14x84

(4) 1965-1980 $100

Grey porcelain strip with Citgo white square/red triangle logo at right end. White "CITIES SERVICE OIL CO." across remainder of sign face.

SECONDARY SIGNS

MEN 10x10

(3) 1965-1972 $25

Blue porcelain sign with white "Men" across bottom.

LADIES 10x10

(3) 1965-1972 $25

Blue porcelain sign with white "Ladies" across bottom.

LINCOLN OIL CO. - FLINT, MI

Lincoln was a southern Michigan marketer, operating more than 50 stations when they were merged into Cities Service in 1930.

No collectible size Lincoln signs known at this time.

ELWOOD OIL CO. - ELWOOD, IN

Indiana jobber brand converted to Cities Service in 1931.

No collectible size Elwood/Red Parrot signs known at this time.

LOUISIANA OIL REFINING CO. - SHREVEPORT, LA

Originally a Gulf Coast refiner/mr/marketer, Loreco became a part of Cities Service in 1932.

IDENTIFICATION SIGNS

LORECO MOTOR OILS GASOLINE 28x36

(5) 1920s $400

Red porcelain panel with two vertical and two horizontal white lines forming red box in center. White outlined red diamond in box with white "LORECO" in diamond. White "MOTOR OILS" above top line, "GASOLINE" below bottom line.

LORECO OVAL 30x36

(5) 1930-1935 $375

Black/white outlined red porcelain oval sign with black/white outlined red diamond superimposed over white center circle. Black outlined white "LORECO" across diamond with "GASOLINE" arched aorund top, "MOTOR OILS" around bottom.

CURB SIGNS

LORECO ETHYL 30"RD

(5) 1926-1935 $375

Black outlined white porcelain sign with black outlined white center circle. Large Ethyl (EGC) logo in center with red "LORECO" arched around top, "GAS" around bottom.

CREW-LEVICK COMPANY - PHILADELPHIA, PA

Crew-Levick was an early Philadelphia lubricants marketer that merged with Cities Service about 1925.

LUBRICATION/MOTOR OIL SIGNS

CREW LEVICK PERFECTION MOTOR OILS 12x24

(5) 1920-1925 $325

White outlined red tin litho sign with blue band diagonal across center, White "PERFECTION" on diagonal band with white "MOTOR OILS" below and "PENNA.PRODUCTS/CREW LEVICK CO./PHILA" above band. Lower white outline is wider to allow for priveledge panel entry on distributor name.

WARNER-QUINLAN COMPANY, NEW YORK, NY

Founded in New York City before 1900, Warner-Quinlan was marketing throughout New England when purchased by Cities Service in 1933.

No collectible size Warner-Quinlan "Warnoco" or "Mileage" signs known at this time.

LINDSEY-MACMILLAN - MILWAUKEE, WI

Early petroleum marketer primarily involved in lubricants. Operated a small number of gasoline stations in the Milwaukee area in the 1920's. Purchased by Cities Service in the early 1930's.

LUBRICANTS SIGNAGE

DELCO MOTOR OIL 42x60
(5) 1926-1930 $400
Wood-framed black tin litho sign with large yellow "Delco" script logo type across center over yellow "MOTOR OIL." White "20% more lubrication" across top with white "LINDSAY-McMILLAN Co MILWAUKEE" across bottom. Red circle at lower left, with yellow imp on circle. IOMA eagle logo superimposed over imp.

SHOEMAKER COAL CO - AUBURN, NY

Purchased by Cities Service about 1935. Little is known about this company.

No collectible size Shoemaker signs known at this time.

MILTON OIL CO - SEDALIA, MO & PIONEER OIL CO-CARTHAGE - JOPLIN, MO

As best as can be determined, Pioneer Oil Co of Joplin Missouri introduced the "DIXCEL" brand sometime around 1930. Milton Oil Company of Sedalia, Missouri, was apparently a jobber for Dixcel that sometime before WWII acquired the brand from Pioneer, who continued operations under the Pioneer brand. Milton rebranded to "Savex" in the 1950s and continued to operate under that brand after Cities Service purchased Milton in 1962.

IDENTIFICATION SIGNS

PIONEER OIL CO DIXCEL GASOLINE 42"RD
(4) 1930s $650
Black outlined orange porcelain sign with black band across center. White outlined orange "DIXCEL" on black band, with black "PIONEER" arched around top over "OIL CO.," "GASOLINE" around bottom.

PIONEER OIL CO DIXCEL GASOLINE 42"RD
(3) 1940s $650
Black outlined orange porcelain sign with black band across center. White oultined orange "DIXCEL" on black band, with black "PIONEER OIL CO." arched around top, "GASOLINE" around bottom.

MILTON OIL CO. DIXCEL PRODUCTS 42"RD
(2) 1940's $650
Black outlined orange porcelain sign with black band across center. White oultined orange "DIXCEL" on black band, with black "MILTON OIL CO." arched around top, "GASOLINE" around bottom.

JENNEY MANUFACTURING CO. - BOSTON, MA

Originally a mid-19th century dealer in whale oils, coal, kerosene and related products, Jenney entered the New England gasoline market in the early years of this century. By 1920 Jenney had a network of elaborate service stations in Boston and the surrounding area. With close ties to Cities Service for many years, Jenney was merged into Cities Service about 1965 and the Citgo brand replaced Jenney.

IDENTIFICATION SIGNS

JENNEY GASOLINE 24x42
(5) 1925-1935 $450
Rectangular self-framed white porcelain sign with dark outline. "JENNEY/GASOLINE" across sign face.

JENNEY 34x72
(5) 1935-1960 $375
Blue outlined white porcelain rounded rectangle sign with "JENNEY" across center.

PUMP SIGNS

SOLVENIZED JENNEY HY-POWER 9"RD
(5) 1934-1946 $1800
Black outlined light blue porcelain sign with black outlined yellow center circle. Light blue band diagonal across center with black script "Jenney" logotype across center. Black outlined yellow "SOLVENIZED" arched around top, "HY-POWER" around bottom. Black cityscape in yellow center circle.

SOLVENIZED JENNEY HY-POWER 12"RD
(5) 1934-1946 $1600
Black outlined light blue porcelain sign with black outlined yellow center circle. Light blue band diagonal across center with black script "Jenney" logotype across center. Black outlined yellow "SOLVENIZED" arched around top, "HY-POWER" around bottom. Black cityscape in yellow center circle.

JENNEY AERO SOLVENIZED 9"RD
(4) 1934-1946 $2400
Yellow outlined light blue porcelain sign with light blue outlined black half-circles at top and bottom of sign. Yellow "JENNEY" on upper half-circle, "SOLVENIZED" on lower half circle with yellow and black airplane in center. Black outlined white "AERO" superimposed over airplane. Small black "JENNEY MANUFACTURING CO BOSTON" at bottom on yellow outline.

JENNEY AERO SOLVENIZED 12"RD
(4) 1934-1946 $2200

Yellow outlined light blue porcelain sign with light blue outlined black half-circles at top and bottom of sign. Yellow "JENNEY" on upper half-circle, "SOLVENIZED" on lower half circle with yellow and black airplane in center. Black outlined white "AERO" superimposed over airplane. Small black "JENNEY MANUFACTURING CO BOSTON" at bottom on yellow outline.

JENNEY AERO 9x12
(2) 1946-1965 $175
Blue/red outlined white porcelain rounded rectangle shaped sign with small blue "JENNEY" over larger red "AERO" in center. Small white "JENNEY MANUFACTURING CO., BOSTON, MASS" on bottom blue border area.

JENNY SUPER AERO 9x12
(2) 1946-1965 $175
Blue/red outlined white porcelain rounded rectangle shaped sign with small red "JENNEY" over larger blue "SUPER-AERO" in center. Small white "JENNEY MANUFACTURING CO., BOSTON, MASS" on bottom blue border area.

ROYAL OCTANE 9x12
(2) 1957-1965 $150
Blue/red outlined white porcelain rounded rectangle shaped sign with small red "JENNEY" over blue "ROYAL/OCTANE" in center. Small white "JENNEY MANUFACTURING CO., BOSTON,MASS" on bottom blue border area.

NEON SIGNS

Neon was added to standard identification signs.

CLARK OIL AND REFINING CORP. - ST. LOUIS, MO

Clark Oil Co. began as a one station operation by Mr. Emory Clark in 1932 in Milwaukee, Wisconsin. By the early 1930s, Clark had evolved into a chain operation, selling "discount" gas through company operated stations in a number of Wisconsin towns. Significant growth came after World War II, when the operation expanded to include stations from Pennsylvania to Kansas, totalling more than 1350 at the peak, including 122 stations acquired when Clark bought out the Owens Oil Co. of Bloomington, IL in 1968. In 1981 Clark was purchased by St. Louis based Apex Oil and in 1988 again sold, this time to Canadian conglomerate Horsham Corp., which continues to operate the more than 900 Clark stations in 12 states today.

IDENTIFICATION SIGNS:

Super CLARK'S Gasoline 72RD
(5) 1935-1956 $200
Orange porcelain sign with black outlined white band across center. Black ""CLARK'S"" across white band with white script "Super" in upper orange area,, "Gas" below.

PUMP SIGNS

ETHYL SUPER 100 10.5x11.5
(4) 1946-1950 $175
Die-cut porcelain sign with orange circle below and black outlined white rectangle across top. Black ""ETHYL" in rectangle,, with black "SUPER" above large black outlined white "100" on orange circle.

PREMIUM SUPER 100 10.5x11.5
(3) 1950-1956 $175
Same as above, "PREMIUM" replacing "ETHYL."

TARGET: Target is a 1960s Clark secondary brand. No Collectible size Target signs known at this time.

OWENS OIL CO. - BLOOMINGTON, IL: Owens was a Bloomington, IL based discount gasoline chain that operated 122 stations in 10 states when purchased by Clark in 1968. By the early 1980s the Clark brandname had replaced the Owens "Oco-Pep" brand.
No collectible size Owens signs known at this time.

RICH OIL COMPANY - IRONTON, OH

Independent marketer founded in 1947 and supplied by Ashland. Purchased by Ashland Oil in the mid 1980's, the brand was adopted for most Ashland "unbranded" locations, replacing Payless, Red Head, Save More, and Save Mart.

SECONDARY SIGNS

RICH FENCE SIGN
(3) 1990-PRESENT $50
White tin sign with large red/white/blue arrow design across bottom. Blue vertical stripes with red "RICH" across top.

COLONIAL OIL INDUSTRIES - SAVANNAH, GA

Colonial Oil Industries was founded in Savannah, Georgia in 1921 as the American Oil Company. From 1921 until 1928 gasoline was sold under the American brand, but with the expansion of the Pure Oil Co. into Georgia, American became a branded Pure Oil jobber. In 1933 when Pan American Petroleum merged with the American Oil Co. (AMOCO), and replaced the Pan-Am brand in Georgia with the Amoco name, Colonial sold the rights to the American name and became Colonial. In 1950 Colonial sold the Pure branded operation to Pure and concentrated on a growing wholesale/terminal business. The Colonial gasoline brand was introduced in the mid-1950s in an effort to give their dealers a unified brand and about 1964 they began operating a discount service station chain, branded "Interstate" and selling Colonial gasoline. By the early 1980s they had replaced many Interstate stations with "Enmark" convenience stores which they continue to operate today.

IDENTIFICATION SIGNS

COLONIAL SHIELD 72x84
(3) 1956-1980 $350
Red, white, and blue die-cut shield with large eagle at top & red "Colonial" lettering on white band across center.

BUILDING SIGNS AND LETTERS

INTERSTATE 2-PANEL SIGN 24x120
(4) 1964-1980 $100
Self framed rectangular litho aluminum panels with blue outline around white center. Red "INTERSTATE" on white area. Two pieces assemble to form complete sign. Attached to building wings.

AMERICAN: Original marketing name,, used from 1921 until 1928.

SECONDARY SIGNS

AMERICAN GASOLINE W/ ARROW 12x28
(5) 1921-1928 $400
Blue porcelain sign with white outline and red/white/blue shield in center. White ""AMERICAN" to left of shield,, "GASOLINE" to right,, with white arrow pointing to right across top of sign.

CONOCO AND AFFILIATES CONTINENTAL OIL COMPANY - PONCA CITY, OK

Continental Oil, founded in 1875 in Denver, Colorado became an operating unit of the Standard Oil Trust, having been purchased by Standard in 1885. After the 1911 breakup of Standard, Continental was spun off as an independent company and by 1913 was involved in production, refining and marketing. The Continental name survived a 1929 takeover by Marland Oil, and over the course of time the Conoco brandname has appeared in nearly all states of the southeast, midwest, mountain and Pacific coast areas. Conoco expanded it's scope worldwide during the 1950s, including the purchase of several marketing companies, such as Western Oil, Kayo and Douglas, greatly expanding the company's marketing presence throughout the United States. Conoco has been a part of the giant chemical company Dupont since 1982 and markets in 37 states today.

IDENTIFICATION SIGNS:

CONOCO GASOLINE MINUTEMAN 26"RD
(5) 1925-1929 $3000
CONOCO GASOLINE MINUTEMAN 30"RD
(3) 1925-1929 $3000
Blue/white outlined yellow porcelain sign with white outlined blue "CONOCO" arched around top, white outlined red "GASOLINE" around bottom. Multicolor detailed minuteman in center.
CONOCO 42"
(1) 1929-1950 $250
CONOCO 72"
(4) 1929-1950 No Listing
Steel framed die cut green and white outlined red porcelain triangle, point down, with red band across upper area. White "CONOCO" across band. Die cut inside triangle around band.
CONOCO 42"
(4) 1929-1950 $250
Steel framed die cut green and white outlined red porcelain triangle, point down, with red band across upper area. White "CONOCO" across band. White solid area inside triangle around band.
CONOCO 30"
(2) 1950-1970 $175
CONOCO 42"
(4) 1950-1970 $200
CONOCO 72"
(1) 1950-1970 $100
Steel framed die cut white outlined red porcelain triangle, point down, with red band across upper area. White "CONOCO" across band. Solid white inside triangle around band.
CONOCO 72"
(4) 1970-PRESENT $50
White outlined red round ended rectangle with white logotype "conoco" across center.

BUILDING SIGNS & LETTERS:

CONOCO RED PORCELAIN LETTERS 18"
(4) 1946-1970 $75
Red die-cut porcelain letters.

PUMP SIGNS:

CONOCO 8.5x9.5
(1) 1950-1960 $65
Green & white die cut triangle with white "CONOCO" on green band across upper triangle.
SUPER 8.5x9.5
(4) 1950-1954 $75
Red & white die cut triangle with white "SUPER" on red band across upper triangle.
ROYAL 8.5x9.5
(4) 1954-1960 $65
Red & white die cut triangle with white "ROYAL" on red band across upper triangle.
DIESEL 8.5x9.5
(4) 1950-1954 $125
Black & white die cut triangle with white "DIESEL" on black band across upper triangle.
CONOCO NEW TRIANGLE 7.5x8.5
(4) 1946-1970 $50
Red & white die cut triangle with white border and white "CONOCO" on red band across upper triangle.

CURB SIGNS:

CONOCO AMALIE 24"
(5) 1924-1929 No Listing
Blue outlined yellow porcelain sign with large red circle in center. Blue "CONOCO" arched around top, "CONTINENTAL OIL COMPANY" around bottom, with blue/white Amalie logo and red PGCOA seal in center.
POLARINE USE CORRECT GRADE 28"RD
(5) 1920-1924 No Listing
Blue outlined yellow porcelain sign with red/blue outlined white triangle in center. Red script "Polarine" logotype across upper triangle, with blue "USE/CORRECT/GRADE" below. Similar in design to Standard of Indiana signs for their product of the same name.
CONOCO MOTOR OIL 24"RD
(5) 1920-1929 No Listing
Blue/white outlined yellow porcelain sign with blue "CONOCO" arched around top and "MOTOR OIL" around bottom. Blue/white detailed minuteman in center. Blue "Extra life" to left, "for your car" to right.
CONOCO GASOLINE 24"RD
(5) 1918-1926 No Listing
Blue outlined yellow porcelain sign with blue "CONOCO" arched around top and red "GASOLINE" around bottom. Blue/yellow minuteman in center.
CONOCO SUPERIOR GASOLINE 24"RD
(5) 1922-1926 No Listing
Same as above, with red "SUPERIOR" above "GASOLINE."
CONOCO ETHYL 28"RD
(4) 1926-1929 $575
Blue/white outlined yellow porcelain sign with blue outlined white circle in center. White outlined blue "CONOCO" arched

around top, white outlined red "GASOLINE" arched around bottom with Ethyl
(EGC) logo with blue rays in center.
CONOCO ETHYL 28"RD
(3) 1929-1946 $450
Green outlined white porcelain sign with green outline around white center circle. Red "CONOCO" arched aorund top, "GASOLINE" around bottom, with large Ethyl
(EGC) logo with yellow rays in center.
CONOCO BRONZE GASOLINE 30"RD
(5) 1936-1940 $300
White porcelain sign with red band across center. "Bronze" logotype lettering in center band with "CONOCO" arched around top, "GASOLINE" around bottom.
CONOCO GERM PROCESSED 30"
(3) 1934-1940 $250
White porcelain die-cut triangular sign with red/white/green Conoco triangle logo at the bottom. Green "CONOCO" across top above red "GERM/PROCESS." Smaller green "PARAFFIN BASE/MOTOR OIL" below, above logo.

LUBRICATION/MOTOR OIL SIGNS:

CONOCO/AMALIE FLANGE SIGN 18"RD
(5) 1924-1929 No Listing
Blue outlined yellow tin litho circle with yellow flange area. Large red circle in center. Blue "CONOCO" arched around top, "CONTINENTAL OIL COMPANY" around bottom, with blue/white Amalie logo and red PGCOA seal in center.
CONOCO/AMALIE LUBESTER SIGN
(5) 1924-1929 No Listing
Yellow sign with red circle in center. Blue "CONOCO" arched around top, "CONTINENTAL OIL COMPANY" around bottom with black & white "AMALIE" logo & PGCOA seal in center circle.
CONOCO MOTOR OIL MINUTEMAN 24x32
(5) 1920-1929 No Listing
Yellow porcelain sign with blue line border. White outlined blue "CONOCO" arched above "MOTOR/OIL" on sign face. Red "A grade for each type of engine" across bottom. Detailed minuteman at left.
CONOCO SUPER MOTOR OIL 24x30
(1) 1950-1954 $100
White porcelain sign with large red and white Conoco triangle logo at top. Red "Super" above green "MOTOR OIL" below logo.

SECONDARY SIGNS:

CONOCO CREDIT CARDS HONORED 24x24
(5) 1950's $75
White tin litho sign with fine black outline. Red and white Conoco triangle logo at top with black "CREDIT" to left, "CARDS" to right and "HONORED" below.
YOUR MILEAGE MERCHANT 8x42
(3) 1934-1950 $60
Red outlined white embossed tin-litho sign with green script "Your Mileage Merchant" across sign.
MEN REST ROOM 8x10
(5) 1936-1950 $35
White porcelain flange sign with green "MEN/REST ROOM/SANITARY SEAT COVERS" on sign face.
LADIES REST ROOM 8x10
(5) 1936-1950 $35
White porcelain flange sign with green "LADIES/REST ROOM/SANITARY SEAT COVERS" on sign face.
CONOCO "MEN" DIECUT 8x10
(3) 1950-1965 $50
White outlined red die-cut oval with red and white Conoco triangle logo die-cut at upper left. White "MEN" in center of oval.
CONOCO "WOMEN" DIECUT 8x10
(3) 1950-1965 $50
Same as above, "WOMEN" replacing "MEN," color uncertain.
GREEN CONOCO "MEN" FLANGE 5x10
(1) 1965-1970 $25
Green tin litho sign with white area curved through center. Red and white Conoco logo at top edge of white area, with green "MEN" below.
GREEN CONOCO "WOMEN" FLANGE 5x10
(1) 1965-1970 $25
Green tin litho sign with white area curved through center. Red and white Conoco logo at top edge of white area, with green "WOMEN" below.
BLUE CONOCO "MEN" FLANGE 5x10
(1) 1965-1970 $25
Blue tin litho sign with white area curved through center. Red and white Conoco logo at top edge of white area, with blue "MEN" below.
BLUE CONOCO "WOMEN" FLANGE 5x10
(1) 1965-1970 $25
Blue tin litho sign with white area curved through center. Red and white Conoco logo at top edge of white area, with blue "WOMEN" below.
CONOCO MEN NEW 4x10
(2) 1970-PRESENT $20
White tin litho sign with red Conoco oval logo to left and "MEN" to right. Black "man" symbol in center.
CONOCO WOMEN NEW 4x10
(2) 1970-PRESENT $20
White tin litho sign with red Conoco oval logo to left and "WOMEN" to right. Black "woman" symbol in center.

BULK PLANT SIGNAGE

CONTINENTAL OIL COMPANY 14x144
(3) 1929-1950 $125
White porcelain strip with green "CONTINENTAL OIL COMPANY" across center. Red/white/green Conoco triangle logo at either end.

TRUCK SIGNS:

CONOCO TRIANGLE ON CIRCLE 9"
(1) 1929-1950 $200
White porcelain sign with red/white/green Conoco triangle logo in center.
CONOCO LETTERING W/DROPSHADE 9"
(1) 1929-1950 $225
White porcelain die-cut individual lettering with thick green drop shading.

MARLAND OIL COMPANY - PONCA CITY, OK

Marland traces its early history to the Oklahoma oil fields of the early 1910s. By 1920 marketing had began under the famous Red Triangle logo and by 1922 there were Marland service stations in 11 states. Marland moved into eastern markets in the late 1920s with the purchase of Prudential Refining of BNaltimore,MD. In 1929 Marland purchased Continental but chose to retain the Continental Oil Co. name for the new corporation. The new Conoco logo was to be the famous Marland triangle rebranded with the name "CONOCO."

IDENTIFICATION SIGNS

MARLAND OILS 30"RD
(5) 1920-1925 No Listing
White tin litho sign with red bar and triangle logo in center. White "MARLAND OILS" across bar.

MARLAND OILS TRIANGLE 42"
(4) 1925-1929 No Listing
White outlined red die cut triangle, point down, with red bar across upper area. White "MARLAND OILS" on bar.

SECONDARY SIGNS

MARLAND GASOLINE 14x29
(4) 1920-1929 $450
Black outlined cream colored sign with red/whte/green Marland Oils triangle to left. Black "MARLAND/GASOLINE/LOOK FOR THE RED TRIANGLE" to right.

MARLAND HIGH TEST GASOLINE 18x30
(5) 1920-1925 $350
White outlined black tin-litho sign with red and white Marland Oils triangle in center. White "MARLAND" across top, "GASOLINE" across bottom with "High" to left and "Grade" to right of triangle.

MARLAND HIGH GRADE MOTOR OIL 12x20
(5) 1925-1930 $350
White outlined black tin-litho flange sign with red and white Marland Oils triangle in center. White "MARLAND" across top, "MOTOR OIL" across bottom with "High" to left and "Grade" to right of triangle.

NATIONAL OIL COMPANY - RICHMOND, VA:

Founded in Richmond in 1896, National Oil entered gasoline marketing in the 1920's in Virginia & Eastern North Carolina. They were purchased by and rebranded to Conoco in 1937, with the National stations and bulk plants being the bulk of the Conoco marketing that was sold to Cities Service when Conoco withdrew from the east in 1950.

IDENTIFICATION SIGNS

NATIONAL OVAL 24x96
(5) 1932-1937 No Listing
Green border area around white/green outlined white rectangular sign with rounded ends. Black outlined green "NATIONAL" across center with black outlined white "NATIONAL OIL COMPANY,INC." across top in border area.

PUMP SIGNS

NATIONAL 3x12
(5) 1932-1937 $200
Red/green outlined white porcelain round end rectangle sign. Green outlined red "NATIONAL" across white center of sign. Small green outlined white "NATIONAL OIL COMPANY, INC" on upper red outline area.

CURB SIGNS

NATIONAL 16"RD
(5) 1925-1937 No Listing
Red outlined white porcelain sign with white/green rings around white inner circle. Green outlined red "NATIONAL" across cente with green outlined white "NATIONAL OIL COMPANY, INC" on upper red outline area.

WESTERN OIL AND FUEL - MINNEAPOLIS, MN

Western Oil and Fuel was a Minneapolis based marketing operation, founded in 1926, that introduced the famous Husky brand in the mid 1930s and operated the Husky stations into the late 1950s. The Husky brandname eventually went to H. Earl Clack, Inc. of Cody, Wyoming and Western continued to market under the Mileage brand. Mileage was a brandname used by a company known as Zenith Gas Pool of Duluth, mn. Western purchased Zenith in 1949 and continued to use the mileage name. Conoco bought out Western in 1958 and about that time introduced "Western" as another gasoline trademark. By 1970 Conoco had converted all the Mileage and DS locations to the Conoco brand, but the Western stations, modeled after the competing Holiday and SuperAmerica convenience stores, remained "Western" up through the late 1970s when they were rebranded "JET."

IDENTIFICATION SIGNS

HUSKY GASOLINE 30"RD
(5) 1930-1935 No Listing
White outlined blue porcelain sign with detailed multi-color orange and yellow Northern Lights scene at top. Extremely detailed blue and white dog's head at center of lights. Yellow "HUSKY/GASOLINE" in blue area below, with orange "GETS YOU THRU" arched around bottom.

HUSKY GASOLINE 28x30
(5) 1930-1935 No Listing
Blue outlined keyhole shaped white porcelain sign with detailed multi-color orange and yellow Northern Lights scene at top. Extremely detailed blue and white dog's head at center of lights. Yellow "HUSKY" over white "GASOLINE" in blue area below circle.

HUSKY SERVICE 48x60
(5) 1935-1946 No Listing
Blue outlined porcelain die-cut shield. Lower area is blue with red outlined yellow "HUSKY" logotype over white "SERVICE." Upper area is red/yellow/white sunrise scene with detailed blue and white husky dog jumping left to right.

HUSKY 72"RD
(5) 1946-1960 $700
Red and blue round porcelain sign with white stripe separating upper blue area from lower red area.Yellow "HUSKY" in upper blue area, with detailed black and white jumping husky in center.

MILEAGE - ZENITH GAS POOL - DULUTH, MN:

Early wholesale terminal operator who introduced the Mileage brandname in the late 1920's. Purchased by Western Oil & Fuel in 1949. Western retained the Mileage brand when phasing out Husky and it continued in use until Conoco replaced the Mileage & DS brands in 1970.

LUBRICATION/MOTOR OIL SIGNS

MILEAGE OILS & GREASES 12x20
(5) 1929-1935 $250
White tin litho signs with blue wheel design at upper center

with blue "Motor Oils/& Greases" below. Red outlined black "Mileage" logotype across center.

SECONDARY SIGNS

MILEAGE GASOLINE 12x20
(5) 1929-1935 $250
White tin litho signs with blue wheel design at upper center with blue "Gasoline" below. Red outlined black "Mileage" logotype across center.

BETTER MILEAGE GASOLINE 72"RD
(5) 1949-1959 $100
Blue porcelain sign with white band across center. Blue outlined red "MILEAGE" across center white band with white "BETTER" in upper area, "GASOLINE" below.

DIRECT SERVICE OIL CO. - MINNEAPOLIS, MN

Minneapolis marketer purchased by Western Oil and Fuel in 1956.

IDENTIFICATION SIGNS

DS OVAL 42x84
(5) 1946-1959 $100
Black/white outlined red porcelain oval with white "DS" with black dropshading in center.

MALCO REFINING - ROSWELL, NEW MEXICO

Malco was a Roswell, New Mexico refiner marketer operating stations in New Mexico under the Malco and Numex brands and under the Valley brand in Arizona. Conoco purchased Malco in 1959 but was forced to sell on anti-trust grounds in 1964. Malco continues to operate as an independent.

SECONDARY SIGNS

MALCO CREDIT CARDS HONORED 20x32
(5) 1950's $225
White porcelain oval with yellow outlined black "MALCO" across sign above black "CREDIT CARDS/ACCEPTED."

VALLEY & NUMEX

Malco secondarys brand from the 1950s.
No collectible size Valley/Numex size known at this time.

DOUGLAS OIL CO. - LOS ANGELES, CA

Douglas Oil dates back to about 1927 and merged with Krieger in 1946. In 1961 Conoco bought Douglas and operated the nearly 300 Douglas stations until 1971 when they were sold off and rebranded, mostly Texaco.

IDENTIFICATION SIGNS

AVIATION TESTED DOUGLAS GASOLINE 42x84
(5) 1930s No Listing
White porcelain can-type sign with die-cut r red/white/blue winged heart logo at upper left. Blue band across cente with white "DOUGLAS" on band. Red acript "Aviation Tested" across top, "GASOLINE" below.

PUMP SIGNS

DOUGLAS AVIATION TESTED GASOLINE 12x14
(4) 1935-1950 $600
White outlined blue porcelain sign with white outlined red Douglas heart logo in center with blue/white wing to upper right. White "DOUGLAS" above heart, with red script "Aviation" to left, "Tested" to right. White "GASOLINE" across bottom.

DOUGLAS REGULAR 12.5x18
(5) 1935-1940 $550
White outlined blue porcelain tombstone shaped sign with white outlined red Douglas heart logo in center with blue/white wing to upper right. White "DOUGLAS" above heart, with red script "Aviation" to left, "Tested" to right. Small white "GASOLINE" below heart, with white script "Regular" across bottom.

DIE CUT HEART 10x22
(5) 1960's $225
White outlined red die-cut heart with blue/white wing positioned to upper right of heart.

DIE CUT HEART 10x22
(5) 1960's $225
White outlined red die-cut heart with blue/white wing positioned to upper left of heart.

HEART ON PANEL 11x16
(5) 1960's $150
White porcelain sign with red heart to lower left. Blue/white wing design positioned to upper right of heart.

DOUGLAS BLEND GASOLINE 10x16
(5) 1960's $150
Blue outlined white tin litho sign with blue vertical panel at left. Red heart with blue/white wing on blue panel. Blue "DOUGLAS" across top, "GAOSLINE" across bottom with red "BLEND" in center.

KRIEGLER

California independent that merged with Douglas in 1946.
No collectible size Kreigler signs known at this time.

F.P.KENDALL AND CO. - CHATTANOOGA, TN

F.P.Kendall and Company opened their first Kayo station in Chattanooga, Tennessee in 1940. The chain expanded throughout the south and had approximately 170 stations in seven states when purchased by Conoco in 1959. Kayo was operated as a separate brand throughout the 1960s and 70s and about 1980 the nearly 900 Kayo stations rebranded "JET." Conoco began replacing the Jet brand in 1990.

IDENTIFICATION SIGNS

KAYO SELLS GAS FOR LESS 36x60
(5) 1946-1959 $200
White outlined porcelain sign with white outlined red "KAYO" logotype at top. Blue and red alternating "SELLS/GAS/FOR/LESS" below.

CO-OP AFFILIATES
FARMLAND INDUSTRIES - KANSAS CITY, MO

Founded in 1928 as Consumers Co-operative Association, Marketing "Union Certified" products, Farmland, commonly refered to as "Double Circle" in reference to their logo, supplies products to farm co-ops throughout the midwest. They continue to market through co-op's under the "CO-OP" brandname and at retail under the "AMPRIDE" brand.

IDENTIFICATION SIGNS

CO-OP 42"RD
(4) 1937-1950 $175
Steel framed black/white outlined red porcelain sign with white "CO-OP with black dropshading across center.

CO-OP DOUBLE CIRCLE 24x48
(1) 1958-PRESENT $100
Red and blue outlined white porcelain die cut double circle logo with black "CO-OP" across center.
CO-OP DOUBLE CIRCLE 48x96
(1) 1958-PRESENT $100
Pole framed red and blue outlined white porcelain die cut double circle logo with black "CO-OP" across center.

LUBRICATION/MOTOR OIL SIGNS

CO-OP MOTOR OIL 10x18
(1) 1958-1965 $25
White rectangular tin litho sign with blue band across bottom. Red/blue/black CO-OP double circle logo above "MOTOR OIL" in center.

SECONDARY SIGNS

WE USE UNION CERTIFIED 12x17.5
(5) 1928-1931 $125
Blue outlined orange embossed tin litho sign with blue "WE USE" across top and "MOTOR OIL GASOLINE" across bottom. White outlined blue "UNION" arched above white outlined blue "CERTIFIED" in center of sign face.

FALCON

A postwar attempt at retail markets by Consumers Co-Operative Association (CCA).
No collectible size Falcon signs known to exist.

FARMERS UNION CENTRAL EXCHANGE - ST. PAUL, MN

Farmers Union Central Exchange is a combination of several farm co-operative organizations that operate across the North Central states. The CENEX brandname was introduced about 1974 to establish a common identity for the various companies. In 1988 Farmers Union Central Exchange merged with Land O'Lakes and all petroleum products are marketed under the CENEX brandname today.

IDENTIFICATION SIGNS

CENEX 72x72
(3) 1973-PRESENT $125
Aluminum framed can type porcelain white porcelain sign with large red and white "CENEX" logo covering most of sign face.

BUILDING SIGNS & LETTERS

CENEX PRIVILEDGE PANEL VARIOUS
(3) 1973-1989 $75
Red aluminum litho panel with red and white CENEX logo and white lettering with the name of the local co-operative across center. Custom manufactured for building signage at each location.
CENEX PRIVILEDGE PANEL VARIOUS
(3) 1989-PRESENT $75
White aluminum litho panel with red/black/white stripe across bottom. Red and white CENEX logo and black lettering with the name of the local co-operative across center. Custom manufactured for building signage at each location.
CENEX/LAND O LAKES PRIVILEDGE PANEL VARIOUS
(3) 1989-PRESENT $75
White aluminum litho panel with black & white stripe across top and bottom. Red and white CENEX logo, Green and white Land O Lakes logo and black lettering with the name of the local co-operative across center. Custom manufactured for building signage at each location.

MIDLAND COOPERATIVES - ST. PAUL, MN

Midland, founded in 1926, sold out to Land O'Lakes in 1985 and Land O'Lakes merged with Farmers Union Central Exchange
(CENEX) in 1988. All gasoline has been sold under the Cenex brand since that time.

IDENTIFICATION SIGNS

CO-OP 42"RD
(5) 1930-1946 $350
Green/white outlined orange porcelain sign with green outlined white center circle. White outlined green "CO-OP" arched aroudn top, "PRODUCTS" around bottom with "CO-OP" across center. Green outlined white "shaking hands" extend between the "Os" in CO-OP.
CO-OP 42"RD
(5) 1930-1946 $350
Same as above, with "GASOLINE" arched around top, "MOTOR OIL" around bottom.
MIDLAND 72x72
(5) 1959-1985 $100
Framed black outlined white porcelain can lype sign with red/white/black Midland "M" logo in center and "MIDLAND" in band across bottom.

FARMERS EDUCATIONAL AND COOPERATIVE STATE UNION (FELCO)-OMAHA, NE:

Marketed petroleum products under the Statex & Felco brands from the 1930's until the 1970's. Merged with Farmers Union Central Exchange in the 1970's.
No collectible size Statex/Felco peetroleum signs known at this time.

SOUTHERN STATES - RICHMOND, VA

Founded as Virginia Seed Service in 1923, Southern States began marketing petroleum products in the 1930s and continues today.

BUILDING SIGNS & LETTERS

SOUTHERN STATES INDEPENDENT DEALER 12x18
(4) 1992-PRESENT $35
Blue tin litho sign with red "SS" logo and white "Suothern/States" across upper area. "Independent/Dealer" in grey area below.

SECONDARY SIGNS

SS DIE CUT 12x12
(1) 1962-1992 $35
Red outlined white tin litho sign with black "double S" logo in center.
SS ON SQUARE VARIOUS
(1) 1962-1992 $50
Square white tin litho sign with red and black "double S" logo in center.

FARMERS CENTRAL EXCHANGE - RALEIGH, NC:

Merged with Southern States about 1980.

IDENTIFICATION SIGNS

FCX TRIANGLE 72x72
(5) 1950-1971 $125
Blue/white outlined red porcelain triangle with white FCX logotype lettering in center.

PUMP SIGNS

FCX TRIANGLE 8x8
(5) 1955-1965 $50
Blue white outlined red tin ltiho sign with white FCX logotype lettering in center,

TENNESSEE CO-OPERATIVES - LaVERGNE, TN

No collectible size Tennessee CO-OP petroleum related signs known at this time.

GROWMARK INDUSTRIES - BLOOMINGTON, IL

Founded as Illinois Farm Supply in the 1920s, gasoline was marketed under the "MAGIC ALADDIN" brand until after WW2. The "FS" brand was introduced about 1960 and continues in use today.

IDENTIFICATION SIGNS

ALADDIN GASOLINE 42"RD
(5) 1930s $450
White porcelain sign with red/white/blue IAA seal in center. White lettering in blue area of seal "ILLINOIS FARM SUPPLY COMPANY" with "PETROLEUM/SERVICE/PRODUCTS" in center. Red "ALADDIN" arched around top, "GASOLINE" around bottom.

ILLINOIS FARM SUPPLY GASOLINE 48"RD
(5) 1930's $400
White porcelain sign with red/white/blue IAA seal in center. White lettering in blue area of seal "ILLINOIS FARM SUPPLY COMPANY" with "PETROLEUM/SERVICE/PRODUCTS" in center. Red "GASOLINE" arched around top, "MOTOR OIL" around bottom.

CURB SIGNS

BLUE SEAL MOTOR OIL IAA SEAL 36"RD
(5) 1940's $350
Similar to above, with red outlined black "BLUE SEAL" arched around top, "MOTOR OIL" around bottom.

MISSOURI FARMERS ASSOCIATION - COLUMBIA, MO

MFA has marketed petroleum products since the 1930s, and continues to market through farm co-operatives and some retail locations in Missouri and Western Tennessee.

IDENTIFICATION SIGNS

MFA OIL COMPANY 42"RD
(5) 1930'S $350
Blue porcelain sign with white center circle. Red/white/blue MFA shield logo in center with "OIL COMPANY" underneath. White "GASOLINE-KEROSENE" arched around upper blue area, "MOTOR OILS-GREASES" around bottom.

MFA SHIELD SMALL 36x42
(5) 1930's $325
Die cut porcelain shield with three white stars in upper blue band and red and white vertical stripes below. Blue outlined white oval across stripes with blue "M-F-A" across oval. Sides of shield are very rounded

MFA SHIELD LARGE 72x72
(3) 1950-PRESENT $275
Die cut porcelain shield with three white stars in upper blue band and red and white vertical stripes below. Blue outlined white oval across stripes with blue "M-F-A" across oval. Sides of shield are less rounded than earlier sign.

CENTRAL CO-OPERATIVE WHOLESALE - SUPERIOR, WISCONSIN

IDENTIFICATION SIGNS

CO-OP OVAL 28x48
(5) 1935-1950 $175
Steel framed black outliend red oval with black outline white "CO-OP" across center.

CO-OPERATIVE GLF (Grange League Farm) - ITHACA, NY

Founded in the 1920s, GLF served farmers in New York and Pennsylvania as well as parts of New England. In 1964, GLF merged with Eastern States Farm Supply to form Agway and continues to market petroleum products under that name throughout the northeast.

IDENTIFICATION SIGNS

GLF QUALITY 24"
(5) 1940's $275
Red porcelain sign with white "omega" outline and large white "GLF" over smaller script "Quality" in center.

GLF QUALITY 24"
(5) 1940's $275
White porcelain sign with red "omega" outline and large red "GLF" over smaller script "Quality" in center.

LANDMARK CO-OPERATIVES - COLUMBUS, OHIO

SECONDARY SIGNS

QUALITY CO-OP PRODUCTS SEAL 16"RD
(5) 1950's $150
Blue/white outlined white porcelain sign with blue seven point co-op seal in center. White bar and circle on seal, with blue "QUALITY" arched around top, "PRODUCTS" around bottom and red "CO-OP" across center band.

INDIANA FARM BUREAU - INDIANAPOLIS, IN

SECONDARY SIGNS

FARM BUREAU CO-OP 16"RD
(5) 1950's $150
White porcelain sign with red and black eight point co-op seal in center. White "FARM/BUREAU" in upper red area, with white "CO-OP" below.

CROWN CENTRAL PETROLEUM - BALTIMORE, MD

Crown Central Petroleum traces its earliest history to the east Texas oil fields where in 1917 oil exploration got underway. By 1900, a Houston refinery was in operation providing high quality lubricants. In 1930, Crown's headquarters shifted to

Texaco 10.5"x12.5" small porcelain keyhole sign. *Dunbar's Gallery.*

Texaco 42" porcelain sign. *Dunbar's Gallery.*

Indian porcelain curved pump plate.

Texaco early cardboard sign. *Leo Mathieu.*

Texaco Filling Station early 42" sign. *Tom Kleeberg.*

Texaco Sea Chief embossed tin pump sign. *Mark McKeown.*

Standard Oil early porcelain No-Smoking sign.

Standard Oil 30" porcelain sign. *Mark McKeown.*

Standard Kerosene early tin flange sign.

Standard of Nebraska 30" porcelain sign. *Mark McKeown.*

Sinclair Opaline Motor Oil 4' porcelain sign. *Tom Kleeberg.*

Sinclair porcelain rest room sign.

Sinclair miscellaneous porcelain signs. (Sinco is 8"). *Bruce Bednar.*

Sinclair Opaline small porcelain sign. *Mark McKeown.*

Baltimore, Maryland under new ownership and in 1933 retail marketing in the east was added to a chain of service stations already operated in Houston. Throughout the 1930s and 40s jobbers were actively sought in markets from Pennsylvania to Alabama. In the mid-1960s Crown became involved in direct retail marketing with the addition of the "Peoples" stations of Energy Oil Co. So successful was direct retailing that by the early 1970s jobber accounts were abandoned in favor of all company operated stations. Crown also entered the convenience store market in the early 1980s with the purchase of Sunoco's "Fast Fare" chain in the southeast. Crown products are currently available at over 600 locations in the southeast.

IDENTIFICATION SIGNS

CROWN OIL/GAS 48"round
(5) 1930s-1945 $350
Framed round orange porcelain sign. Black band arched across center with white "CROWN" on band. White "GAS * OIL" below.

CROWN "wing" 54x84
(3) 1945-1966 $250
Pressed tin/lithographed diecut wing shaped signs assembled back to back. Removable formed tin star. Blue outline around orange logo, blue outlined white "CROWN" across center.

CROWN "wing" 32x72
(5) 1950s-1966 $350
Porcelain diecut sign with formed upper "frame." Removable star. Orange with blue outlined white "CROWN" across center.

CROWN "wing" ON SHIELD 48x84
(4) 1966-1973 $100
Pressed tin/lithographed shield shaped sign assembled can style. Blue outlined white with Crown wing logo in center.

BUILDING SIGNS

CROWN "wing" 54x84
(3) 1956-1966 $200
Single sided version of identification sign listed above. Assembled on wooden frame & mounted on building pylons.

"COMPLETE CAR SERVICE" 16"
(5) 1946-1960 $75
Orange porcelain lettering.

"COMPLETE CAR SERVICE" 16"
(5) 1956-1973 $50
Orange plasic lettering.

ILLUMINATED SIGNS

CROWN "wing" 54x84
(5) 1952-1966 $125
Formed plastic freeform sign faces for internally lit sign.

PUMP SIGNS

CROWN REGULAR 9x12
(4) 1940-1956 $75
Black tin litho sign with orange "CROWN OIL/GAS" logo above white "REGULAR"

CROWNZOL PREMIUM 9x12
(5) 1940-1956 $75
Orange tin litho sign with white "CROWNZOL" logo above white "PREMIUM."

CROWNZOL SUPER PREMIUM 12x24
(5) 1955-1956 $75
Blue outlined gold tin litho sign. Orange "CROWNZOL/SUPER/PREMIUM" over blue "Higher Octane/than Maximum/Requirement for/All NEW Cars/U.S. or Foreign."

(CROWN GOLD)"King of all Premiums" 10x16
(4) 1956-1973 $35
Blue tin litho sign with gold outlined white lettering "KING/OF ALL/PREMIUMS/ULTRA HIGH OCTANE.

(CROWN SILVER)"Sterling Performance" 10x16
(4) 1956-1973 $35
Orange tin litho sign with black outlined white lettering "Sterling/Performance/AT REGULAR/PRICE."

CROWN WING ON BLUE OVAL 4x7
(3) 1960-1966 $50
Blue porcelain oval with orange/white/blue porcelain wing logo. Used as blanking plate over site glass openings.

OIL RACK SIGNS

CROWN MOTOR OIL 14x18
(4) 1946-1956 $75
White tin litho sign with large orange/black "CROWN MOTOR OIL" logo.

CROWN PREMIUM MOTOR OIL 14x18
(5) 1946-1956 $75
Blue tin litho sign with large Crown "wing" logo over "PREMIUM."

CROWN PREMIUM MOTOR OIL 14x18
(5) 1956-1962 $75
Formed gold tin litho embossed signs assembled back-to-back. Orange wing logo at top over blue "V." Blue outlined orange "PREMIUM" below.

LUBRICATION/MOTOR OIL SIGNS

CROWN PREMIUM MOTOR OIL "CAN" 24x36
(3) 1956-1962 $65
Gold tin litho die-cut sign-image of large oil can.

SECONDARY SIGNS

CROWN CREDIT CHECKS HONORED 12x20
(4) 1940s $100
Blue tin litho flanged sign with separate flange mount. Orange wing logo to upper left, with black outlined "CHECKS" banner to right. White "HONORED HERE" below.

CROWN CREDIT CARDS HONORED 12x20
(4) 1950s $100
Blue tin litho flanged sign with large orange wing logo above white "CREDIT CARDS HONORED."

HIGHEST OCTANE AT LOWEST PRICE - CROWN GOLD
(5) 1956-1960 $150
Blue tin litho sign with gold lettering.

TRUCK SIGNS

CROWN TIN WING ON BACKGROUND 16x36
(5) 1950s $200
Blue tin litho sign with orange and white Crown wing logo in center.

THRIFT

Was a 1950s Crown secondary brand.
No collectible size Thrift signs known to exist.

ENERGY OIL COMPANY - ANNANDALE, VA

Founded as Peoples Self-Service Gas Stations, Inc., in Annandale, Virginia in the late 1940s, Peoples operated a chain of retail gasoline stations in Virginia, Maryland and Washington,

DC. By the mid-1960s, the corporate name had become "Energy Oil Company" and over 20 stations were being operated. In 1972 Crown, who had supplied Peoples for years, purchased the chain and rebranded the stations Crown. Peoples then became a Crown secondary brand, appearing on a few remaining jobber-operated stations in North Carolina. The name was re-introduced for use on some company operated stations Crown purchased from Ashland in 1985, but these were rebranded Crown in 1987 and the Peoples brand is no longer in use.

No collectible size Peoples signs known to exist.

DERBY OIL CO. - WICHITA, KS

Derby Oil Co. was a mid-continent refiner/marketer founded in Wichita, Kansas in 1920. In the late 1930s, Derby was instrumental in the creation of Sovereign Service, a marketing organization for independent dealers supported by a number of mid-continent refiners. After WW2, Derby concentrated on an expanding chain of retail service stations as well as numerous dealer operated units and a large jobber/marketer arrangement with Piester. In 1955 Colorado Interstrate Gas Company purchased Derby and with it's large oil production system allowed for greater expansion of the Derby refinery and marketing network. In the early 1970s Colorado Interstate Gas became a part of the Coastal Corporation and in 1988 Derby stations, along with others in the Coastal organization were rebranded Coastal.

IDENTIFICATION SIGNS

DERBY GASOLINE 30"RD
(3) 1930-1950 $300

White outlined blue porcelain sign with white outlined red center circle. White "DERBY" arched around top, "GASOLINE" around bottom with blue detailed white star in center. Blue "D.O.CO." on star.

DERBY 48"RD
(4) 1950-1988 $250

DERBY 72"RD
(2) 1950-1988 $150

Blue/white outlined red poircelain sign with white star in center. Blue band across star, with white "DERBY" in band.

CURB SIGNS

DERBY PRODUCTS 30"RD
(3) 1930-1950 $325

White outlined blue porcelain sign with white outlined red center circle. White "DERBY" arched around top, "PRODUCTS" around bottom with blue detailed white star in center.

DERBY'S FLEXGAS ETHYL 30"RD
(4) 1930-1940 $325

Red porcelain sign with black outlined white center circle. White "DERBY'S" arched around top, "FLEXGAS" around bottom and Ethyl
(EGC) logo in center circle.

SOVEREIGN SERVICE

Sovereign Service was a franchise program allowing independent dealers supplied by the various mid-continent refiners that participated in the program to be branded "Sovereign Service." Dealers were held to high standards of operation, supported by cooperative advertising of the programs' features (independent dealers only, clean rest rooms, 24 hour service, etc.). Derby was a leading participant in the program, along with Kanotex/Apco, Vickers, Rock Island, Eason, Elreco, Yale, Shallow Water and perhaps others. The program was initiated in the late 1930s (about 1937) and survived as late as 1959. Globes from the individual brands are listed with that company's listing in this book.

IDENTIFICATION SIGNS

SOVEREIGN SERVICE SHIELD 42"
(4) 1937-1959 $275

SOVEREIGN SERVICE SHIELD 72"
(3) 1937-1959 $200

Steel framed maroon outlined yellow die-cut porcelain shield. Smaller maroon shield at top with yellow "SS" in center. Maroon "SOVEREIGN/SERVICE" diagonal across sign face.

CURB SIGNS

SOVEREIGN SERVICE SHIELD 30"
(3) 1937-1959 $350

Steel framed maroon outlined yellow die-cut porcelain shield. Smaller maroon shield at top with yellow "SS" in center. Maroon "SOVEREIGN/SERVICE" diagonal across sign face.

DIXIE DISTRIBUTORS, INC. - ANN ARBOR, MI

Dixie Distributors, Inc., was a distributor association put together by a group of Michigan based oil jobbers in Ann Arbor, Michigan in 1926. Almost immediately independent jobbers from all around the country applied for a Dixie franchise and joined the association in order to be able to market their products under the Dixie name. The organization expanded greatly until World War II, when gasoline shortages forced many independents, including many of the Dixie affiliates out of business. Marketing efforts by the majoro after WW II were so much greater than an independent organization could hope to compete with and many former Dixie jobbers eventually were branded by the majors. Nevertheless, Dixie stations survived until the 1960s or later in some areas, and the organization survives today.

IDENTIFICATION SIGNS

DIXIE OILS GASOLINE 42x72
(4) 1930's $300

White outlined blue porcelain oval with white oval in center. White "DIXIE" arched around top, "GASOLINE" around bottom, with blue outlined yellow diamond in center. Red "OILS" on diamond.

PUMP SIGNS

DIXIE EGC ETHYL 12"RD
(4) 1949-1960 $100

White outline blue sign with white circle in center. White "DIXIE" arched around top, "ETHYL" around bottom with Ethyl (EGC) logo in center.

NEON

DIXIE OILS GAS 6' 42x72
(4) 1930's $300

White outlined blue porcelain oval with white oval in center. White "DIXIE" arched around top, "GAS" around bottom, with blue outlined yellow diamond in center. Red "OILS" on diamond. Punched for neon tubing.

CURB SIGNS

DIXIE OILS GASOLINE POWER TO PASS 24x30
(4) 1930's $375
Keyhole shaped tin litho embossed sign. Blue sign with Dixie Oils Gasoline logo at top over white "POWER TO PASS"

DIXIE OILS GASOLINE POWER TO PASS 24x36
(4) 1940's $300
Steel framed porcelain sign. Large Dixie Oils Gasoline logo at top over blue "POWER TO PASS"

SECONDARY SIGNS

DIXIE OILS GASOLINE POWER TO PASS 36x60
(4) 1930's $275
Wood framed yellow tin litho sign. Large Dixie Oils Gasoline logo at top over blue "POWER TO PASS."

DIXIE OILS GASOLINE 48x72
(4) 1940's $275
Die cut tin litho sign with reflective white areas. Large Dixie Oils Gasoline logo at top, extending above yellow rectangle. Blue stripes across yellow area interrupted by log. Blue "THE POWER TO PASS, THAT'S DIXIE GAS" across bottom.

THE POWER TO PASS THAT'S DIXIE GAS 30x48
(5) 1930's $275
White tin litho sign with Dixie Oils Gasoline logo to lower left. Blue "the Power to Pass" across top, with blue "that's/DIXIE/GAS" to right of logo.

MC NUTT OIL & REFINING CO/EL PASO - EL PASO, TX: Large Dixie branded jobber that later marketed under the El Paso brand.

IDENTIFICATION SIGNS

DIXIE DIAGONAL ON LARGE OVAL 42x72
(5) 1950's $150
Blue outlined white porcelain oval with rede "DIXIE" diagonal across center.

PUMP SIGNS

EL PASO DIE CUT TORCH 10x12
(4) 1958-1964 $275
Die-cut porcelain sign. Blue outlined white rectangle across blue torch shaft with red fame. Red "EL PASO" on rectangle.

DIXIE 390 10x12
(4) 1958-1964 $175
Red outlined whit eporcelain rectangular sign with blue outlined red oval in center. White "DIXIE" in oval. Red "390" across bottom of sign.

DX AND AFFILIATES MID-CONTINENT PETROLEUM - TULSA, OK

Mid-Continent was organized in 1916 from the various holdings of oil pioneer Josh Cosden and others. Gasoline marketing under the Diamond brandname began about 1920 and by the mid-1920s, Mid-Continent operated a large number of stations throughout the Central United States. A super-premium fuel "Diamond D-X" was introduced about 1933 and by 1940 stations were branded as "Diamond D-X" stations. After WW II the brandname became simply "D-X." In 1955 Mid-Continent merged with Sunray Oil Co. and the corporate name became Sunray DX Oil Company, revised in 1962 to DX-Sunray Oil Company. In 1968 DX-Sunray merged with Philadelphia based Sun Oil Company and many dual logo Sunoco/DX items were introduced at this time. DX globes had been phased out by the time of the merger. Shortly after the merger the Sunoco brandname replaced DX in states where Sunoco already marketed. The DX brand was retained in other areas until after 1980 when Sunoco began gradually replacing DX. As of this writing only small areas of DX branding remain, primarily jobber operated stations that just haven't yet been converted.

IDENTIFICATION SIGNS

DIAMOND GASOLINE MOTOR OIL 36x60
(5) 1925-1933 $225
White porcelainsign with black corners and white band across bottom. Red/white "DIAMOND" logo in center with black "GASOLINE MOTOR OIL" across lower band.

DIAMOND 40x72
(5)1933-1946 $150
Steel framed black/tan outlined red porcelain flat-diamond shaped sign with tan diamond in center. Black "DIAMOND" across center of sign.

D-X 40x72
(2)1946-1957 $150
Steel framed black/tan outlined red porcelain flat-diamond shaped sign with tan diamond in center. Black "D-X" across center of sign.

D-X 52x108
(5)1946-1957 $100
Steel framed black/tan outlined red porcelain flat-diamond shaped sign with tan diamond in center. Black "D-X" across center of sign.

DX 52x108
(2)1957- $100
Porcelain can type sign with flat-diamond shaped white porcelain faces. Red triangles top and bottom of face forming white band across center. Large blue "DX" on center band.

BUILDING LETTERING & SIGNAGE

DIAMOND SERVICE STATION 12x144
(4)1930-1946 $125
White outlined black embossed porcelain sign with white "Diamond Service Station" across sign.

DX OLD DIAMOND 36x60
(3)1946-1957 $250
Red porcelain channel forming flat-diamond frame, hollow in center. Black porcelain channel "D-X" over hollow center.

DX NEW DIAMOND 36x60
(3)1957-1970 $200
White porcelain channel type sign with formed red triangles to and bottom of flat-diamond face. Blue formed "DX" on white center.

PUMP SIGNS

DX 7x12
(5)1957-1974 No Listing
Die-cut white porcelain flat-diamond sign with red triangles top and bottom with blue "DX" on center band.

DX 12x14
(4)1957-1974 $35

DX 14x24
(1)1957-1974 $35

DX 18x24

(2)1957-1974 $35
Red porcelain pump panel with white "flat-diamond" DX logo with red triangles and blue "DX" at top of panel.

DX
Gold anodized aluminum panel with litho white flat-diamond "DX" logo with red triangles and blue "DX" at top of panel.

DX BORON 14x24
(3)1957-1960 $35
DX BORON 18x24
(4)1957-1960 $35
Gold anodized aluminum panel with litho white flat-diamond "DX" logo with red triangles and blue "DX" at top of panel. Red "BORON" below.

DX SUPER BORON 14x24
(1)1960-1971 $35
DX SUPER BORON 18x24
(2)1960-1971 $35
Gold anodized aluminum panel with litho white flat-diamond "DX" logo with red triangles and blue "DX" at top of panel. Red "SUPER/BORON" below.

DX MARINE 14x24
(4)1957-1974 $250
Blue porcelain pump panel with white "flat-diamond" DX logo with red triangles and blue "DX" at top of panel. White "MARINE" and wave line below, with white sppedboat to either side and red "GASOLINE" below

NEON SIGNS

DX small can-type neon over porcelain
(5)1946-1957 $500
Black/tan outlined red porcelain flat-diamond shaped sign faces assembled can-type. Tan diamond in center. Black "D-X" across center of sign. Neon outline and "D-X."

CURB SIGNS

NEVRNOX ETHYL
(EGC) 30"RD
(3) 1926-1935 $375
Red outlined cream color sign with black outlined red "NevrNox" arched across top. Ethyl
(EGC)logo with yellow rays below.

SUNRAY DX w/OLD LOGO 25"
(4)1955-1958 No Listing
Octagon shaped die-cut sign. Black outlined orange porcelain sign with green band across bottom. Red outlined cream diamond with black "D-X" on lower green band with white "PETROLEUM PRODUCTS" below diamond. Yellow sun and rays on upper orange area with black "SUNRAY" arched around bottom of sun.

SUNRAY DX w/NEW LOGO 25"
(5)1957-1962 No Listing
Octagon shaped die-cut sign. Red outlined white porcelain sign. Red outlined tan sun and rays at top with white outlined dark blue "SUNRAY" arched around bottom of sun. Red, white, and blue "DX" flat diamond logo below with red "PETROLEUM PRODUCTS" arched under logo.

LUBRICATION/MOTOR OIL SIGNS

DIAMOND MOTOR OIL 20x28
(2) 1930-1935 $100
Cream colored octagon shaped porcelain sign with red diamond around cream diamond and red solid diamond in center. Black "DIAMOND" across center diamond with "MOTOR/OIL" below.

DIAMOND 760 MOTOR OIL 32x60 $150
Cream colored porcelain die-cut sign with red and black "DIAMOND" logo on upper sign face. Large black outline "760" below diamond with cream "MOTOR OIL" on red band across bottom.

REFILL HERE DIAMOND 760 32x60
(5) 1930s $150
Black outlined cream tinlitho sign with red and black "DIAMOND" logo in center. Black outline "760" at bottom of logo over black "MOTOR OIL." Cream "REFILL HERE" across top.

DIAMOND 760 MOTOR OIL 22x30
(4) 1930s $150
Cream tin litho sign with black band across top and bottom. Red and cream flat diamond logo in center with black "DIAMOND" across center. Black outlined "760" below, with cream "MOTOR OIL" across bottom band.

DX MOTOR OIL STRIP 12x72
(3) 1946-1958 $75
Black outlined cream self-framed porcelain sign with red & black DX diamond logo at left. Black "MOTOR OIL" across strip.

DX MOTOR OIL 24x30
(3) 1946-1957 $75
Die cut tin litho sign. Cut out to enclose red and tan flat diamond logo at top with black "D-X" across center. Black "MOTOR OIL" on tan area below.

WE RECOMMEND DX MOTOR OIL 24x30
(2)1946-1957 $75
Tan tin litho sign with red and black D-X logo in center. Black "WE RECOMMEND" across top, "MOTOR OIL" across bottom.

SECONDARY SIGNS

DIAMOND CREDIT CARDS HONORED 13x18
(5)1935-1946 $100
Black porcelain sign with cream band across upper sign face. Red flat-diamond logo outline with black "DIAMOND" across center in upper cream area with cream "CREDIT CARDS/ HONORED HERE" on black area below.

DIAMOND NEVRNOX GASOLENE 24x32
(4) 1926-1930 $125
Red tin litho sign with tan flat diamond in center. Small red flat diamond at top with black "DIAMOND" across center. Black "NevrNox" logotype across center over red band. Tan "GASOLENE" on band. Black "STOPS CARBON" below.

FLANGE SIGNS

DIAMOND GAS/OILS 14x20
(5) 1910s $225
White outlined red tin litho sign with white diamond in center. White/red outlined black "DIAMOND" across center diamond with black outlined white "GAS" across top, "OILS" across bottom.

BULK PLANT SIGNAGE

DX SUNRAY OIL COMPANY 12x144
(3) 1955-1958 $125
White outlined black embossed porcelain sign with white "D-X Sunray Oil Company" across sign.

SUNRAY OIL

Sunray, a mid-continent area producer/refiner dating back to the 1920s, was historically very involved with production and unbranded marketing with very little attention paid to branded "Sunray" marketing. Gasoline was sold under the company name through a small number of jobbers and due to the limited branded marketing, the variety of signs available is limited and all are rare. Sunray Oil merged with Mid-continent Petroleum in 1955 and the "DX" brand was added to Sunray logos at existing Sunray stations. About 1962 those few stations became simply "DX" stations.

PUMP SIGNS

SUNRAY DX PETROLEUM PRODUCTS 8.5x8.5
(5) 1955-1958 No Listing
Octagon shaped die-cut sign. Black outlined orange porcelain sign with green band across bottom. Red outlined cream diamond with black "D-X" on lower green band with white "PETROLEUM PRODUCTS" below diamond. Yellow sun and rays on upper orange area with black "SUNRAY" arched around bottom of sun.

CURB SIGNS

SUNRAY NATURAL POWER OILS 25x25
(4) 1950-1955 No Listing
Octagon shpaed die cut porcelain sign. Black outlined orange porcelain sign with green band across bottom. Yellow sun and rays on upper orange area with black "SUNRAY" arched around bottom of sun above black "NATURAL POWER." White outlined black "OILS" on green band.

SECONDARY SIGNS

SUNRAY NATURAL POWER OILS 7x7
(4) 1950-1955 No Listing
Octagon shaped die-cut porcelain sign. Black outlined orange porcelain sign with green band across bottom. Yellow sun and rays on upper orange area with black "SUNRAY" arched around bottom of sun above black "NATURAL POWER." White outlined black "OILS" on green band.

PREMIER - LONGVIEW, TX

Independent Texas refiner with stations in Texas, Oklahoma, Arkansas and Louisiana. In 1964 the company merged with Sunray DX. Some stations were spun off to Vickers, while DX retained the Premier, also introducting the "Freeway" name as another secondary brand in conjunction with Premier. Some unsuccessful DX stations were later converted to Premier or Freeway and the brands survived at a small number of locations until DX rebranded Sunoco in the 1980s.

PUMP SIGNS

SKY RANGER 10"RD
(4) 1950-1964 $225
Blue/white/red outlined white porcelain sign with blue "Sky/Ranger" diagonal across sign face. Red airplane with trail below

FREEWAY NEVR-NOX ETHYL 12x13.5
(2) 1964-1970 $50
White porcelain sign with red lower half. Blue "Freeway" logotype across top with white "NEVR-NOX/ETHYL" on red area below.

CURB SIGNS

SKY RANGER AVIATION OIL 26x29
(3) 1946-1960 $450
Blue outlined white porcelain tombstone shaped sign with large red circle on upper sign face. Blue outlined white "Sky/Ranger" on circle with white airplane and trail above. Blue "AVIATION OIL" on lower sign face below circle.

SKY RANGER OIL/PREMIER RANGER 26x29
(3) 1946-1960 $450
Blue outlined white porcelain tombstone shaped sign with large red circle on upper sign face. Blue outlined white "Sky/Ranger" on circle with white airplane and trail above. Blue "AVIATION OIL" on lower sign face below circle. Reverse side is blue outlined white with large red circle on upper sign face. Blue outlined white "Ranger" on circle with white "PREMIER" above and "100% PARAFFIN BASE" below. Blue "MOTOR OIL" on lower sign face below circle.

BARNSDALL REFINING CO. - BARNSDALE, OK

One of the oldest oil refiners in the business, Barnsdall began retail gasoline marketing in 1921. At one time one of the largest mid-continent marketers, Barnsdall was reorganized during the depression as Barnsdall Refining and in 1940 the Bareco brandname replaced the Barnsdall brand. In 1952 Bareco closed their last refinery to concentrate on wax production and the gasoline marketing was sold off to Mid-Continent Petroleum. Mid-Continent eventually phased out the Bareco brand after the merger with Sunray.

IDENTIFICATION SIGNS

BARNSDALL BE SQUARE PRODUCTS 48"RD
(4) 1922-1935 $550
Blue outlined white porcelain sign with blue outlined square in center. Large blue "B" in center square with small "BE" above and "SQUARE" below. Blue outlined red "BARNSDALL" arched around top, "PRODUCTS" around bottom.

BARNSDALL "B" DEPENDABLE PRODUCTS 48x48
(4) 1935-1942 $500
Blue outlined white porcelain sign with blue outlined square in center. Large blue "B" in center square. Red "BARNSDALL" across upper sign face with red "Dependable Products" across bottom.

BUILDING SIGNS AND LETTERS

THE WORLD'S FIRST REFINER BARNSDALL 48x96
(5) 1921-1930 No Listing
Elaborate die-cut multi-color porcelain sign. Blue outlined white cloumns form frame around lower blue area with white cloud at bottom. Blue outlined red "BARNSDALL" on lower cloud area. World globe in center extending above blue area with red/blue outlined square with blue "B" in center superimposed over globe. White "The World's" to left of globe on blue area, Red "First" and white "Refiner" to right. White band arched above globe with blue squares & "B" at each end and red "BE SQUARE TO YOUR MOTOR" around top.

PUMP SIGNS

BARNSDALL BE SQUARE PRODUCTS 8"RD
(4) 1922-1935 $500
Blue outlined white porcelain sign with blue outlined square in center. Large blue "B" in center square with small "BE" above and "SQUARE" below. Blue outlined red "BARNSDALL" arched around top, "PRODUCTS" around bottom.

CURB SIGNS

BE SQUARE MOTOR OIL 30"RD
(4) 1925-1935 $375
Blue outlined white porcelain sign with blue outlined square in center. Large blue "B" in center square with small "BE" above and "SQUARE" below. Blue outlined red "BE SQUARE" arched around top, "MOTOR OIL" around bottom.

BARNSDALL SUPER GAS ETHYL 30"RD
(3) 1926-1935 $350
Blue outlined white porcelain sign with blue "BARNSDALL" arched around top over red "SUPER-GAS." Ethyl (EGC) logo with yellow rays at bottom.

SECONDARY SIGNS

DEPENDABLE BARNSDALL PRODUCTS 12x20
(5) 1920s $200
Red outlined blue porcelain flange sign with white band diagonal across center. Red outlined blue "BARNSDALL" across band, with white outlined red "DEPENDABLE" above band and "PRODUCTS" below. White square in lower right corner with blue "Square B" logo.

BE SQUARE ARROW 6x24
(5) 1930s $225
Die cut red porcelain arrow with white "BE SQUARE TO YOUR MOTOR" across arrow.

TRUCK SIGNS

BE SQUARE TO YOUR MOTOR 3"RD
(4) 1930s $225
Red porcelain sign with white center circle. Blue square with blue "B" in center. White "BE SQUARE TO" arched around top, "YOUR MOTOR" around bottom.

MONARCH MANUFACTURING CO. - COUNCIL BLUFFS, IA

Owner of the famous "MonaMotor" trademark,promoted through their own radio station "K.O.I.L.," Monarch began gasoline marketing about 1920 and merged into Barnsdall about 1930.

IDENTIFICATION SIGNS

MONAMOTOR 24x42
(5) 1925-1930 No Listing
Black outlined red porcelain sign with black outlined white flat diamond covering most of red area and black band across bottom. Black "MonaMotor" across center of diamond over "Oil," white "100%/Paraffin Base" in lower left corner, "Tune in/KOIL" in lower right. White "BEST BECAUSE PUREST" in balck band across bottom.

LUBRICATION/MOTOR OIL SIGNS

MONA MOTOR OIL 10x28
(4) 1925-1930 No Listing
Red tin litho sign with white diamond shaped band across center. Black underlined "MonaMotorOil" across band.

SECONDARY SIGNS

EXPERT CAR WASHING 10x28
(5) 1928-1930 $75
Blue/black outlined white tin litho sign with black "EXPERT/CAR WASHING" across top. Smaller "A Careful High Grade/Inspected Job" below with outlined MonaMotor logo in lower right corner.

BATTERY SERVICE 10x28
(5) 1928-1930 $75
Blue/black outlined white tin litho sign with black "BATTERY SERVICE" across top. Smaller "OUR BATTERIES/Last Longer Cost Less" below with outlined MonaMotor logo in lower right corner.

TIRE SERVICE TUBE VULCANIZING 10x28
(5) 1928-1930 $75
Blue/black outlined white tin litho sign with black "TIRE SERVICE" across top. Smaller "TUBE/Vulcanizing" below with outlined MonaMotor logo in lower right corner.

ACCURATE INFORMATION AND ROAD MAPS 10x28
(5) 1928-1930 $75
Blue/black outlined white tin litho sign with black "ACCURATE INFORMATION/and ROAD MAPS" across top. Smaller "for Tourists" below with outlined MonaMotor logo in lower right corner.

HAWKEYE OIL COMPANY

Iowa marketer operating stations under the Hawkeye and Parco (Not affiliated with Producers and Refiners Parco brand) brandnames. Purchased by Mid-Continent about 1930.
No collectible size Hawkeye or Parco signs known at this time

ERICKSON & AFFILIATES - MINNEAPOLIS,MN

Founded in Barron, WI in 1928, Erickson was originally a wholesale grocery company. They entered the gasoline business in 1931, opening a small discount type station in Barron. After World War II Erickson moved their headquarters to Minneapolis and greatly expaned their operating, opening stations under the Erickson and Life brands. Also about this time family members began a small refining operation in nearby St.Paul known as Northwestern Refining. In 1950 other family members began operating a second Erickson chain, calling their company Erickson Oil Products and basing it in Hudson, WI. Being involved in the grocery business gave Erickson the opportunity to introduce the "Holiday Stationstore" concept in 1959 and began a gradual conversion of their operation over to the new convenience store method of operation. Billed as "America's Most Unusual Service Stations," Holiday Stationstores offered everything from automotive products to sporting goods to Holiday brand bicycles. Based on the success of the Holiday operation, those family members operating Northwestern Refining began a second convenience store operation they called SuperAmerica. Suddenly the entire petroleum industry was watching this Minnesota marketer. Ashland Oil purchased the Northwestern Refining/SuperAmerica operation in 1970, and it continues to lead the convenience store industry today. By the 1970's virtually all of the Erickson locations had become Holiday Stationstores, and the company operates approximately 300 Holiday stations today, along with Holiday supermarkets throughout the upper midwest.

IDENTIFICATION SIGNS

ERICKSON OVAL 42x84
(5) 1946-1970 $200
Yellow porcelain oval with red script "Erickson" across center.

HOLIDAY - WORLD WIDE, INC. - MINNEAPOLIS, MN:
Erickson Oil Company became "World Wide, Inc." in the early 1960's and continues top operate under that corporate heading today.

SECONDARY SIGNS
HOLIDAY CREDIT CARDS 12x16
(5) 1960-1972 $35
White rectangular tin litho sign with red outline. Blue "Holiday" logotype across center, with red "USE YOUR" above and "CREDIT CARD" below.

ERICKSON OIL PRODUCTS - HUDSON, WI:
Founded in 1950 by Erickson family members, Erickson Oil Products operates supermarkets under the "Erickson's More 4" banner and convenience stores known as Erickson's Freedom Valu-Centers. Stations are located throughout the upper midwest.
No collectible size signs known at this time.

EXXON AND AFFILIATES STANDARD OIL COMPANY OF NEW JERSEY - NEW YORK, NY
Esso is, of course, Standard Oil of New Jersey, largest of the Standard companies to emerge from the 1911 breakup of Standard Oil. Standard (NJ) began marketing in several eastern seaboard states and in Tennessee, Louisiana, and Arkansas through Standard of Louisiana, a subsidary assigned to them with the 1911 breakup. The Esso brand was first introduced in 1926 as a Ethyl grade gasoline and in 1933 replaced Standard as the company's brandname. Esso marketing expanded to include stations in 19 states with marketing after WW2 added throughout the country under various brandnames detailed below. In 1959 all domestic marketing was grouped under the subsidary Humble Oil and Refining, formerly an independent company that Standard purchased in 1919. Under Humble, the Esso brandname was retained in the east and south. Enco was introduced to replace the various other regional brands and Humble was used in Ohio when Sohio objected to both Esso and Enco. In 1972 the brandname Exxon was introduced to replace Esso, Enco and Humble, although Enco remained in use in the upper midwest area until that area was abandoned several years later. Exxon continues to market through over 11,000 stations in 40 states.

STANDARD & ESSO IDENTIFICATION SIGNS
"STANDARD" GASOLINE 18x36
(3) $125
Red porcelain sign with white "STANDARD"/GASOLINE" across sign face.
"STANDARD" OIL COMPANY'S GASOLINE
(5) 1915-1917 $450
Red porcelain sign with white "STANDARD OIL COMPANY'S" across the top above with "S" logo with ""STANDARD" in upper space, "GASOLINE" in lower space and red "MOTOR" on center stroke of "S."
STANDARD MOTOR GASOLINE 30"RD
(2) 1917-1924 $300
Blue outlined white porcelain sign with large red "S" in center with red crescents on wither side. Blue outlined white "STANDARD/MOTOR/GASOLINE" on stroke of red "S" with blue "MOTOR" in upper "S" space, "OIL" in lower.
GASOLINE/STANDARD/MOTOR OIL 30"RD
(3) 1924-1926 $250
White porcelain sign with red line bar & circle logo in center. Blue "STANDARD" in center bar with red "GASOLINE" arched around top, "MOTOR OIL" around bottom.
STANDARD/SCRIPT ESSO 42"RD
(2) 1926-1933 $250
One of the few oil industry signs with two different faces. One side is white porcelain sigh red line bar & circle logo design. Blue "STANDARD" in bar area across center. Reverse has wide red border and thin blue/white line around white center circle. Blue script "Esso" diagonal across center.
STANDARD SIGN FOR GLOBE INSERT 36x84
(4) 1926-1933 No Listing
Cany type sign with white outlined red porcelain colonial shield shaped faces. White "STANDARD" across sign face. Cutout at top for globe lens to show through, having been mounted in frame attached to internal sign structure.
STANDARD W/ PRIVILEDGE PANEL 42x84
(5) 1930-1933 $300
Steel framed porcelain sign with white outlined red panel across bottom. White "STANDARD" across lower panel. Upper panel is white outlined blue with white "DEALER NAME" across panel
STANDARD ESSO STATION "MINOR SIGN"
(5) 1933-1938 $300
STANDARD ESSO STATION "MAJOR SIGN" 42x84
(5) 1933-1938 $100
Dark blue outlined white oval porcelain sign with red "Esso" logotype across center. Blue "STANDARD" above logotype, "STATION" below.
"DEALER" PANEL 8x18
(5) 1935-1938 $40
White porcelain sgin with blue outlined white "DEALER" across sign. Designed as a conversion panel for above "major sign" , bolted over "STATION" when the station network converted to dealer operated stations.
STANDARD ESSO DEALER "MINOR SIGN"
(2) 1933-1946 $300
STANDARD ESSO DEALER "MAJOR SIGN" 42x84
(3) 1933-1946 $100
Dark blue outlined white oval porcelain sign with red "Esso" logotype across center. Blue "STANDARD" above logotype, "DEALER" below.
ESSO 42x84
(1) 1946-1965 $150
Dark blue outlined white oval porcelain sign with red "Esso" logotype across center.
ESSO
ESSO 42x84
(4) 1965-1972 $100
White porcelain sign with wide light blue border. Red "Esso" logotype across center.

BUILDING SIGNS AND LETTERS
ESSO SERVICE 16x48
(5) 1933-1946 $75

Sinclair Opaline small porcelain sign. *Mark McKeown.*

Sinclair Track Stop porcelain sign. *Neil Stockwell.*

Sinclair Aircraft 42" porcelain sign.

"Shell" Air small porcelain sign.

Shell credit card porcelain sign. *Jim Hollabaugh.*

Shell miscellaneous porcelain signs. *Jim Hollabaugh.*

Shell Motor Oil metal flange. *Jim Hollabaugh.*

Golden Shell porcelain sign.

Unusual Shell porcelain Locomotive sign. *Bruce Miller.*

Shell Green Streak porcelain pump plate. *Bruce Miller.*

Magnolia Magnolene 30" porcelain sign. *Bruce Bednar.*

Union 76 porcelain 24" sign.

Col-tex 30" porcelain sign. *Bruce Bednar.*

White porcelain panel with blue "Esso Service" across center.
ESSO SERVICENTER BLUE PORCELAIN 14"
(4) 1946-1961 $200
Blue die-cut domed porcelain lettering "Esso Servicenter."
HAPPY MOTORING BLUE PORCELAIN
(5) 1946-1961 $200
Blue die-cut domed porcelain lettering "Happy Motoring."

PUMP SIGNS

"STANDARD" GASOLINE 8x16
(5) 1918-1923 $225
Blue outlined white curved porcelain sign with blue outlined red "STANDARD.
(in quotes)/GASOLINE" across center.
USE STANDARD MOTOR OIL 10x10
(5) 1918-1923 $225
Blue outlined white porcelain sign with small blue "Use" above blue "STANDARD/MOTOR/OIL."

NEON SIGNS

STANDARD ESSO DEALER NEON 42x84
(5) 1935-1946 No Listing
Can-type neon over porcelain sign. Each face is dark blue outlined white oval with red "Esso" logotype across center. Blue "STANDARD" above logotype, "DEALER" below. Neon outline and "Esso."

CURB SIGNS

ESSO AVIATION PRODUCTS/STAND CLEAR 24"RD
(4) 1946-1960 $650
Red porcelain sign with blue lower half. Blue and white wings on blue outlined white oval with red "Esso" logo on upper red area. White "AVIATION/PRODUCTS" on lower blue area. Reverse side of sign is solid red with white "STAND/CLEAR" above winged Esso oval logo.
ESSOLUBE MOTOR OIL 24"RD
(4) 1933-1940 $250
White porcelain sign with red "Essolube" over blue "MOTOR/OIL" and red dot.
ASK FOR STANDARD MOTOR OIL 24"RD
(5) 1924-1933 $300
White porcelain sign with blue outlined "Ask for/STANDARD/MOTOR OIL" on upper sign face above small red/white/ blue "STANDARD" bar & circle.
STANDARD MOTOR OIL/GASOLINE 30"RD
(3) 1924-1933 $250
Blue outlined white porcelain sign with blue outlined "STANDARD/MOTOR OIL" on upper sign face above small red/white/ blue "STANDARD" bar & circle. Reverse is white outlined red with white "STANDARD/GASOLINE" on upper sign face above small white circle with red/blue "STANDARD" bar & circle.
ESSO OVAL 18x26
(5) 1946-1965 No Listing
Dark blue outlined white oval porcelain sign with red "Esso" logotype across center.

LUBRICATION/MOTOR OIL SIGNS

WE SELL STANDARD MOTOR OIL 18x36
(4) 1924-1930 $300
Blue outlined white porcelain sign with blue outlined red "STANDARD/MOTOR OIL" across center of sign face. Small red and blue Standard "Bar & Circle" logo to left of "MOTOR OIL." BLue "We Sell" across top with white "MAKES MOTORS RUN SMOOTHER AND COSTS RUN LOWER" on blue band across bottom of sign.
POLARINE FOR MOTOR CARS 10x26
(4) 1918-1924 $300
Blue/red outlined blue porcelain flange sign with white script "Polarine" logotype across sign face. Blue "THE STANDARD OIL" on white underline of "Polarine" with white "FOR MOTOR CARS" across bottom of sign.
ESSO MOTOR OIL UNEXCELLED 30x60
(4) 1934-1940 $150
Wood framed white tin litho sign with blue band across bottom. Red "Esso/MOTOR OIL" above band with white "UNEXCELLED" on blue band.
ESSO MOTOR OIL UNEXCELLED W/OVAL 30x60
(3) 1940-1952 $150
Wood framed tin litho sign with red band across top, white band across center and blue band across bottom. Blue outlined white "Esso" oval on upper red band with blue "MOTOR OIL" on center band and white "UNEXCELLED" on lower blue band.

SECONDARY SIGNS

AUTHORIZED STANDARD STATION 16x16 $225
Steel framed white porcelain sign with red line bar & circle logo in center. Blue "STANDARD" on center bar, with blue "AUTHORIZED" across top, "STATION" across bottom.
STANDARD NO SMOKING 6x30
(2) $350
Red/white outlined red porcelain sign with white "NO SMOKING" across center. White circle at either end with red bar & circle logo in center. Blue "STANDARD" across center of bar.

ESSO CREDIT CARDS HONORED
"BUY AT THE ESSO SIGN" 14x18
(4) $300
Blue porccelain sign with white band across center. Small white oval in upper blue area with red "Esso" in center and blue "BUY AT THE" above "Esso," "SIGN" below. Red "CREDIT" in center white band with white "CARDS HONORED" on lower blue area.
ESSO CREDIT CARDS HONORED BLUE 14x18
(5) $350
Blue porccelain sign with white band across center. Small white oval in upper blue area with red "Esso" in center. Red "CREDIT" in center white band with white "CARDS HONORED" on lower blue area.
ESSO CREDIT CARDS HONORED RED $125
Red porccelain sign with white band across center and blue band below. Blue outlined white oval in upper blue area with red "Esso" in center. Blue "CREDIT" in center white band with white "CARDS HONORED" on lower blue area.
SANITIZED STERILSEAT REST ROOMS 14x18
(4) 1946-1955 $100
Rectangular porcelain sign with red upper half and blue lower half. White circle with blue plus sign in center. White "STERILESEAT" across plus sign.
(Refers to a system devised about 1940 of using UV light to sterilize toilet seats) White "STERILIZED" in upper red area, "REST ROOMS" in lower blue area.

HUMBLE TOP RATED REST ROOMS 30x30
(3) 1961-1972 $40
Die-cut pentagon shaped white tin litho sign. Blue triangles top and bottom forming white band across center with vertical lines at top of sign. White "HUMBLE" in upper triangle, "REST/ROOMS" in lower with red "TOP RATED" across center band. Used at Esso and Enco locations as well as Humble branded stations.
MEN FLANGE 4x14
(3) 1936-1965 $35
White porcelain flange sign with blue lines across bottom below blue "MEN."
LADIES FLANGE 4x14
(3) 1936-1965 $35
White porcelain flange sign with blue lines across bottom below blue "LADIES."
NUSO CRANK-CASE FLUSHING SERVICE 24x24
(5) 1925-1930 No Listing
Red outlined blue porcelain flange sign with red diamond covering most of sign face. Blue outlined white "NUSO/CRANKCASE" on upper diamond above white "FLUSHING" and blue outlined white "SERVICE" on lower diamond. White "WE EXTEND" across top, "STANDARD OIL CO.N.J." across bottom.
ESSO SPREADER BAR PANEL $50
White porcelain sign with blue outlined Esso oval with red logotype "Esso" in center. Used as centerpiece of red/white/red "spreader bar" strip sign above pump islands.

BULK PLANT SIGNAGE

STANDARD OIL COMPANY WHOLESALE ONLY 12x24
(5) 1924-1933 No Listing
White porcelain sign iwht red line bar & circle logo at top. Blue "STANDARD" across center bar. Blue "STANDARD OIL COMPANY/OF NEW JERSEY/WHOLESALE ONLY" below logo.
ESSO HEATING OILS 20x36
(5) 1946-1960 $225
Porcelain sign with red band across top, white band across center and blue band across bottom. Blue outlined white oval logo with red "Esso" centered on upper red band. Red "HEATING OILS" across center white band with white "AUTHORIZED DEALER" on bottom blue band.
ESSO OIL BURNERS 20x36
(5) 1946-1960 $225
Porcelain sign with red band across top, white band across center and blue band across bottom. Blue outlined white oval logo with red "Esso" centered on upper red band. Red "OIL BURNERS" across center white band with white "AUTHORIZED DEALER" on bottom blue band.
ESSO STANDARD WHOLESALE PLANT 12x60
(4) 1946-1959 $200
Steel framed white porcelain sign with blue "ESSO STANDARD OIL COMPANY" across top above blue "WHOLESALE PLANT."
ESSO PRIVILEDGE PANEL
(2) 1965-1972 No Listing
White porcelain sign with red stripe area to right. Esso logo at left with jobber name below logo and stripe. Custom made for each bulk plant location.
ESSOMARINE OILS & GREASES 18x30
(4) 1935-1940 $550
Red porcelain sign with blue band across lower sign face. Blue outlined white oval covers most of sign with red "Essomarine" across center of oval above blue "O LS & GREASES."

TRUCK SIGNS

ESSOLUBE 10x48
(4) 1930-1933 $250
White self framed porcelain panel with black outlined red "Essolube" across panel.
STANDARD 10x48
(4) 1930-1933 $250
White self framed porcelain panel with black outlined red "STANDARD"(in quotes) across panel.
SONJ NO ACCIDENTS 7"RD
(5) 1920-1923 No Listing
White outlined black porcelain sign with white outlined red band across center. Black outlined white "NO ACCIDENTS" on center band. Black and white line drawing of Mack "Bulldog" truck, c1920, in black outlined white circle at bottom. White circle with red Standard Oil Company "S" logo at top. Red and white scrollwork around sign elements. White band across bottom with mounting holes. License plate attachment awarded to safety concious drivers.

ADVANCE SIGNS

STANDARD MOTOR GASOLINE DIRECTIONAL SIGN 60x60
(5) 1918-1923 $600
Blue outlined white diamond shaped porcelain sign with red outlined blue trapezoids in upper point of diamond. White town names & distances on blue trapezoid. Blue outlined red "STANDARD/MOTOR/GASOLINE" above blue outlined red and white "SONJ" circle at bottom. Placed at intersections giving directions and distances to nearest towns where Standard products were available.

EXXON

The Exxon brandname was introduced in the fall of 1972 replacing Esso in 19 eastern states, Humble in Ohio and Enco elsewhere. When Exxon was introduced the Enco brandname remained in use in a limited number of stations in the Chicago marketing area that had already been scheduled for abandonment about 1976.

IDENTIFICATION SIGNS

EXXON
(1) 1972-1983 $50
Aluminum framed white porcelain sign with blue band across bottom. Red "Exxon" logotype across above band.

BUILDING SIGNS AND LETTERS

EXXON ON RED PANEL
(3) 1987-PRESENT $25
Black/white outlined red porcelain sign with white "Exxon" logotype to upper left.
PRIVILEDGE PANEL
(3) 1987-PRESENT $25
Black/white outlined red porcelain sign with white dealer or store name to upper left. Custom made "priviledge panel" for retail locations.

PUMP SIGNS

EXXON TRIMLINE PANEL
(3) 1987-PRESENT $35

Black/white outlined blue porcelain Gilbarco Trimline pump door with white "Exxon" logotype to upper left.

EXXON TOKHEIM PANEL
(3) 1987-PRESENT $35
Black/white outlined blue porcelain Tokheim 1250 pump door with white "Exxon" logotype to upper left.

SECONDARY SIGNS

EXXON SPREADER BAR PANEL
(2) 1972-1987 $25
White porcelain sign with red Exxon logotype over blue bar in center. Used as centerpiece of red/white/red "spreader bar" strip sign above pump islands.

BULK PLANT SIGNAGE

EXXON PRIVILEDGE PANEL
(1) 1972-1987 $100
White porcelain sign with red stripe area to right. Exxon logo at left with jobber name below logo and stripe. Custom made for each bulk plant location.

STANDARD OIL CO. OF LOUISIANA

Marketing affiliate assigned to Standard NJ at the 1911 breakup of Standard Oil. Stanocola operated stations in Tennessee, Arkansas and Louisiana under the Stanocola brand until 1923 when the "STANDARD" bar and circle image was introduced throughout the marketing territory.

IDENTIFICATION SIGNS

STANOCOLA GASOLINE 18x36
(3) 1918-1923 $200
Red porcelain sign with white "STANOCOLA/GASOLINE" across sign face.

CURB SIGNS

STANOCOLA PETROLEUM PRODUCTS 18"RD
(4) 1918-1923 $650
Black outlined red porcelain sign with black outline around orange center circle. Green/black/red/white shield in center circle with "STANOCOLA/PETROLEUM/PRODUCTS" on shield. White "STANDARD OIL COMPANY" arched around top, "OF LOUISIANA" around bottom.

LUBRICATION/MOTOR OIL SIGNS

STANOCOLA POLARINE 18x18
(4) 1918-1923 $350
Thick green border area around white flange porcelain sign. Green outlined red "STANOCOLA/Polarine" on upper sign face above "The Standard/Motor Oil." Small multi-color Stanocola shield and cirlce on lower right corner.

*WE SELL STANOCOLA POLARINE 18x36
(4) 1918-1923 $450
Black outlined red porcelain sign with black outlined white "STANOCOLA/Polarine" across center. Black "We Sell The "STANDARD" Motor Oil" across top and green "MAKES MOTORS RUN SMOOTHER AND COSTS RUN LOWER" on black band across bottom of sign.

SECONDARY SIGNS

WE SELL NOTHING BUT STANOCOLA 12x20
(5) 1918-1923 $450
White/red outlined white porcelain sign with red band across upper sign face. White "We Sell Nothing But/STANOCOLA PRODUCTS" on upper red band with green "STANOCOLA GASOLINE" over "STANOCOLA Polarine" and other small lettering on lower white area of sign.

HUMBLE OIL AND REFINING - HOUSTON, TX

Houston based producer/refiner/marketer purchased by Standard (NJ) in 1919. The Humble brand was used in Texas and New Mexico from 1919 until 1961. In 1959 all United States marketing came under the subsidary "Humble Oil and Refining" and "Humble" was added to stations of each brand then in use, although each retained its own regional brandname (Esso, Carter, Oklahoma, etc.) In 1961 Enco replaced Humble as the station brand in Texas and Oklahoma. The Humble brand was introduced in Ohio when both Esso and Enco were blocked by Sohio and remained in use in Ohio until converted to Exxon in 1972.

IDENTIFICATION SIGNS

HUMBLE OILS W/ GEAR 42"RD
(4) 1919-1933 $400
Blue outlined white porcelain sign with blue gear-tooth circle in center. Red "HUMBLE" arched around top, "GASOLINE" around bottom with red "HUMBLE/OILS" across center of gear. White "HUMBLE OIL & REFINING CO." arched around top of gear, "HOUSTON, TEXAS" around bottom.

HUMBLE 42x84
(4) 1933-1945 $250
Blue/white outlined red porcelain oval with white "HUMBLE" across center.

HUMBLE DEALER 48x96
(4) 1945-1972 $200
Blue outlined white porcelain sign with red "HUMBLE" across center above smaller red "DEALER."

BUILDING SIGNS AND LETTERS

CERAMIC HUMBLE GEAR 24"RD
(5) 1919-1933 No Listing
Blue outlined white cast ceramic sign with blue gear-tooth circle in center. Red "HUMBLE" arched around top, "GASOLINE" around bottom with red "HUMBLE/OILS" across center of gear. White "HUMBLE OIL & REFINING CO." arched around top of gear, "HOUSTON, TEXAS" around bottom. Inset in building structure.

PUMP SIGNS

HUMBLE "CONTINOUSLY IMPROVED" 10x18
(4) 1939-1945 $100
Blue/white outlined white porcelain sign with thick blue/thin red outline dwhite circle at top. Red "Humble" in circle. Blue "continuously/improved" below circle with blue outlined red oval logo at bottom. White "HUMBLE" in oval.

ESSO EXTRA "EXTRA QUALITY... 10x18
(4) 1939-1945 $100
Blue/white outlined white porcelain sign with thick red/thin blue outlined white circle at top. Blue "Esso/EXTRA" in circle. Blue "EXTRA QUALITY/EXTRA/PERFORMANCE" below circle.

HUMBLE "CONTINOUSLY IMPROVED" 10x18
(2) 1945-1956 $100
Blue/white outlined white porcelain sign with thick red/thin blue outlined white circle at top. Red "HUMBLE" in circle. Blue "continuously/improved" below circle with blue outlined white oval logo at bottom. Red "HUMBLE" in oval.

ESSO EXTRA "EXTRA QUALITY... 10x18

(2) 1945-1956 $100
Blue/white outlined white porcelain sign with thick red/thin blue outlined white circle at top. Blue "Esso/EXTRA" in circle. Blue "EXTRA QUALITY/EXTRA/PERFORMANCE" below circle with blue outlined white oval logo at bottom. Red "HUMBLE" in oval.

NEON SIGNS

HUMBLE 42x84
(4) 1933-1945 No Listing
Blue/white outlined red porcelain oval with white "HUMBLE" across center. Punched for neon "HUMBLE" across sign. Faces assembled back-to-back over can-type frame.

SECONDARY SIGNAGE

WE DONT SMOKE ON DRIVE WAY 10x16
(4) 1946-1950 $125
White/green outlined white porcelain sign wtih green and white attendants head with "HUMBLE" logo on cap at left. Green "For Your Safety" across top with "WE DON'T/SMOKE ON/DRIVE WAY" to right of attendant.

BULK PLANT SIGNAGE

DISTRIBUTOR ESSO PETROLEUM PRODUCTS 20x34
(4) 1946-1961 $100
White porcelain sign with blue outlined white oval in center. Red "Esso" logotype in oval, with blue "DISTRIBUTORS" above logo, "PETROLEUM PRODUCTS" below.

TRUCK SIGNS

HUMBLE OILS WITH GEAR 10"RD
(4) 1919-1933 $350
Blue outlined white porcelain sign with blue gear-tooth circle in center. Red "HUMBLE" arched around top, "GASOLINE" around bottom with red "HUMBLE/OILS" across center of gear. White "HUMBLE OIL & REFINING CO." arched around top of gear, "HOUSTON, TEXAS" around bottom.

HUMBLE SMALL RED DOOR SIGN
Blue/white outlined red porcelain oval with white "HUMBLE" across center.

HUMBLE SMALL DOOR SIGN $300
Blue outlined white porcelain oval sign with red "HUMBLE" across center.

HUMBLE LARGE DOOR SIGN $275
Blue outlined white porcelain oval sign with red "HUMBLE" across center.

HUMBLE SPLIT BACK DOOR SIGN $350
Blue outlined white porcelain oval sign with red "HUMBLE" across center. Two halves split vertically for installation on rear doors of tank truck.

HUMBLE OHIO DOOR SIGN 7.5x10
(4) 1961-1965 $125
Blue outlined white porcelain rounded oval sign with red "HUMBLE" across center.

TEXAS PACIFIC COAL AND OIL - FT. WORTH, TX

Independent Texas marketer sold out to and rebranded Humble in 1945.

IDENTIFICATION SIGNS

TEXAS PACIFIC GASOLINE/MOTOR OIL 42"RD
(5) 1930-1945 No Listing
Red/black otulined white porcelain sign with red/white/black outline around white center circle. Black outlined red teepee in center with white outlined black "TP" across center. Black "PRODUCTS" below. Black "TEXAS PACIFIC" arched around top, "GASOLINE MOTOR OIL" around bottom.

BUILDING SIGNS AND LETTERS

TEXAS PACIFIC COAL AND OIL CO PANEL 12x96
(5) 1930-1945 No Listing
White outlined black embossed porcelain sign with white "TEXAS PACIFIC COAL AND OIL CO." across panel

CURB SIGNS

THURMAN MOTOR OIL 24"RD
(5) 1930s No listing
White porcelain sign with black "THURMAN" arched across center above "MOTOR/OIL." White lettering on black at bottom and red design at top of sign.

TP AERO MOTOR LUBRICATING OIL 16x30
(5) 1930s No Listing
Red and black rectangular porcelain sign with black and white winged teepee at top. Black "TP" on teepee. Black and white "Aero Motor Lubricating Oil" across center with winged TP logo at bottom.

TEXAS PACIFIC ETHYL GASOLINE 30"RD
(5) 1926-1945 $350
Red/black outlined white porcelain sign with white center circle. Large Ethyl
(EGC) logo in center with "TEXAS PACIFIC" arched around top, "GASOLINE" around bottom.

COLONIAL BEACON OIL COMPANY - BOSTON, MA

Formed in a late 1920s merger between Boston area based Colonial Oil Company and Beacon Oil. Colonial Beacon, an innovative marketer with stations throughout New England, sold out to Standard (NJ) in 1930. The Colonial name, as a subsidary, survived until the end of WW2, although products were branded Esso in 1933. Colonial influenced the entire industry with the introduction of service bays at stations and the marketing of tires, batteries and automotive accessories (known as TBA in the trade) through their stations. Colonial was instrumental in the formation of the Atlas Supply Company to supply the various Standard companies that participated in the program a TBA line.

IDENTIFICATION SIGNS

COLONIAL SERVICE STATION 36x42
(5) 1930-1933 $300
Steel framed red porcelain sign with cream colored band across lower sign face. White "Service Station" on upper red area with black outlined green "Colonial" across lower cream area.

COLONIAL ESSO DEALER MINOR
(4) 1933-1946 $325

COLONIAL ESSO DEALER MAJOR 42x84
(5) 1933-1946 $125
Dark blue outlined white oval porcelain sign with red "Esso" logotype across center. Blue "COLONIAL" above logotype, "DEALER" below.

BUILDING SIGNS AND LETTERS

BEACON OIL COMPANY PANEL 16x48
(5) 1920-1925 $600

Red porcelain sign with green outlined cream "Beacon Oil/Company" across sign face.
COLONIAL BEACON OIL COMPANY 16x64
(5) 1925-1930 $600
Red porcelain sign with green outlined cream "Colonial Beacon/Oil Company" across sign face.
COLONIAL GAS PANEL 24x36
(4) 1925-1930 $600
Green outlined cream porcelain sign with black outlined green "Colonial/Gas" across sign face.

PUMP SIGNS

BEACON OILS 9x10
(4) 1926-1930 $450
Red outlined die-cut shield shaped porcelain sign with green upper half and white lower half. Red/white/green beacon with basket in center with white outlined red "BEACON" in green area above balloon and green outlined red "OILS" in white area below.
COLONIAL ETHYL 9x10
(5) 1926-1930 $525
Red outlined die-cut shield shaped porcelain sign with green upper half and white lower half. Black outlined white circle in center with Ethyl
(EGC) logo. White outlined red "COLONIAL" in green area above balloon and green outlined red "GAS" in white area below.

CURB SIGNS

BEACON OILS SWIRL W/ BEACON 24"RD
(5) 1926-1930 $625
Green porcelain sign with black and green balloon with basket in center. Series of cream swirlled rays out from beacon in center. Black "BEACON" arched around top, "OIL" around bottom.
BEACON PENN MOTOR OIL 24SQ $600
Yellow/black/white outlined green porcelain sign with white shaped area in center. White vertical stripes on lower shield area with black band diagonal across striped area. Yellow "PENN" on black band with black outlined yellow "BEACON" across top and black "MOTOR OIL" across bottom.
COLONIAL ETHYL 25.5x30
(4) 1926-1930 $550
Black outlined white porcelain sign with thick yellow border area. Large green triangle with white rays in center. White outlined black "Colonial" across upper triangle with Ethyl.
(EGC) logo below.
ESSO ETHYL 25.5x30
(3) 1930-1933 $500
Red/blue outlined white square porcelain sign with cut corners, forming octagon. Blue script logotype "Esso" across sign face. Small Ethyl
(EGC) logo in lower right corner.

LUBRICATION/MOTOR OIL SIGNS

GIVE COLONIAL LUBRICATION SERVICE 32x32
(4) 1926-1930 $650
Green outlined white porcelain sign with green ribbon/red seal desing in center. Black and white minuteman soldier to wither side of seal at bottom. Small white "WE ARE/EQUIPPED AND TRAINED/TO GIVE" over larger "Colonial/LUBRICATION/SERVICE" on seal.

LUBESTER SIGNS

BEACON W/ BALOON "LIGHT" 6"RD
(5) 1926-1930 No Listing
Black outlined green tin litho sign with baloon and seirls design. Other details not available.

SECONDARY SIGNS

COLONIAL CHARGE ACCOUNTS 10x24
(5) 1928-1931 $300
Black/green outlined white porcelain sign with black outlined green "Colonial/Charge Accounts/Accepted Here" across sign face.

ATLAS TIRE COMPANY: Colonial Beacon Oil Company had been an early leader in introducting automotive accessory sales through their service stations. Prior to this time, major oil companies had sold gasoline, motor oils and lubricants, and little else. Shortly after Colonial Beacon was purchased by Standard Oil of New Jersey, Esso embarked on a program to introduce this type of merchandising to their stations. To create a brand for their TBA (tires, batteries, accessories) that could be sold and warrantied nationwide, Esso joined with other Standard affiliates (under the mistaken reasoning that they were not competing in each other's territory - and probably never would) Standard of Indiana, California, Kentucky, and Ohio to form Atlas Tire Company. The brand was introduced nationwide at the stations from the above companies as a oil industry TBA brand that would be backed nationwide. The operation continued intact until the early 1990s when convenience store marketing came to dominate the industry. Exxon and Amoco withdrew from the plan, leaving Chevron and BP (formerly Sohio) selling Atlas products today.

NOTE: Atlas signage that also bears the brandnames or logo from affiliated companies alongside the Atlas brand are listed with that brand under "SECONDARY SIGNAGE."

SECONDARY SIGNS

ATLAS TIRES BATTERIES ACCESSORIES 30x72
(2) 1931-1960 $75
Black wood framed red tin litho sign with white outlined blue diagonal stripes in upper left and lower right corners. White "ATLAS/TIRES - BATTERIES/ACCESSORIES" across sign face.
ATLAS TIRES 12x72
(2) 1931-1960 $100
Black self framed red tin litho sign with white outlined blue diagonal stripes in upper left and lower right corners. White "ATLAS" vertical down sign above "TIRES" across sign face.

PENNSYLVANIA LUBRICATING COMPANY "PENOLA": Pennsylvania Lubricating Company was the lubrications manufacturing arm of Standard Oil of New Jersey. While most products were sold under the various Esso brandnames, some export items and some lubricants made for distribution in parts of the country where use of Esso was forbidden were branded Penola. This sign is thought to be from before Standard purchased Pennsylvania Lubricating Company, and carries the brandname Penotol.
PENOTOL 30"RD
(4) 1926-1930 No Listing
Yellow porcelain sign with black outlined red circle at top.

Black and white "Minuteman" sillouette in circle. Black outlined red "PENOTOL/MOTOR OIL" across sign center with white outlined black "extra quality" inbetween. Black and white PGCOA seal at bottom with black "PENNSYLVANIA LUBRICAING CO" arched around bottom.

POWERINE OIL CO. - DENVER, CO

Denver based refiner/marketer. Purchased by Standard (NJ) in 1938, the Powerine name was retained until the end of WW2 when it was replaced by Oval-E.

IDENTIFICATION SIGNS

THAT GOOD POWERINE GAS 20x28
(2) 1925-1930 $300
Red porcelain sign with black outlined white band with rounded areas top and bottom across center. Black outlined red "POWERINE" across band with black "THAT GOOD" arched around top of white area, "GAS" below.

LUBRICATION/MOTOR OIL SIGNS

POWERLUBE MOTOR OIL SMOOTH AS THE/TREAD OF A TIGER W/ SEAL 20x28 No Listing
White outlined blue porcelain sign with white outlined yellow "Power-lube" logotype across upper sign face with detailed yellow and black tiger below. White outlined yellow "MOTOR/OIL" to right of tiger. Blue outlined yellow bar across bottom with PGCOA seal in center. Blue "SMOOTH AS THE" to left of seal, "TREAD OF A TIGER" to right.

POWERLUBE MOTOR OIL REFLECTIVE 25x48
(5) 1920s No Listing
Blue/yellow outlined blue reflective tin litho sign with yellow "Power-lube" logotype across upper sign face with detailed yellow and black tiger below. White outlined yellow "MOTOR/OIL" to right of tiger. Blue outlined yellow bar across bottom with PGCOA seal in center. Small yellow band just below tiger with blue "SMOOTH AS THE TREAD OF A TIGER" across band.

NORTHWEST REFINING CO. - CUT BANK, MT

Marketed Grizzly gasoline. Retail marketing purchased by Standard in 1942 and rebranded Carter in 1945.

IDENTIFICATION SIGNS

GRIZZLY VERTICAL OVAL 42x66
(5) 19401945 No Listing
Blue outlined white vertical oval with large standing brown bear in center. Blue "GRIZZLY" across below bear.

GRIZZLY GASOLINE 72"RD
(5) 1936-1945 No Listing
White porcelain sign with large detailed standing brown bear in center. Blue "GRIZZLY" arched around top, "GASOLINE" around bottom.

GRIZZLY DUBBS CRACKED GASOLINE 42"RD
(5) 1936-1945 No Listing
Blue outlined white porcelain sign with large detailed standing brown bear in center. Blue "GRIZZLY" arched around top, "GASOLINE" around bottom. Red script "Dubbs Cracked" to lower left of bear.

PUMP SIGNS

GRIZZLY DUBBS CRACKED GASOLINE 12"RD
(5) 1936-1945 $
Blue outlined white tin sign with large detailed standing brown bear in center. Blue "GRIZZLY" arched around top, "GASOLINE" around bottom. Red script "Dubbs Cracked" to lower left of bear.

NORDAL 12"RD
(5) 1936-1945 $75
Blue outlined white tin sign with blue "Nordal" logotype across center.

CURB SIGNS

GRIZZLY GASOLINE DUBBS CRACKED 24x36
(5) 1930-1936 $1800
Yellow porcelain tombstone shaped sign with large brown bear in center, standing on black mound. Red outlined black "GRIZZLY/GASOLINE" arched around top with red script "Dubbs Cracked" to lower left of bear. Yellow "PERFECT/POWERFUL PERFORMANCE" above red script "Watch your Miles" on black mound at bottom.

YALE OIL CO. - BILLINGS, MT:

Used the name "Hancock" in conjunction with Litening in the 1920's, later the Litening brand was used without Hancock. Also marketed Wildfire gasoline. Purchased by Standard (NJ) in 1946 and rebranded Carter.

SECONDARY SIGNS

HANCOCK LITENING GASOLINE 12x24
(5) 1920's $225
Green outlined black porcelain sign with green "HANCOCK" above orange "LITENING" and yellow "GASOLINE" on lower sign face. Orange "100% PURE PENNSYLVANIA" over yellow "MOTOR & TRACTOR/OILS" with PGCOA seal to either side on lower sign face.

CHANGE TO LITENING 12x24
(5) 1930's $225
Wood framed yellow tin litho sign with black "CHANGE TO" over black outlined red "LITENING" across sign face.

OVAL E

Name introduced in 1946 to replace Powerine. Changed to Carter in 1950.

IDENTIFICATION SIGNS

OVAL E "MAJOR SIGN" 42x84
(5) 1946-1950 $150
Dark blue outlined white oval porcelain sign with red logotype "E" in center.

CARTER OIL CO. - TULSA, OK

Long a Standard production subsidiary, the Carter brandname was introduced in 1945 to replace Grizzly and Litening. In 1950 Carter replaced Oval-E to bring all of Standard's western marketing under the Carter brand. The Carter brand was replaced by Enco in 1961.

IDENTIFICATION SIGNS

CARTER "MINOR SIGN" 30x60
(5) 1946-1961 $200
Dark blue outlined white oval porcelain sign with red "Carter" logotype across center.

CARTER "MAJOR SIGN" 42x84
(3) 1946-1961 $100
Dark blue outlined white oval porcelain sign with red "Carter" logotype across center.

BUILDING SIGNS AND LETTERS

CARTER/SERVICE/CARTER $150

White porcelain sign with blue outlined white oval at either end. Red "Carter" in oval with blue "SERVICE" across center.

SECONDARY SIGNS

CARTER CREDIT CARDS HONORED $450

White porcelain sign with blue outlined Carter oval with red logotype "Carter" in center. Used as centerpiece of red/white/red "spreader bar" strip sign above pump islands.

ENCO

Introduced in 1960, first in the Oklahoma/Pate marketing territory as a replacement for those brandnames and expanded to Carter and Humble territories in 1961, Enco
(from the slogan America's Leading ENergy COmpany), was to be Standard's primary brandname in all areas where Esso could not be used. Replaced by Exxon in 1972 except in select areas where Standard planned to withdraw from marketing. Used as late as 1980 in those areas.

IDENTIFICATION SIGNS

ENCO "MAJOR SIGN" 42x84
(1) 1946-1965 $125

Dark blue outlined white oval porcelain sign with red "Enco" logotype across center.

ENCO "MINOR SIGN" 30x60
(5) 1965-1972 $125

ESSO "MAJOR SIGN" 42x84
(4) 1965-1972 $100

White porcelain sign with wide light blue border. Red "Enco" logotype across center.

SECONDARY SIGNS

ENCO PANEL FOR ISLAND SIGN 17.5x36
(2) 1965-1972 $50

White porcelain sign with blue outlined Enco oval with red logotype "Enco" in center. Used as centerpiece of red/white/red "spreader bar" strip sign above pump islands.

BULK PLANT SIGNAGE

ENCO PRIVILEDGE PANEL
(1) 1965-1972 No Listing

White porcelain sign with red stripe area to right. Enco logo at left with jobber name below logo and stripe. Custom made for each bulk plant location.

PATE OIL CO - MILWAUKEE, WI

Founded in Milwaukee in 1933, Pate operated over 50 stations when purchased by Standard in 1956. The Pate brand was replaced by Enco in 1960.

BUILDING SIGNS AND LETTERS

PATE AIR GLIDE GASOLINE 32x96
(5) 1946-1956 No Listing

Blue porcelain can type sign with yellow logotype "AIR GLIDE" across center of sign. White "GASOLINE" with yellow lines below. Blue outlined white oval with red "PATE" added to top of signs from 1956-1959.

PATE LETTERING 16"
(5) 1946-1959 $100

Blue porcelain lettering "PATE."

NEON SIGNS

PATE "HELLO NEIGHBOR" NEON SHIELD 48x96
(5) 1946-1959 $1200

Can-type neon over porcelainsign. Upper section is die-cut blue/white outlined red shield with white outlined blue band diagonal across center. White "PATE" on band. White outlined blue lower section with white "HELLO NEIGHBOR" slogan across sign.

OKLAHOMA OIL COMPANY - CHICAGO, IL

Oklahoma was a Chicago area discount marketer that Standard (NJ) purchased in 1956. Almost immediately another Chicago discounter, Perfect Power, was purchased and rebranded Oklahoma. The Gaseteria/Hoosier Pete operation was added in 1957 and also rebranded Oklahoma. The Enco name was first introduced to replace Oklahoma in 1960.

PUMP SIGNS

OKLAHOMA REGULAR 12"RD
(5) 1950-1956 $225

White porcelain sign with blue half-circles top and bottom forming white band across center. Blue band arched around bottom. Blue "OKLAHOMA" arched around upper white area, red "REGULAR" in center band with white "GASOLINE" on lower blue band. Other red lettering in half circles.

GASETERIA OIL CO. - INDIANAPOLIS, IN

Founded prior to WW2 in Indianapolis, Gaseteria operated the Bonded stations throughout Indiana and Kentucky. In the mid-1950s Bonded opened staions in Ohio. They also purchased the remnants of Johnson Oil Refining/Brilliant Bronze, as well as Hoosier Petroleum in 1956. In 1957 Standard purchased Gaseteria and rebranded the operation Oklahoma and the Ohio locations were spun off as an independent company, based in Springfield, Ohio.

PUMP SIGNS

BONDED 68 10"RD
(5) 1938-1950 $250

White porcelain sign with gold metal/blue ribbon design in center. White "Bonded" above red "68" in center of ribbon.

BONDED 78 10"RD
(5) 1938-1950 $250

White porcelain sign with gold metal/blue ribbon design in center. White "Bonded" above red "68" in center of ribbon.

JOHNSON OIL REFINING CO. - CHICAGO, IL

Mid-Continent refiner/marketer that marketed gasoline under the Johnson, Jo-Re-Co, and Brilliant Bronze brandnames. Gaseteria purchased Johnson in 1956 and sold the refinery operation to Deep Rock. Rebranded Oklahoma with the 1957 purchase of Gaseteria by Standard (NJ).

IDENTIFICATION SIGNS

JOHNSON TIME TELLS GASOLENE 42"RD
(3) 1925-1946 $2000

Black outlined orange porcelain sign with thick black ring around orange center circle. Orange/black/white hourglass with black and white wings in center with black "TIME" "TELLS" on wings. White "JOHNSON" arched around top, "GASOLENE" around bottom.

BRILLIANT BRONZE 72"RD
(5) 1946-1956 $225

Black outlined white circle in center. Lower half of circle is orange with black/white winged hourglass design. Black "brilliant/bronze" in upper half of circle.

PUMP SIGNS

BRILLIANT BRONZE 10x12
(4) 1946-1956 $750
White/orange outlined white porcelain sign with black outlined white circle in center. Lower half of circle is orange with black/white winged hourglass design. Black "brilliant/bronze" in upper half of circle.

CURB SIGNS

JOHNSON WINGED 70 GASOLINE 30x30
(5) 1925-1930 $1200
White outlined black tombstone shaped sign with circular sign face and black areas at the bottom. Orange across top and bottom of circular area with black "JOHNSON" in upper area and "ANTI-KNOCK" below. White "Winged/70/GASOLINE" across black area in center. Orange and black wings in center.
JOHNSON ETHYL/TIME TELLS MOTOR OIL 30"RD
(4) 1926-1940 $1600
JOHNSON ETHYL/TIME TELLS MOTOR OIL 42"RD
(3) 1926-1940 $1500
JOHNSON ETHYL/TIME TELLS MOTOR OIL 48"RD
(3) 1926-1940 $1200
Black outlined orange porcelain sign with thick black ring around orange/black outlined white center circle. Ethyl (EGC) logo in center circle with black and white wings in center with black "TIME" "TELLS" on wings. White "JOHNSON" arched around top, "GASOLENE" around bottom. Reverse is black outlined orange with thick black ring around orange center circle. Orange/black/white hourglass with black and white wings in center with black "TIME" "TELLS" on wings. White "JOHNSON" arched around top, "GASOLENE" around bottom.
BRILLIANT PENN MOTOR OIL 20x32
(5) 1940s $1000
Black steel framed yellow porcelcelain sign picturing red/black/yellow "BRILLIANT PENN" motor oil quart can.

HOOSIER PETROLEUM CORP. - INDIANAPOLIS, IN

Indiana marketer purchased by Gaseteria in 1956. Originally marketed under the Vortex brand and later as Hoosier Pete.
No Collectible Signs Known At This Time

RYAN'S

A Hoosier Pete affiliate.
No Collectible Signs Known At This Time

SENCO, INC. - ATLANTA, GA

Formed after WW2 as Clark's Oil Company with signage similar to Milwaukee based Clark Oil, the company rebranded Senco about 1956 and was purchased by Standard (NJ) in 1962 and rebranded Esso.

IDENTIFICATION SIGNS

Super CLARK'S Gasoline 72RD
(5) 1946-1956 $100
Orange porcelain sign with black outlined white band across center. Black ""CLARK'S"" across white band with white script "Super" in upper orange area,, "Gas" below.

SOUTHERN OIL CO. - MIAMI, FL: Marketed under the "Deem" brand through about a dozen stations in Florida. Purchased by Standard (NJ) in 1962 and rebranded Esso.
No Collectible Signs Known At This Time

SUN-FLASH OIL CO. - COLUMBUS, OH

Columbus based discount operation founded in the 1930s. They were operating about 30 stations when purchased by Standard (NJ) in 1961. Locations were rebranded Humble.

NEON SIGNS

SUN FLASH 42x96
(5) 1935-1940 $125
Die-cut porcelain faces assembled can-type back-to-back. Black outlined and accented yellow porcelain faces with black "SUN/FLASH" across sign. Faces punched for neon "SUN/FLASH."

SECONDARY SIGNS

SUN FLASH 12x72
(5) 1935-1940 $150
Wood framed tin litho vertical panel sign. Black band across top and bottom with white center. Black "SUNFLASH" on vertical yellow stripe down center superimposed over black sun. White "GASOLINE" in upper black stripe, "100% PENNA/ MOTOR OIL" on lower.

ALERT OIL COMPANY - WILMINGTON, DE

Alert was formed in the late 1960s as an Esso secondary brand.
No Collectible Signs Known At This Time

IMPERIAL OIL LIMITED: Standard's Canadian affiliate, Imperial was assigned to Standard of New Jersey with the breakup of Standard Oil. They are a fully integrated company and continue to operate across Canada today.

IDENTIFICATION SIGNS

IMPERIAL GASOLINE 30x72
(5) 1925-1933 No Listing
Red/white outlined blue porcelain sign with white outlined red circle projecting above upper edge. White "IMPERIAL" logotype on circle, with white "GASOLINE" across blue band below.
IMPERIAL DEALER 30x60
(5) 1934-1946 $250
IMPERIAL DEALER 42x84
(4) 1934-1946 $200
Blue outlined white oval with three red stars across top. Red "IMPERIAL" across center with red "DEALER" below."
IMPERIAL ESSO DEALER 30x60
(3) 1946-1965 $150
IMPERIAL ESSO DEALER 42x84
(2) 1946-1965 $100
Thick blue outlined white oval with red "Esso" logotype across center. Blue "IMPERIAL" across top, "DEALER" below."
IMPERIAL ESSO PRODUCTS 30x60
(4) 194601965 $150
Thick blue outlined white oval with red "Esso" logotype across center. Blue "IMPERIAL" across top, "PRODUCTS" below."

CURB SIGNS

IMPERIAL MARVELUBE 24"RD
(5) 1925-1933 $225

White outlined red porcelain sign with white outlined red circle at top. White "IMPERIAL" logotype across circle with white "Marvelube" below. Other small lettering on sign is white.

FINA AND AFFILIATES

AMERICAN PETROFINA - DALLAS, TX

American Petrofina is the Dallas, Texas based subsidary of Belgium's PetrofinaSA. American Petrofina has been formed almost entirely by mergers and buyouts, beginning with the purchase of Panhandle Oil Corp. of Wichita Falls, Texas in October 1956. Subsequent purchases, where an entire company's marketing was purchased and converted, are listed below. Fina currently operates stations throughout the mid-continent and southeast.

IDENTIFICATION SIGNS

FINA SHIELD 48"
(3) 1956-PRESENT $125
Blue/white outlined red die cut porcelain shield with white circle in center. Blue FINA across white center circle.

FINA SHIELD 72"
(1) 1956-PRESENT $100
Porcelain framed can type blue/white outlined red die cut porcelain shield with white circle in center. Blue FINA across white center circle.

FINA INSERT 32x60
(2) 1974-PRESENT $35
White FRP insert for use over BP porcelain signs in those area where Fina absorbed BP marketing. Red white and blue Fina shield over red band at bottom.

PUMP SIGNS

FINA SHIELD 10"
(4) 1956-1970 $75
Blue/white outlined red die cut porcelain shield with white circle in center. Blue FINA across white center circle. Possible for Canadian use only

PANHANDLE OIL CO. - WICHITA FALLS, TX

Operated stations in Texas until selling out to and rebranding Fina in 1956.

IDENTIFICATION SIGNS

PANHANDLE GASOLINE TOMBSTONE 30x40
(5) 1926-1935 $1500
Outlined inverted tombstone sign with Panhandle rounded triangle logo in lower area. White "PANHANDLE/REFINING/COMPANY" on triangle, with script "Panhandle/GASOLINE" above logo.

PANHANDLE PRODUCING/REFINING 42"
(5) 1930-1946 $1200
Die-cut porcelain Panhandle rounded triangle logo with horns across center. White "PANHANDLE" across top, with "PRODUCING" along left side and "REFINING" along right side.

PANHANDLE 42"
(5) 1946-1956 No Listing
Die-cut porcelain Panhandle rounded triangle logo with horns across center. White "PANHANDLE" across top with small star at bottom.

CURB SIGNS

PANHANDLE ETHYL GASOLINE 30"RD
(5) 1926-1940 No Listing
White porcelain sign with large Ethyl (EGC) logo in center. "PANHANDLE" arched around top, "GASOLINE" around bottom.

AMERICAN LIBERTY - DALLAS, TX:

Formed in the 1930s from oil holdings of the Murchison family, sold out to Fina in October 1957.

IDENTIFICATION SIGNS

AMLICO COMPASS 72"RD
(5) 1946-1957 $150
Black outlined white porcelain sign with green/gray compass in center. Red "AMLICO" across white center band, with black "AMERICAN LIBERTY" arched around top, "OIL COMPANY" around bottom.

PUMP SIGNS

AMLICO REGULAR GASOLINE 10"RD
(5) 1950-1957 $35
Green outlined white circle with gray center circle and green band across center. White "REGULAR" in center band, with gray "AMLICO" arched around top, "GASOLINE" around bottom.

AMLICO PREMIUM GASOLINE 10"RD
(5) 1950-1957 $35
Green outlined gray circle with white center circle and green band across center. White "PREMIUM" in center band, with white "AMLICO" arched around top, "GASOLINE" around bottom.

PETRO-ATLAS - EL DORADO, KS

Petro Atlas is a consolidation of the old Eldorado Refining and K-T Refining. Petro-Atlas sold out to Fina in March 1958.

IDENTIFICATION SIGNS

OILS/KT/GASOLINE 48"RD
(5) 1930-1940 $225
Black outlined white porcelain sign with black outleind red center circle. White "KT" in center circle, with "OILS" arched around top. "GASOLINE" around bottom.

ELDORADO REFINING ELRECO 48"RD
(5) 1940-1950 $200
White outlined black porcelain sign with white outlined red circle in center. Black/white outlined band across circle with black "ELRECO" on band. White "EL DORADO REFINING" arched around top.

ELRECO OVAL 42x72
(5) 1950-1958 $100
Black outlined red porcelain oval with black outlined white band across center. Black "ELRECO" on band.

LUBRICATION/MOTOR OIL SIGNS

FLEET THE K-T OIL CORPORATION
MOTOR OIL GREASE $125
White outlined red tin litho sign with black and white bands across bottom. White "FLEET" with black car and airplane in upper red area, with white "THE K-T OIL COMPANY" on black band and red "MOTOR OIL-GREASE" on white band.

PETROLEUM MARKETING CO. - TULSA, OK

Pemco is a Tulsa based independent that rebranded Fina about 1960.
No collectible size Pemco signs known at this time.

Sohio Motor Oil porcelain sign. *Jim Hollabaugh.*

Sohio dealer porcelain sign. *Jim Hollabaugh.*

Caldwell & Taylor's Benzol 10"x14" porcelain flange sign.

Signal Ethyl 12" porcelain pump plate. *Bruce Bednar.*

Miscellaneous porcelain pump signs. *Bruce Bednar.*

Richfield 48" porcelain sign. *Jim Hollabaugh.*

Sterling porcelain pump plate 12"x18."

Sterling small tin credit card sign.

Sterling porcelain pump plate.

Pure Oil Steamship rare 30" porcelain sign. *Dunbar's Gallery.*

White Star 30" scarce porcelain sign. *C. Hook.*

Mobil Gargoyle porcelain sign 24"x19." *Dunbar's Gallery.*

Mobilgas embossed small porcelain building sign. *Dunbar's Gallery.*

Rocolene 7" Lubester sign. *Dunbar's Gallery.*

GOLDEN RULE - ST. LOUIS, MO

Golden Rule is a St Louis, Missouri based independent that branded Fina about 1960.

No collectible size Golden Rule signs known at this time.

COSDEN PETROLEUM - BIG SPRING, TX

Founded about 1925 by Josh Cosden after he lost controlling interest in Mid-Continent Petroleum which he had founded in 1916. Cosden was operating a Big Spring, TX refinery and petrochemical operation and marketing through stations in six states when they merged with American Petrofina in April, 1963. Cosden stations were rebranded Fina shortly after the merger.

IDENTIFICATION SIGNS

COSDEN LIQUID GAS 42"RD
(5) 1926-1946 $650

Green/white outlined orange porcelain sign with green outlined white bar & circle across center. White "COSDEN" with black dropshading arched around top, "OILS" around bottom. Green "LIQUID GAS" across center.

COSDEN GREEN/RED 72"RD
(3) 1946-1960 $75

Red/white outlined green porcelain sign with red/white outlined red band diagonal across center. White "COSDEN" in band.

COSDEN RED/WHITE 72"RD
(4) 1960-1963 $75

Black outlined white porcelain sign with red band diagonal across center. White "COSDEN" in band.

PUMP SIGNS

COSDEN HIGHER OCTANE 10"RD
(2) 1946-1960 $150

Red/white outlined green porcelain sign with red/white outlined red band diagonal across center. White "COSDEN" in band, with white "HIGHER" above and "OCTANE" below.

COSDEN ETHYL 10"RD
(2) 1946-1960 $150

White outlined red porcelain sign with white center circle. Ethyl (EC) logo in center, with white "COSDEN" arched around top, "GASOLINE" around bottom.

CURB SIGNS

COSDEN LIQUID GAS HORSEPOWER PLUS 30"RD
(5) 1935-1940 No Listing

Detailed porcelain sign with black "background" arched around green center circle. Orange and blue horse heads on black background, with black outlined white "COSDEN/LIQUID GAS" diagonal across upper sign face. Smaller white "HORSEPOWER+PLUS" above "SPECIAL" below.

COSDEN POLICEMAN DIE CUT No Listing

Detailed die-cut policeman in blue and white uniform pointing to round Cosden Higher Octane sign in Red, white and green.

COL-TEX

As best as can be determined, Col-Tex was sold off to Cosden after Apco purchased and rebranded Kanotex, greatly expanding the Apco brand from it's original Oklahoma marketing territory. When Apco became the primary brandname, Col-Tex stations were either converted to Apco or sold off with the Col-Tex brand to Cosden, who operated the Col-Tex stations as a secondary brand until 1963.

IDENTIFICATION SIGNS

COL-TEX 72"RD
(3) 1953-1963 $75

Black outlined white porcelain sign with red/white outlined red band diagonal across center. White "COL-TEX" in band.

PUMP SIGNS

COL-TEX 8"RD
(3) 1953-1963 $100

Black outlined white porcelain sign with red/white outlined red banband diagonal across center. White "COL-TEX" in band.

ONYX

Was a Texas independent that sold out to Cosden and beacame a Cosden secondary brand.

No collectible size Onyx signs known at this time.

WIDES OIL COMPANY - MURPHYSBORO, IL

Wides is a Murphysboro, Illinois based independent bought by Fina about 1972 and converted to the Fina brand after 1977.

IDENTIFICATION SIGNS

WIDES/SAVE WITH WIDES/GAS FOR LESS 72"RD
(4) 1950-1970 $150

Red outlined white porcelain sign with red circle in center. White band diagonal across cricle with blue "WIDES" in band. Blue "SAVE WITH WIDES" arched around top, "GAS FOR LESS" around bottom.

LIBERTY OIL CO. - MT. VERNON, IL: Discount marketer dating from the 1930's that was purchased by Fina before 1980.

IDENTIFICATION SIGNS

LIBERTY 42x84
(5) 1950-1970 $350

Blue porcelain oval sign with white band across center. Detailed blue and white statue of liberty in center, partially covered by center band. Red "LIBERTY" on band.

CANADIAN PETROFINA: The first entry into the North American continent by Belgian Petrofina SA came with the establishment of Canadian Petrofina in the early 1950's.

PUMP SIGNS

FINA SHIELD 12"
(3) 1960'S $100

Blue/white outlined red die cut porcelain shield with white circle in center. Blue FINA across white center circle.

FINA HY-POWER 11x13.5
(2) 1960s $100

Blue/white outlined red die cut porcelain shield with band arched across top and blue band across bottom. White circle in center. Blue FINA across white center circle, with white "HY-POWER" in upper band, "GASOLINE" below.

FLYING A/TYDOL & GETTY

TIDEWATER OIL COMPANY - NEW YORK, NY Founded in New York in 1887, Tidewater is best known for their world famous motor oil brand "Veedol." Tidewater entered the gasoline market about 1915 and by 1920 was marketing all along the eastern seaboard under the Tydol brand. During the 1930s control of Tidewater fell into the hands of Standard Oil of New Jersey who set up Mission Corporation as a subsidary to operate Tidewater and Skelly. J. Paul Getty purchased control of Mission in 1937 and merged Tidewater with California's Associated Oil Company in 1938. Associated had introduced the Flying A brand in 1932 for it's premium gasoline and with the creation of Tidewater Associated Oil Company, stations in the west became Associated Flying A stations and stations in the east became Tydol Flying A stations. In 1956 the Tydol and Associated designations were dropped and the brand became simply Flying A. In 1966 stations in the former Associated territory, including the secondary Seaside operation were sold to Phillips Petroleum and rebranded Phillips 66. Also in 1966 the Tidewater name was dropped in favor of Getty Oil Company, Getty having completed the purchase of Mission in 1951 and having operated Tidewater Associated and Skelly as separate subsidaries since that time. The Getty brand first appeared in 1970, replacing Flying A at stations in the northeast. In 1983 Getty also began replacing the Skelly brand with Getty, however before the changeover could be completed Texaco purchased Getty Oil and the Getty marketing in the northeast was sold off to New England independent Power-Test. The former Skelly locations were rebranded Texaco or sold off and Power Test, renamed Getty Oil Marketing and based in Jericho, New York continues to operate Getty brand stations from Maine to Virginia.

IDENTIFICATION SIGNS

TYDOL GASOLINE 42"RD
(1) 1925-1942 $225
Orange band and black outline around white porcelain sign. Large black "TYDOL" (large "T") above "GASOLINE" across sign face.

PUMP SIGNS

TYDOL MONOGRAM LOGO ON BLACK 13"SQ
(3) 1925-1942 No Listing
Curved black porcelain sign with orange band and black outline around white center circle. Large black "TYDOL" (large "T") across sign face, with black "T-W-O" monogram below.

TYDOL LOGO ON BLACK 13"SQ
(3) 1925-1942 $225
Black porcelain sign with orange band and black outline around white center circle. Large black "TYDOL" (large "T") across sign face.

TYDOL ETHYL 13"SQ
(3) 1925-1942 $225
Black porcelain sign with orange band and black outline around white center circle. Large black "TYDOL" (large "T") across sign face over Ethyl (EGC) logo below.

CURB SIGNS

TIDE WATER APPROVED DEALER 16x18
(5) 1920-1925 $350
Orange/black outlined white porcelain sign with black "TIDE WATER/APPROVED/DEALER" across center.

TYDOL WITH ETHYL 30"RD
(2) 1926-1935 $325
Orange ring around black outlined white center circle. Black "TYDOL" across upper white area with lage Ethyl (EGC) logo with orange rays below.

ASK FOR/VEEDOL/MOTOR OILS/GREASES 24"RD
(4) 1920-1925 $350
Black porcelain sign with orange "VEEDOL" across center. White script "Ask for" across top, with orange "Motor Oils/Greases" across bottom.

VEEDOL THE FILM OF PROTECTION 24"RD
(5) 1925-1930 $350
Black porcelain sign with orange "VEEDOL" across center. White script "Ask for" across top, with orange "Motor Oils Greases/THE/FILM OF PROTECTION" across bottom.

LUBRICANTS SIGNAGE

VEEDOL/MOTOR/OILS/GREASES 10x14
(4) 1920s $125
Black/orange outlined black porcelain sign with orange "VEEDOL/Motor/Oils/Greases" across sign.

VEEDOL GREASING SERVICE LASTING LUBRICATION 30x72
(5) 1920s $150
Black self framed tin litho sign with orange outlined white band across bottom. Orange "VEEDOL/GREASING SERVICE" across upper sign with black "LASTING LUBRICATION" on white band.

DRIVE IN FOR VEEDOL LUBRICATION 30x72
(5) 1920s $150
Black self framed tin litho sign with two orange lines across top and across bottom forming black band. Orange "VEEDOL/COMPLETE LUBRICATION" in center with white "DRIVE IN FOR" on upper band and "MOTOR TRANSMISSION DIFFERENTIAL CHASSIS" on lower band. Small PGCOA seal in lower left corner. Orange outlined white band diagonal above "VEEDOL" with black "100% PENNSYLVANIA/SUPREME QUALITY" on band.

VEEDOL MOTOR OILS VERTICAL 15x72
(5) 1920S $125
Wood framed black tin litho sign with orange "VEEDOL" vertical down center. White "Motor Oils" across below "VEEDOL," with orange outlined white band across bottom. Black "100% PENNSYLVANIA/SUPREME QUALITY" in band.

TOMBSTONE SIGNS

ASK FOR VEEDOL MOTOR OILS 22x28
(3) $200
Black/orange outlined black porcelain sign with white "Ask for" across top above orange "VEEDOL/Motor Oils" across center. Orange outlined white band across lower sign face with black "100% PENNSYLVANIA/SUPREME QUALITY" on band.

BUY VEEDOL MOTOR OILS 22x28
(2) $200
Black porcelain tombstone shaped sign with black outlined orange ring around white circle on upper sign face. Orange script "Buy" at top above black "VEEDOL/MOTOR OILS" in circle. White "100% PENNSYLVANIA/AT IT'S FINEST" below circle.

BUY VEEDOL MOTOR OILS 22x28
(2) $200
White porcelain tombstone shaped sign with black outlined orange ring around white circle on upper sign face. Orange script "Buy" at top above black "VEEDOL/MOTOR OILS" in

circle. Black "100% PENNSYLVANIA/AT IT'S FINEST" below circle.

OIL RACK SIGNS

VEEDOL/MOTOR OIL/100% PENNSYLVANIA 12x22
(5) 1930s $125
Black tin litho sign with white outlined orange "VEEDOL" over orange "MOTOR OIL" across top. White "100% PENNSYLVANIA/AT ITS FINEST" below.

SECONDARY SIGNAGE

TIDEWATER ASSOCIATED CREDIT CARDS 12x16
(5) 1939-1942 $350
Orange/black outlined white tin litho sign with black "TIDE WATER/ASSOCIATED/CREDIT CARDS/HONORED HERE" across sign face.

TRUCK SIGNAGE

TIDEWATER ASSOCIATED TYDOL/VEEDOL 7.5x11
(5) 1939-1942 $400
Black/white outlined black porcelain sign with white "TIDE WATER/ASSOCIATED" across top. Orange/white/green "TYDOL/VEEDOL" logo at bottom center with white "OIL" to left and "CO" to right.

ASSOCIATED OIL COMPANY - SAN FRANCISCO, CA

Founded in 1901, Associated was a west coast integrated oil company that was marketing gasoline by 1915. In 1932 they introduced the Flying A brand as a premium grade gasoline and after the 1938 merger with Tidewater, Flying A became the company's primary brandname. Also operated stations under the Seaside brand as a secondary operation. Flying A stations in the Associated territory, as well as Seaside, was sold to Phillips in 1966.

IDENTIFICATION SIGNS

"GAS" (FLYING A) 30x60
(4) 1932-1940 $650
Green/white outlined red porcelain sign with white "GAS" across center. Small white wings to either side of "A" that is slightly taller than other letters.

BUILDING LETTERING & SIGNAGE

ASSOCIATED PANEL VARIOUS
(4) 1915-1956 $300
Green/white outlined red porcelain panel with white "ASSOCIATED" across center.

DIECUT A 24x42
(4) 1940s $750
Black outlined red "A" with black outlined and detailed red wings.

PUMP SIGNS

FLYING A GASOLINE 10"SQ
(2) 1950-1956 $125
White/red outlined white porcelain sign with black outlined red "A" and wings with black details in center. Red "FLYING" across top, "GASOLINE" across bottom.

FLYING A ETHYL GASOLINE 10"SQ
(2) 1950-1956 $150
Green/white outlined red porcelain sign with black outlined white "A" and wings with black details in center. White "FLYING" across top, white outlined black "ETHYL" above white "GASOLINE" across bottom.

FLYING A DIECUT 9x13.5
(4) 1946-1950 $225
White outlined red porcelain sign with black outlined white "A" in center with black outlined and detailed wings projecting to either side.

FLYING A DIECUT EMBOSSED 9x13.5
(4) 1946-1950 $225
White outlined red embossed porcelain sign with black outlined white "A" in center with black outlined and detailed wings projecting to either side.

RED A ON WHITE ON RECTANGLE 9x15
(4) 1940s $200
Red porcelain sign with white oval in center. Red "A" with red wings/white details in center of oval.

NEON SIGNS

"GAS" (FLYING A) 30x60
(4) 1932-1940 No Listing
Green/white outlined red porcelain sign faces assembled on can-type frame with white "GAS" across center. Small white wings to either side of "A" that is slightly taller than other letters.

CURB SIGNS

ASSOCIATED GASOLINE 24"RD
(5) $2800

ASSOCIATED GASOLINE 30"RD
(4) $2500
Green/white outlined red porcelain sign with white pitcher with green details in center. Green "MORE MILES/TO THE/GALLON" on pitcher. White "ASSOCIATED" arched aorund top, "GASOLINE" around bottom.

ASSOCIATED GASOLINE 30"RD
(4) $2200
White/green/white outlined red porcelain sign with white pitcher with green details in center. Green "MORE MILES/TO THE/GALLON" on pitcher. White "ASSOCIATED" arched aorund top, "GASOLINE" around bottom.

USE ASSOCIATED GASOLINE 30"RD
(4) $2200
Blue outlined red porcelain sign with white pitcher with blue details in lower center. Blue "MORE MILES/TO THE/GALLON" on pitcher. White "USE ASSOCIATED GASOLINE" arched around top of sign.

LUBRICANTS SIGNAGE

CYCOL HEXAGON FLANGED SIGN 20x28
(5) 1925-1935 $850
Green/yellow outlined red hexagonal red porcelain flange sign. Yellow "CYCOL" across above center with black "MOTOR OIL" below. Smaller black lettering across center.

SECONDARY SIGNAGE

FLYING A OUR CREED 8x11
(4) 1932-1935 $450
Red/yellow/green outlined white porcelain sign with red circle at top. White einged "A" in circle with white "FLYING" above and "ASSOCIATED OIL COMPANY" arched around bottom. Green "OUR CREED/FRIENDLINESS/CLEANLINESS/HONEST VALUES/SUSTAINED/QUALITY/& SERVICE" below circle on sign

ASSOCIATED OUR CREED 8x11
(4) 1935-1950 $450

Red/yellow/green outlined white porcelain sign with red outlined white circle at top. Green attendant figure in circle with green "SMILING" arched aorund top, "ASSOCIATED SERVICE MEN" arched around bottom. Green "OUR CREED/FRIENDLINESS/ CLEANLINESS /HONEST VALUES/SUSTAINED/QUALITY/& SERVICE" below circle on sign

CERTIFIED CLEAN REST ROOMS 30"RD
(4) 1940-1946 $375
Green outlined white porcelain sign with red script "Certified" across top with red circle/white triangle Associated logo to right. Green "CLEAN/COMFORT/STATIONS" below.

MEN 3x10
(4) 1940-1956 $150
Red outlined white porcelain die cut sign with red outlined white circle with red script "Certified" across top with red circle/white triangle Associated logo to right. Green "CLEAN/COMFORT/STATIONS" below. Green "MEN" to right.

WOMEN 3x10
(4) 1940-1956 $150
Red outlined white porcelain die cut sign with red outlined white circle with red script "Certified" across top with red circle/white triangle Associated logo to right. Green "CLEAN/ COMFORT/STATIONS" below. Green "WOMEN" to right.

TIDEWATER ASSOCIATED OIL COMPANY - NEW YORK, NY: Tidewater and Associated merged in 1938. Signs listed below are from the post-merger era.

IDENTIFICATION SIGNS

TYDOL FLYING A RED/BLK 48"RD
(2) 1941-1950 $325
TYDOL FLYING A RED/BLK 72"RD
(1) 1941-1950 $150
Black outlined cream porcelain sign with red over black band across lower sign face. Black/white outlined red "A" with black/white wings on band. Black "TYDOL" on upper sign face.

TYDOL FLYING A GREEN 48"RD
(2) 1950-1956 $325
TYDOL FLYING A GREEN 72"RD
(1) 1950-1956 $150
Black outlined cream porcelain sign with green band across lower sign face. Black/white outlined red "A" with black/white wings on green band. Black "TYDOL" on upper sign face.

FLYING A SERVICE 72"RD
(1) 1956-1970 $125
Red porcelain sign with black outlined white "A" with wings in center. Black outlined white "FLYING" across top, "GASOLINE" arched around bottom.

FLYING A SERVICE EMBOSSED 48"RD
(3) 1956-1970 $850
FLYING A SERVICE EMBOSSED 72"RD
(4) 1956-1970 $750
Red embossed porcelain sign with embossed black outlined white "A" with wings in center. Embossed black outlined white "FLYING" across top, "GASOLINE" arched around bottom.

FLYING A EMBOSSED 48"RD
(3) 1956-1970 $850
FLYING A EMBOSSED 72"RD
(4) 1956-1970 $750
Red embossed porcelain sign with embossed black outlined white "A" with wings in center.

BUILDING LETTERING & SIGNAGE

FLYING A 16"RD
(4) 1956-1970 $350
Red embossed tin sign with embossed black outlined white "A" with wings in center.

EMBOSSED WINGED V PORCELAIN 20x36
(5) 1941-1956 $450
Embossed porcelain black outlined white "V" with black outlined white wings. Each wing is a separate piece.

EMBOSSED WINGED A PORCELAIN 20x36
(5) 1956-1970 $450
Embossed porcelain black outlined white "A" with black outlined white wings. Each wing is a separate piece.

TYDOL 12"
(5) 1946-1956 $100
Black porcelain lettering "TYDOL."

VEEDOL MOTOR OIL 12"
(5) 1946-1956 $100
Black porcelain lettering "VEEDOL MOTOR OIL."

PUMP SIGNS

TYDOL FLYING A 10"RD
(2) 1950-1956 $125
Red band and black outline around beige porcelain sign with green band across lower sign face. Black outlined red "A" in center of band with black outlined white wings to either side. Large black "TYDOL" over small italic "FLYING" above band.

TYDOL ETHYL FLYING A 10"RD
(2) 1950-1956 $125
White band and black outline around red porcelain sign. Black outlined white "A" with wings at bottom with black outlined white "TYDOL/ETHYL" above "A."

FLYING A GASOLINE 10"RD
(1) 1956-1960 $100
Red outlined white porcelain sign with black outlined red "A" in center with red wings detailed in black. Red "FLYING" across top, "GASOLINE" arched around bottom.

FLYING A SUPEREXTRA 10"RD
(1) 1956-1960 $100
White outlined red porcelain sign with black outlined white "A" in center with white wings detailed in black. White "FLYING" across top, "GASOLINE" arched around bottom.

CURB SIGNS

USE VEEDOL 30"RD
(5) 1967-1970 $175
Red tin litho sign with white "USE/VEEDOL" across top. White "THE WORLD"S MOST FAMOUS/MOTOR OIL/100% PENNSYLVANIA" below.

VEEDOL 30"RD
(5) 1970-1975 $50
Red tin litho sign with white band across center. Orange "VEEDOL" with red "O" enclosing atomic logo across white band. White "THE WORLD"S MOST FAMOUS/MOTOR OIL/100% PENNSYLVANIA" below.

TOMBSTONE SIGNS

VEEDOL "V" MOTOR OILS RED/BLK 22x28

(2) 1941-1945 $225
Red porcelain sign with black outlined tan circle on upper sign face. Red over black band across lower circle with black outlined red "V" with black and white wings on band. Black "VEEDOL" above band, small black "MOTOR OILS" below. Tan "100% PENNSYLVANIA" in red area below circle.
VEEDOL "V" MOTOR OILS RED/BLK 22x28
(2) 1945-1950 $225
Red porcelain sign with black outlined white circle on upper sign face. Red over black band across lower circle with black outlined red "V" with black and white wings on band. Black "VEEDOL" above band, small black "MOTOR OILS" below. White "100% PENNSYLVANIA" in red area below circle.
VEEDOL "A" MOTOR OILS RED/BLK 22x28
(2) 1941-1950 $225
Red porcelain sign with black outlined white circle on upper sign face. Red over black band across lower circle with black outlined red "A" with black and white wings on band. Black "VEEDOL" above band, small black "MOTOR OILS" below. White "100% PENNSYLVANIA" in red area below circle.
VEEDOL "A" MOTOR OIL RED/GR 22x28
(2) 1950-1956 $225
Red porcelain sign with black outlined white circle on upper sign face. Red over green band across lower circle with black outlined red "A" with black and white wings on band. Black "VEEDOL" above band, small black "MOTOR OILS" below. White "100%/ PENNSYLVANIA" in red area below circle.

LUBRICANTS SIGNAGE

VEEDOL "V" MOTOR OIL 18"RD
(3) 1946-1950 $300
Red/black outlined cream tin litho sign with red/black band across bottom. Black/white outlined red "V" on band, with black outlined white wings to either side. Black "VEEDOL" above band, "MOTOR OIL" below.
VEEDOL "A" 10-30 CAN FLANGE 10x16
(3) 1960-1965 $75
White tin litho flange sign in the design of the 1960-1970 oil can. Gold bedallion around upper blue area with white "VEEDOL" across blue. Gold circle as "medal" on medallion with blue "A" with white wings on circle. Blue outlined gold "10-30" below logo with blue lettering across bottom.
CHANGE FOR THE BETTER 10x16
(4) 1960-1965 $75
White tin litho flange sign in the design of the 1960-1970 oil can. Gold bedallion around upper blue area with white "For/Distinguished/Service" across blue. Gold circle as "medal" on medallion with blue "A" with white wings on circle. Red band below, with white "CHANGE/for the BETTER" across band.

SECONDARY SIGNAGE

TIDEWATER ASSOCIATED CREDIT CARDS 12x18
(4) 1946-1956 $200
Gold tin litho sign with black lines forming gold band across top. Black outlined of circle on right side of band wiht black outlined red "A" with wings in circle. Black "friendliness lives here..." on band with black "TIDE WATER ASSOCIATED" just below. Black outlined red "CREDIT CARDS" across cente rwith black "WELCOME" below.

BULK PLANT SIGNAGE

DISTRIBUTOR/TIDEWATER OIL/PRODUCTS 8x16
(5) 1955-1970 $125
Red porcelain sign with white "DISTRIBUTOR/TIDEWATER OIL/ PRODUCTS" across sign.

GETTY OIL COMPANY - NEW YORK, NY:

In 1951 J.Paul Getty completed his purchase of Misson Corporation, owner of Tidewater-Associated Oil Company. His Pacific Western Oil Company became Gatty Oil in 1956. In 1966 Tidewater Oil was renamed Getty Oil & in 1970 the Flying A brand was replaced with brandname Getty. Texaco purchased Getty in 1983 and sold off remaining branded marketing in the northeast to Power Test Oil Company. Power Test was rerenamed Getty Oil marketing and continues to brand stations from Maine to Virginia.
No collectible size Getty signs known at this time.

SEASIDE OIL COMPANY - SANTA BARBARA, CA

California refiner/marketer purchased by Tidewater Associated about the end of WW2. Seaside remained a Flying A secondary brand until the western marketing was purchased by Phillips in 1966. Phillips used the Seaside brand for a short time after the purchase but by 1970 the stations were rebranded Phillips 66.

IDENTIFICATION SIGNS

SEASIDE GASOLINE W/ GULL TRIANGLE 38x38
(4) 1935-1960 $650
Orange triangular porcelain sign with blue band along right side. Blue outlined yellow triangle superimposed over orange/blue area to center rightr. Blue "Seaside/Gasoline" logotype across top of yellow area with blue outlined soaring gull below.
SEASIDE GASOLINE W/ WAVES TRIANGLE 38x38
(5) 1946-1965 $650
Orange triangular porcelain sign with blue band along right side. Blue/yellow outlined yellow triangle superimposed over orange/blue area to center right with blue area below, separated by "waves." Blue "Seaside" across yellow area with white "Gasoline" logotype below, above white outlined soaring gull below.

NEON SIGNS

SEASIDE GASOLINE W/ GULL TRIANGLE 64x64
(5) 1946-1965 No Listing
Orange can type triangular porcelain sign with blue band along right side. Blue/yellow outlined yellow triangle superimposed over orange/blue area to center right with blue area below, separated by "waves." Blue "Seaside" across yellow area with white "Gasoline" logotype below, above white outlined soaring gull below. Neon outline and "Seaside."

SECONDARY SIGNS

STOP YOUR MOTOR NO SMOKING 6x36
(5) 1955-1965 $350
Yellow/red outlined yellow porcelain sign with red/yellow/ blue Seaside logo with waves in center. Blue "STOP YOUR MOTOR" to left, "NO SMOKING" to right.

MOHAWK OIL COMPANY - LOS ANGELES, CA:

West Coast refiner/marketer with close ties to Associated & Tidewater Associated Oil Company.

IDENTIFICATION SIGNS

MOHAWK GASOLINE 30"RD
(3) 1925-1960 $3200
Black outlined red porcelain sign with black outlined white center circle. Black outlined white "MOHAWK" arched around top, "GASOLINE" around bottom with red and black Indian head in center.

MOHAWK GASOLINE 42"RD
(3) 1925-1960 3000
Black outlined white porcelain sign with black outlined white center circle. Black outlined red "MOHAWK" arched around top, "GASOLINE" around bottom with red and black Indian head in center.

PUMP SIGNS

MOHAWK GASOLINE TAB SIGN 12"RD
(5) 1920-1925 No Listing
Black outlined red porcelain sign with black outlined white center circle. Black outlined white "MOHAWK" arched around top, "GASOLINE" around bottom with red and black Indian head in center. White tab for fastening to cylinder rods.

NEON SIGNS

MOHAWK GASOLINE 48"RD
(3) 1925-1960 No Listing
Black outlined red porcelain sign with black outlined white center circle. Black outlined white "MOHAWK" arched around top, "GASOLINE" around bottom with red and black Indian head in center. Punched for neon installation, faces are assembled back to back.

PATHFINDER PETROLEUM - SOUTH PASADENA, CA:

West Coast marketer purchased by Tidewater Associated about 1960.

SECONDARY SIGNS

PATHFINDER 30x30
(5) No Listing
Die cut tin litho sign with upper area forming half circle with rectangular panel below. Blue half circle with yellow "PATHFINDER" with multi-color figure in center. Blue outlined yellow panel across bottom with blue "GASOLINE/LEADS THE WAY" on panel.

NORTHWESTERN OIL COMPANY - SUPERIOR, WI:

Originally marketing under the Northwestern Vimamite and Noco brands, Northwestern was purchased by Tidewater Oil in 1931 to serve as a company owned jobber for Tydol produts in Wisconsin & Minnesota. Formed the basis of Tydol's entry into the midwest, where they branded stations until selling them to Sunray-DX in 1957.
No collectible size Northwestern signs known at this time.

GULF AND AFFILIATES

GULF OIL CO. - PITTSBURGH, PA

The Gulf Oil Company was founded in 1901 at the Spindletop, Texas oil fields. Very early in the company's history they established a marketing presence throughout the south and east that was to dominate the company's direction for the next eighty years. Gulf is credited with having opened the first "built for that purpose, offstreet gasoline station in the world," having opened a small retail station in Pittsburgh in December 1913. Other stations pre-date this, but they were converted from other uses. Gulf continued to market in 28 southern and eastern states until the 1950s when several purchases of regional chains made Gulf a coast-to-coast marketer. The gas shortages of the 1970s forced Gulf to withdraw from many areas and the early 1980s saw Gulf in serious financial trouble. In 1984 Chevron (Standard of California) purchased Gulf Oil, converting stations in several states to Chevron and selling the southeast marketing to BP and the northeast marketing to Cumberland Farms. Cumberland continues to operate stations under the Gulf brand.

IDENTIFICATION SIGNS

GULF 30"RD
(2) 1920-1936 $325
White outlined orange porcelain sign with blue "GULF" across center. Light blue and white dropshading on letters.

GULF 42"RD
(1) 1920-1936 $300
Pipe framed white outlined orange porcelain sign with blue "GULF" across center. Light blue and white dropshading on letters.

GULF DEALER 66"RD
(1) 1936-1954 $150
Steel framed blue/white outlined orange porcelain sign with blue "GULF" across center. Blue and white lines form drop shading for letters. Blue "DEALER" with dropshading below.

GULF DEALER 66"RD
(1) 1954-1960 $150
Same as above, without blue and white stripes in dropshading.

GULF DEALER 72"RD
(1) 1960-1963 $100
Pole framed blue/white outlined orange porcelain sign with blue "GULF" with white dropshading across center. Blue "DEALER" across bottom.

GULF NEW DISC 72x78
(1) 1963-PRESENT $75
Pole framed can type blue/white outlined orange porcelain sign with white rounded center band Blue "Gulf" on center band.

ILLUMINATED SIGNS

THAT GOOD GULF GASOLINE 66"RD
(5) 1920-1936 No Listing
Framed orange tin sign with white opal glass lettering "THAT GOOD/GULF/GASOLINE.

GULF DEALER 72"RD
(4) 1954-1963 $150
Formed orange plexiglass face for internally illuminated signs with laminated blue lettering and white dropshading.

BUILDING LETTERING & SIGNAGE

GULF BLUE PORCELAIN 10"
(3) 1936-1963 $100

GULF BLUE PORCELAIN 12"
(1) 1936-1963 $100

GULF BLUE PORCELAIN 18"
(1) 1936-1963 $75

GULF BLUE PORCELAIN 24"
(1) 1946-1963 $50
GULFLEX 12"
(1) 1950-1963 $50
LUBRICATION 12"
(4) 1936-1950 $50
WASHING 12"
(1) 1936-1963 $50
TIRES 12"
(2) 1936-1963 $50
BATTERIES 12"
(4) 1936-1963 $50
ACCESSOIRES 12"
(4) 1936-1963 $50
All are blue porcelain formed letters mounted on white porcelain rails. All of the above were also found on orange porcelain rails.

PUMP SIGNS

THAT GOOD GULF GASOLINE 12"RD
(4) 1932-1936 $125
White outlined blue porcelain sign with orange "bar and circle" across center. Blue "GOOD GULF" on center bar. Blue "THAT" at top, "GASOLINE" at bottom.
THAT GOOD GULF GASOLINE 10.5"RD
(1) 1946-1950 $100
Blue outlined white porcelain sign with orange "bar and circle" across center. Blue "GOOD GULF" on center bar. Blue "THAT" at top, "GASOLINE" at bottom.
GULF NO NOX KNOCKPROOF 10.5"RD
(1) 1946-1950 $100
White outlined blue porcelain sign with white "GULF/NO-NOX" on upper sign face above white band across bottom. Orange "KNOCKPROOF" on band.
GULF DIESELECT 10.5"RD
(4) 1946-1950 $200
Orange/white outlined blue porcelain sign with white band across lower sign face. Gulf logo at top, with orange "DIESELECT/ FUEL" across band.
GULF DIESEL 10.5"RD
(4) 1946-1950 $200
Orange/white outlined blue porcelain sign with white band across lower sign face. Gulf logo at top, with orange "DIESEL/ FUEL" across band.
DIESEL #1 10.5"RD
(4) 1950-1960 $50
Blue outlined white porcelain sign with blue "1" in center. Orange "GULF" at top, "DIESEL" below.
DIESEL #2 10.5"RD
(3) 1950-1960 $50
Same as above, "2" replacing "1."
GULF MARINE WHITE GASOLINE 10.5"RD
(4) 1946-1950 $300
Blue outlined white porcelain sign with blue band across center. Blue "GULF" in upper white area with "GASOLINE" below. White "MARINE/WHITE" in center band.
GOOD GULF (INTERLOCKING) 10.5"RD
(1) 1950-1956 $75
Blue outlined whit eporcelain sign with Orange "Good" over blue "Gulf." Interlocking "G's" in Good and Gulf.
GULF NO NOX 8.5x10.5
(1) 1950-1956 $75
Blue white outlined blue porcelain sign with white "GULF/NO-NOX" across center.
GULF DIESELECT 8.5x10.5
(3) 1950-1956 $100
Rectangular porcelain sign with white/blue outliend white upper half and blue/white outlined blue lower half. Blue "GULF" at top, white "DIESELECT" at bottom.
GULF MARINE WHITE 8.5x10.5
(3) 1950-1956 $125
Blue outlined white porcelain rectangular sign with blue outlined orange "GULF/MARINE/WHITE"
All of the following porcelain signs listed as 8.5" high and 11" wide are trapezoidal in shape.
GOOD GULF 8.5x11.5
(1) 1956-1963 $50
Blue outlined porcelain sign with white upper half and blue lower half. Orange "Good" on white area, white "Gulf" on blue.
GULF SUPER NO NOX 8.5x11.5
(2) 1956-1959 $50
Blue outlined porcelain sign with blue upper half and white lower half. White "Gulf" on blue area, orange "SUPER/No-nox" on white area.
GULF NO NOX 8.5x11.5
(1) 1959-1962 $50
Blue outlined porcelain sign with blue upper half and white lower half. White "Gulf" on blue area, orange "No-nox" on white area.
GULF NEW NO NOX 11.5x17.5
(2) 1962-1963 $75
White porcelain sign with blue outline shield design. Three blue stars above and below blue "NEW" over orange "NO-NOX."
GULF CREST 11.5x17.5
(5) 1956-1962 $350
Gold porcelain shield shaped sign with white square in center. Blue "Gulf" over orange "Crest" in square.
GULFTANE 8.5x11.5
(2) 1962-1963 $50
White outlined porcelain sign with orange stripe across top, white across center, blue across bottom. Blue "Gulftane" across center white stripe.
DIESEL 1 FUEL 8.5x11.5
(5) 1960-1963 $250
Blue outlined white porcelain sgin with blue "Diesel" across top, "Fuel" below with blue outlined orange circle in center. White "1" in orange circle.
DIESEL 2 FUEL 8.5x11.5
(5) 1960-1963 $250
Blue outlined white porcelain sgin with blue "Diesel" across top, "Fuel" below with blue outlined orange circle in center. White "2" in orange circle.
GULF KEROSENE 10.5"RD
(2) 1956-1963 $100
Blue outlined white porcelain sign with blue italic "Gulf" at top over orange italic "Kerosene"
GOOD GULF 8.5x11.5

(1) 1963-1971 $45
Orange outlined white porcelain sign with orange "Good" across top above Gulf new disc logo.
NO NOX 8.5x11.5
(1) 1963-1971 $45
Blue outlined white porcelain sign with Gulf new disc logo at top above blue "No-nox."
GULFTANE LARGE LOGO 8.5x11.5
(2) 1963-1971 $45
Blue outlined white porcelain sign with orange band across top and blue band across bottom. Large Gulf new disc at left in center extending most of the way into color bands, with blue "tane" to right.
GULFTANE SMALL LOGO 8.5x11.5
(2) 1963-1971 $35
Blue outlined white porcelain sign with orange band across top and blue band across bottom. Small Gulf new disc at left in center, with blue "tane" to right.
GULF DIESEL 8.5x11.5
(1) 1963-1971 $75
Blue outlined white porcelain sign with Gulf new disc logo at top above orange "DIESEL."
GULF DIESELECT 8.5x11.5
(1) 1963-1971 $100
White outlined blue porcelain sign with Gulf new disc logo at top above white "DIESELECT."
GULF KEROSENE 8.5x11.5
(1) 1963-1971 $100
Blue outlined white porcelain sign with Gulf new disc logo at top above orange "KEROSENE."
GULF MARINE 8.5x11.5
(1) 1963-1971 $100
Blue outlined white porcelain sign with Gulf new disc logo at top above blue "MARINE."
GULF MARINE WHITE 8.5x11.5
(1) 1963-1971 $125
Orange outlined white porcelain sign with Gulf new disc logo at top above orange "MARINE" over small blue "WHITE."
GULF SUPERMARINE 8.5x11.5
(1) 1963-1971 $150
Blue outlined white porcelain sign with Gulf new disc logo at top above orange "SUPER" over blue "MARINE."
AND GULFPRIDE OUTBOARD MOTOR OIL 8.5x11.5
(1) 1963-1971 $150
Blue outlined white porcelain sign with blue "AND/Gulf-pride/OUTBOARD/MOTOR OIL covering sign. For use in conjunction with other Marine product pump signs.
MICRONIC FILTRATION 4.5x10
(5) 1940's $100
Blue outlined white porcelain sign with blue "MICRONIC/FILTRATION" above orange "FOR YOUR PROTECTION" on sign.
REGULAR 3x7.5
(4) 1946-1956 $75
Orange outlined white porcelain strip with orange "REGULAR" across center. For use in Florida and North Carolina where grade label had to be posted in addition to brandname.
PREMIUM 3x7.5
(4) 1946-1956 $75
Blue outlined white porcelain strip with blue "PREMIUM" across center. For use in Florida and North Carolina where grade label had to be posted in addition to brandname.

NEON SIGNS

GULF 30" LETTER SET 30"
(2) 1946-1963 $425
Orange porcelain faces with blue channel porcelain returns. Four individual letters mounted on angle iron for rooftop mounting

CURB SIGNS

THAT GOOD GULF GASOLINE 30"RD
(5) 1926-1930 $375
Blue outlined white porcelain sign with large blue/orange/white Gulf logo covering most of center of sign. Blue "THAT GOOD" arched around top, "GASOLINE" around bottom.
NO NOX ETHYL 30"RD
(5) 1926-1930 $375
White/blue outlined orange porcelain sign with white outlined blue rounded side square covering most of center of sign. White circle in center with Ethyl
(EGC) logo, with white "NO-NOX" in upper blue area, "ETHYL" in lower blue area.
GULF SUPREME MOTOR OIL 32x36
(5) 1926-1930 $375
White porcelain tombstone sign with large blue/orange/white Gulf logo covering entire top of sign face. Blue "SUPREME" with blue/white striped dropshading over blue "MOTOR OIL" below logo on lower white area.
"LIGHTHOUSE" PANELS
From about 1918 until about 1930 Gulf installed throughout their marketing territory a series of roadside highway markers with various warning messages of road hazards ahead. Most common was a marker advising motorists that they were approaching a cross road intersection. Constructed in the shape of lighthouses, the warning message appeared on a round upper panel with changable porcelain panels advertising Gulf products on the pedestals. The following list of signs are descriptions of the panels that are known at this time. All are white porcelain with orange and blue lettering and scenes as described and most have a large orange/white/blue Gulf logo on them. All are 27.5" wide and 60" high with text as listed.
THERE IS/MORE POWER IN/THAT GOOD/GULF(LOGO)/GASOLINE/AT THE SIGN OF/THE ORANGE DISC/GULF REFINING COMPANY
(NO CAR) 27.5x60
(5) 1920-1925 $1100
SUPREME/MOTOR/OIL(ACROSS LOGO)/FOR A/SMOOTH RUNNING/MOTOR/AT THE SIGN OF/THE ORANGE DISC/GULF REFINI9NG CO.
(ORANGE CAR FACING RIGHT) 27.5x60
(4) 1920-1925 $1100
SUPREME/GULF(LOGO)/MOTOR OIL/AT THE/SIGN OF THE/ORANGE DISC/GULF REFINING CO.
(ORANGE CAR FACING RIGHT) 27.5x60
(4) 1920-1925 $1100
THAT GOOD/GULF(LOGO)/GASOLINE/AT THE SIGN OF/THE ORANGE DISC/GULF REFINING CO.
(BLUE CAR FACING RIGHT) 27.5x60
(3) 1925-1930 $1100
GULF(LOGO)/SUPREME/MOTOR/OIL/AT THE SIGN OF/THE

ORANGE DISC/GULF REFINING CO.
(BLUE CAR FACING RIGHT) 27.5x60
(3) 1925-1930 $1100
THAT GOOD/GULF(LOGO)/GAOSLINE/AT THE SIGN OF/THE ORANGE DISC/GULF REFINING CO.
(CLOSED ROADSTER BLUE FACING LEFT) 27.5x60
(3) 1925-1930 $1100
GULF(LOGO)SUPREME/MOTOR/OIL/AT THE SIGN OF/THE ORANGE DISC/GULF REFINING CO.
(CLOSED ROADSTER BLUE FACING LEFT) 27.5x60
(3) 1925-1930 $1100

LUBRICANTS SIGNAGE

GULFPRIDE PANEL SIGN 24x48
(4) 1936-1946 $350
Tin litho embossed sign with blue upper band and white lower band. Orange and white stripes across center with small Gulf orange disc logo in center. White script "Gulfpride" across top, with blue "THE WORLD'S FINEST MOTOR OIL/SUPER-REFINED BY ALCHLOR PROCESS/100% PURE PENNSYLVANIA" across bottom.

GULFLUBE PANEL SIGN 24x48
(4) 1936-1946 $350
Tin litho embossed sign with blue upper band and white lower band. Orange and white stripes across center with small Gulf orange disc logo in center. White "GULFLUBE" across top, with blue "THE HIGH MILEAGE MOTOR OIL/MULTI-SOL PROCESSED" across bottom.

GULFPRIDE FENCE SIGN 24x48
(3) 1946-1952 $350
Self framed tin litho embossed sign with blue upper band and white lower band. Orange and white stripes across center with small Gulf orange disc logo in center. White script "Gulfpride" across top, with blue "THE WORLD'S FINEST MOTOR OIL/SUPER-REFINED BY ALCHLOR PROCESS/SOLD HERE IN SEALED CANS ONLY-NOT SOLD IN BULK" across bottom.

GULFLUBE FENCE SIGN 24x48
(4) 1946-1952 $350
Self framed tin litho embossed sign with blue upper band and white lower band. Orange and white stripes across center with small Gulf orange disc logo in center. White "GULFLUBE" across top, with blue "THE HIGH MILEAGE MOTOR OIL/MULTI-SOL PROCESSED" across bottom.

GULFLEX FENCE SIGN 24x48
(4) 1946-1952 $350
Self framed orange tin litho embossed sign with blue/white otuline and blue band across bottom. Blue "GULFLEX" with blue/white striped dropshading diagonal across sign with small Gulf logo below. Blue "YOUR CAR NEEDS" at top, white "REGISTERED LUBRICATION" on blue band.

GULFPRIDE MARINE FENCE SIGN 24x48
(5) 1946-1952 $500
Self framed blue tin litho embossed sign with white band across bottom. White script "Gulfpride" over "MARINE" in upper blue area. Blue "THE WORLD'S FINEST MARINE OIL/SUPER-REFINED BY ALCHLOR PROCESS" on white band with Gulf logo at either end on color split.

OIL RACK SIGNS

GULFPRIDE OIL RACK SIGN 12x24
(3) 1936-1952 $100
Self framed tin litho embossed sign with blue upper band and white lower band. Orange and white stripes across center with small Gulf orange disc logo in center. White script "Gulfpride" across top, with blue "THE WORLD'S FINEST MOTOR OIL/SUPER-REFINED BY ALCHLOR PROCESS/SOLD HERE IN SEALED CANS ONLY-NOT SOLD IN BULK" across bottom.

GULFLUBE OIL RACK SIGN 12x24
(3) 1932-1952 $100
Self framed tin litho embossed sign with blue upper band and white lower band. Orange and white stripes across center with small Gulf orange disc logo in center. White "GULFLUBE" across top, with blue "THE HIGH MILEAGE MOTOR OIL/MULTI-SOL PROCESSED" across bottom.

GULFPRIDE HD SELECT 10x26
(3) 1952-1959 $75
Blue outlined white tin litho sign with notches near bottom on each side. Blue "Gulfpride" over orange "H.D.SELECT" with small Gulf logo at lower right.

GULFPRIDE SELECT SINGLE G 10x26
(3) 1959-1963 $75
Gold outlined white tin litho sign with purple triangle at either end. Blue "Gulfpride/SELECT" over purple "SINGLE-G" on sign face with small orange "THE MOTOR OIL FOR WIDE RANGE PROTECTION" at top.

GULFPRIDE MARINE HD RACK SIGN 10x26
(4) 1955-1963 $150
Blue tin litho sign with white stripe across lower sign face. White "Gulfpride" in upper blue area, with small Gulf logo at either end of white stripe and blue "MARINE H.D." between.

SECONDARY SIGNAGE

GULF AUTHORIZED DEALER 10x36
(3) 1926-1936 $225
Blue outlined white porcelain sign with blue and white Gulf logo at either end and blue "AUTHORIZED/DEALER" across center.

THAT GOOD GULF GASOLINE PRICE SIGN 12"RD
(4) 1924-1936 $150
Orange outlined white porcelain sign with black "THAT GOOD GULF" arched around top, "GASOLINE" around bottom. Center is designed so that interchangable orange & black price discs can be attached.

NO NOX ETHYL PRICE SIGN 12"RD
(4) 1924-1936 $150
Orange outlined white porcelain sign with blue "NO-NOX" arched around top, "ETHYL" around bottom. Center is designed so that interchangable orange & black price discs can be attached.

KEROSENE PRICE SIGN 10x16
(5) 1920s $100
White outlined blue porcelain sign with white "GULF" across top above KEROSENE. Location for attachment of pricing numbers with white "PER GALLON" below.

GULF 24"RD
(5) 1950's $500
Blue/white outlined orange porcelain sign with blue "GULF" across center. Blue and white lines form drop shading for letters.

GULF 30"RD
(5) 1962-1963 $300

Tire rack sign designed for "Gulf Tire Mart" merchandiser. Blue/white outlined orange porcelain sign with blue "GULF" with white dropshading across center.

GULF NEW DISC 30"RD
(5) 1962-1963 $100
Tire rack sign designed for "Gulf Tire Mart" merchandiser. Blue/white outlined orange porcelain sign with white rounded band across center. Blue "GULF" on white band across center.

ANOTHER GULF STATION 48x96
(5) 1950-1963 $200
White tin litho sign with large orange/blue/white Gulf disc logo at right end. Blue outlined white area to left with blue "ANOTHER GOOD/GULF STATION/WILL OPEN HERE"

GULF CARDS HONORED STRIP 12x60
(5) 1936-1946 $125
White porcelain strip with blue "GULF CARDS HONORED" across sign.

CLEAN REST ROOMS STRIP 12x60
(5) 1936-1946 $125
White porcelain strip with blue "CLEAN REST ROOMS" across sign.

GULF CARDS HONORED 12x18
(2) 1946-1963 $100
Blue outlined white porcelain sign with blue "GULF/CARDS/HONORED" across sign.

CLEAN REST ROOMS 12x18
(2) 1946-1963 $100
Blue outlined white porcelain sign with blue "CLEAN/REST/ROOMS" across sign.

SANITARY REST ROOMS 12x24
(3) 1954-1963 $100
White tin litho sign with black "Sanitary" over black "REST ROOMS."

MEN 6.5x17
(3) 1928-1940 $75
Blue outlined white porcelain sign with blue "MEN" across center.

LADIES 6.5x17
(3) 1928-1940 $75
Blue outlined white porcelain sign with blue "LADIES" across center.

MENS REST ROOMS 6.5x17
(3) 1928-1940 $75
Blue outlined white porcelain sign with blue "MENS/REST ROOM" across sign.

LADIES REST ROOM 6.5x17
(3) 1928-1940 $75
Blue outlined white porcelain sign with blue "LADIES/REST ROOM" across sign.

MEN PROJECTING 5x15
(5) 1946-1950 $50
Blue outlined white porcelain projecting sign with orange base with blue and white stripes. Blue "MEN" across sign.

LADIES PROJECTING 5x15
(5) 1946-1950 $50
Blue outlined white porcelain projecting sign with orange base with blue and white stripes. Blue "LADIES" across sign.

MEN FLANGE 5x15
(1) 1950-1963 $50
White porcelain flange sign with blue "MEN" across sign.

LADIES FLANGE 5x15
(1) 1950-1963 $50
White porcelain flange sign with blue "WOMEN" across sign.

REST ROOMS FLANGE 5x18
(2) 1956-1963 $75
White porcelain flange sign with rounded forward edge with blue "REST ROOMS" above arrow.

MEN 2x8
(1) 1946-1989 $35
White porcelain round end rectangle sign with blue "MEN" across center.

WOMEN 2x8
(1) 1946-1989 $35
White porcelain round end rectangle sign with blue "WOMEN" across center.

GULF TIRES BATTERIES FENCE SIGN 24x48
(3) 1946-1963 $250
Blue outlined white self-framed tin litho embossed sign. Large blue "GULF" across top, over small orange "ACCESSORIES-BATTERIES." Blue "TIRES" across bottom.

EARLY GULF FLANGE SIGNS

SUPREME AUTO OIL 14x24
(4) 1912-1920 $450
Yellow porcelain sign with black "SUPREME/AUTO/OIL" over small "MANUFACTURED BY" over larger "GULF REFINING COMPANY." Black & yellow spoked wheel/radiator "GRC" logo with "SUPREME" at top, "AUTO OIL" at bottom to left side of sign face.

SUPREME AUTO OIL 18x22
(4) 1920-1922 $425
White porcelain sign with black outline and large orange circle in center. Large black lettering SUPREME/AUTO/OIL over smaller MANUFACTURED BY/GULF REFINING COMPANY/PITTSBURGH, PA/FOR SALE HERE

THAT/GOOD GULF/GASOLINE 18x22
(4) 1920-1922 $425
White porcelain sign with black outline and large orange circle in center. Large black lettering THAT/GOOD GULF/ GASOLINE over smaller MANUFACTURED BY/GULF REFINING COMPANY/PITTSBURGH, PA/FOR SALE HERE

SUPREME AUTO OIL-LEAVES LESS CARBON 18x22
(3) 1922-1924 $425
White porcelain sign with oragne circle in center. Blue "SUPREME/AUTO/OIL" with blue/white striped dropshading across circle with blue "LEAVES LESS CARBON" across bottom of circle. Blue "GULF REFINING COMPANY" across bottom of sign.

SUPREME MOTOR OIL-FOR A SMOOTH RUNNING MOTOR 18x22
(3) 1924-1930 $425
White porcelain sign with oragne circle in center. Blue "SUPREME/AUTO/OIL" with blue/white striped dropshading across circle with blue "FOR A SMOOTH RUNNING MOTOR" across bottom of circle. Blue "GULF REFINING COMPANY" across bottom of sign.

Mobilgas horse with attached hanging sign. *Al Alford.*

Socony Aircraft Oils porcelain sign 30"x20." *Bruce Bednar.*

Tulsa 60"x40" porcelain sign. *Dunbar's Gallery.*

Old Dutch 6' porcelain sign. *Lonnie Hop.*

Old Dutch 6' plastic sign. *Lonnie Hop.*

Republic small pump plate sign. *Mark McKeown.*

Marathon 48" porcelain sign. *Dunbar's Gallery.*

Republic 48"x52" shield shape sign. *Dunbar's Gallery.*

Marathon 48" porcelain sign.

Marathon 48" metal flange. *Leo Mathieu.*

Lion 60" metal sign.

Frontier 60" porcelain sign. *Jim Hollabaugh.*

El Paso porcelain pump plate. *Mark McKeown.*

Seaside Gasoline triangular sign. *Al Alford.*

Tydol Flying A 12" porcelain pump sign. *Mark McKeown.*

THAT GOOD GULF GASOLINE 18x22
(3) 1922-1930 $400
White porcelain sign with orange disc Gulf logo in center. Blue "THAT GOOD" across top of logo, "GASOLINE" across bottom with blue "GULF REFINING COMPANY" across bottom of sign.
NO NOX MOTOR FUEL 18x18
(2) 1926-1930 $425
White porcelain sign with blue outlined white circle covering most of sign. Small Gulf logo in center, with blue "NO-NOX" arched around top, "MOTOR FUEL" around bottom. Blue "GULF REFINING COMPANY" across bottom.
NO NOX MOTOR FUEL 18x18
(2) 1926-1930 $425
White porcelain sign with blue outlined white circle covering most of sign. Small Gulf logo in center, with blue "NO-NOX" arched around top, "MOTOR FUEL" around bottom. Blue "GULF REFINING CO OF LA" across bottom.

TRUCK SIGNAGE

GULF 12"
(5) 1936-1950 $225
Blue/white outlined orange porcelain sign with blue "GULF" across center. Blue and white lines form drop shading for letters.

BULK PLANT SIGNAGE

GULF FUEL OIL DEALER 30"RD
(5) 1936-1950 $475
Steel framed blue/white outlined orange porcelain sign with blue "GULF" across center. Blue and white lines form drop shading for letters. Blue "FUEL OIL/DEALER" below.
GULF OIL CORPORATION SINGLE PANEL 12x144
(2) 1936-1950 $200
Blue self framed embossed porcelain strip with white "GULF OIL CORPORATION" across center.
"DISTRIBUTOR" PANEL 8x30
(4) 1936-1950 $50
White outlined blue embossed porcelain rectangle with rounded ends. White "DISTRIBUTOR" across center. For mounting above GULF OIL CORPORATION sign listed above.
"PRODUCTS" PANEL 8x30
(4) 1936-1950 $50
White outlined blue embossed porcelain rectangle with rounded ends. White "PRODUCTS" across center. For mounting below GULF OIL CORPORATION sign listed above.
DISTRIBUTOR GULF OIL CORPORATION PRODUCTS 42x288
(2) 1950-1989 $200
Four blue wood framed porcelain panels assembled with blue panel with white "DISTRIBUTOR" across top, blue panel with white "PRODUCTS" across bottom with two panels making up "GULF OIL CORPORATION" in center.
GULF SOLAR HEAT PANEL 30x42
(2) 1960-1963 $250
White porcelain sign without border. Left side of sign has Gulf Solar Heat logo with orange flame. To right of logo is blue "HEATING OILS." Designed such that white lettered blue porcelain panel with jobber name bolts across bottom for assembly.
GULF SOLAR HEAT PANEL 48x96
(4) 1960-1963 $100
Two piece white porcelain sign with white/blue border. Left panel has Gulf Solar Heat logo with blue outlined orange and white flame. Right panel has blue "HEATING OILS." Designed such that white lettered blue porcelain panel with jobber name bolts across bottom for assembly.
GULF SOLAR HEAT PANEL 30x42
(2) 1960-1963 $200
White porcelain sign with white/blue border. Left side of sign has Gulf Solar Heat logo with blue outlined orange and white flame. To right of logo is blue "HEATING OILS." Designed such that white lettered blue porcelain panel with jobber name bolts across bottom for assembly.
GULFTANE LP GAS BUTANE/PROPANE
(5) 1950s $350
White tin litho sign. Description not available
NO SMOKING 6x16
(3) 1932-1946 $100
White outlined blue porcelain sign with white "NO SMOKING" across center.
NO SMOKING 6x16
(3) 1932-1946 $100
Blue porcelain sign with white "NO SMOKING" across center.

WILSHIRE OIL CO. - LOS ANGELES, CA

Wilshire, a west coast refiner/marketer dating from the 1910s, was purchased by Gulf in 1957. In 1965 the Wilshire brand was replaced by Gulf.

IDENTIFICATION SIGNS

POLY GAS DIECUT PARROT 36x60
(4) 1930-1946 $4000
Detailed die-cut tin litho multi-colored parrot with black/red outlined yellow banner across center below parrot perch. White outlined black "POLLY/GAS" on banner.

SUPERIOR OIL CO. - DES MOINES, IA

Superior was a Des Moines based marketer, datinginging from the mid-1930s. They merged with Carpenter Oil Co. in the early 1960s. About 1968 Gulf purchased Superior and re-branded stations Gulf.

PUMP SIGNS

SUPERIOR 400 REGULAR GASOLINE 11"RD
(5) 1940's $75
Red outlined whit eporcelain sign with red center circle. Blue "SUPERIOR" arched around top, "GASOLINE" around bottom with white "400/REGULAR" in red center circle.

CARPENTER OIL CO. - LINCOLN, NE

Founded in the 1930s
(possibly earlier) as L.L.Coryell and Sons, Carpenter owned the Coryell brand by the early 1950s and operated stations throughout Nebraska. In the early 1960s they merged with Superior Oil of Des Moines, who was bought out by Gulf in 1968.

IDENTIFICATION SIGNS

CORYELL 70 40x60
(4) 1946-1960 $225
White/blue outlined red porcelain sign with white "CORYELL"

arched across top above large "70" with white blocks above, below and to each side.

CORYELL 70 60x96
(4) 1946-1960 $150
Can type porcelain sign with two-panel faces. White/blue outlined red porcelain sign with white "CORYELL" arched across top above large "70" with white blocks above, below and to each side.

PUMP SIGNS

ALWAYS LESS CORYELL 70 HIGHER ANTI-KNOCK 12x12
(5) 1935-1950 $225
Black tin litho sign with white circle in center. Large yellow and black framed red Coryell 70 shield in center with whtie lettering. Black "ALWAYS LESS" above shield, "HIGHER ANTI-KNOCK" below.

BULK PETROLEUM CO. - CHICAGO, IL

Another old Chicago private brand that sold out to a major oil company during the market expansion in the 1950s-60s. Gulf purchased the operation, branded "BULKO" in 1968. Most of the Chicago area stations were converted to Gulf, but some unsuccessful Gulf locations around the country were operated as Bulko stations in the 1970s.

PUMP SIGNS

BULKO GASOLINE YELLOW TRIANGLE 11x12.5
(5) 1935-1955 $250
Blue outlined yellow triangle, point up, with rounded corners. Red "BULKO" with blue dropshading across center with red "GASOLINE" below and detailed red/blue/yellow elephant with red "B" on yellow blanket at top.

BULKO GASOLINE WHITE TRIANGLE 11x12.5
(5) 1946-1955 $250
Black outlined white triangle, point up, with rounded corners. Red "BULKO" across center with red "GASOLINE" below and detailed red/black/white elephant with red "B" on white blanket at top.

SUNSET OIL COMPANY - LOS ANGELES, CA:

Sunset was a California marketer dating from the 1914. In the early 1960's Gulf purchased the marketing operation, selling off the Arizona stations to form an independent company that survived into the 1970s.

CURB SIGNS

SUNSET GASOLINE RAINBOW MOTOR OIL 30"RD
(5) 1930s $2400
Red outlined green porcelain sign with red outlined mulit-color sunset scene in center circle. White outlined yellow "SUNSET" arched around top, "GASOLINE" around bottom with yellow banner just above. Red "RAINBOW MOTOR OIL" on banner.

SUNSET MOTOR OIL 20x26
(5) 1930s $2200
Green porcelain tombstone shaped sign with PGCOA seal superimposed over yellow/red sunset scene in center. White outlined yellow "SUNSET" arched across top, "MOTOR OIL" across bottom.

SECONDARY SIGNS

NO SMOKING STOP YOUR MOTOR 6x30
(5)1930s $600
White porcelain sign with red outlined white rectangle to either end. Yellow outlined green "SUNSET/PRODUCTS" above green "THANK YOU" in center. Red "NO SMOKING" in left rectangle, "STOP YOUR MOTOR" in right.

BRITISH AMERICAN OIL COMPANY - TORONTO, ONTARIO: Founded in 1906, B-A was a fully integrated oil company with operations across Canada. Gulf began purchasing B-A stock after WW2, and in 1968 completed the purchase and formed Gulf Canada. The Gulf brand replaced B-A in 1970. When Gulf was purchased by Chevron in 1984, Gulf Canada was purchased by Petro-Canada and stations were converted to the Petro-Canada brand.

IDENTIFICATION SIGNS

BRITISH AMERICAN GASOLENE 30x60
(5) 1920s No Listing
Can type sign with opal glass letter inserts. White "BRITISH-AMERICAN" above white "GASOLINE" across sign face.

B-A 30x60
(4) 1930-1949 $250
Green/white outlined porcelain sign with red "bowtie" across center and green bands above and below. White "B-A" on bowtie, with white "SERVICE" across top band, "PRODUCTS" across bottom.

BA 42"RD
(4) 1949-1970 $200
Green/white outlined porcelain sign with red left side and green right side separated by curved white line. White "B" on red, "A" on green.

PUMP SIGNS

BA 88 10"RD
(4) 1949-1967 $75
Green porcelain sign with white "88" across center. Black "B-A" across top, "GASOLINE" across bottom.

BA 98 10"RD
(4) 1949-1967 $75
Red porcelain sign with white "98" across center. Black "B-A" across top, "GASOLINE" across bottom.

BUILDING SIGNS & LETTERS

B-A 30x60
(5) 1940-1949 $200
Green/white outlined vertical rectangle sign with red vertical bowtie in center. White "B-A" vertical on bowtie.

ROYALITE OIL COMAPNY, LTD - CALGARY, ALBERTA: Founded in 1920 from the remnants of Calgary Petroleum Products, Royalite operated as a natural gas subsidary of Imperial Oil until 1949. Spun off as independent in 1949, Royalite entered the refining business, and in 1953 began opening service stations in western Canada. In 1962 British American purchased Roylaite, operating it a a subisdary until 1968 when all activities were consolidated under Gulf Canada.

No collectible size Royalite signs known at this time.

AMERADA HESS - NEW YORK, NY

Hess. The only oil company that is better known for its annual promotion than for its products. Chances are, if you collect gasoline memorobilia, the only thing you know about Hess is that they've produced and sold a toy truck, a replica of one of their tankers or other company vehicles, every Christmas since 1964. The company's belief in absolute silence to the public and to the press has prevented much more than information about the trucks to ever be made public. The extensive research for this book has brought up fragmants of information that we will attempt to consolidate here.

Hess began operations as a coal yard/service station in Asbury Park, New Jersey about 1925. They evolved from coal into fuel oil during the depression and from fuel oil into heavy residual oils for boiler fuels. Terminal facilities at Perth Amboy, New Jersey were added after WW2 and the company opened it's first Hess branded retail service stations about 1959. In 1963 the purchased independent refiner Delhi-Taylor, which owned a majority of Billups Eastern as well as Southern Oil out of Ft. Lauderdale, Florida. Southern stations were converted to Hess, and in 1965 Hess purchased the remaining portion of Billups Eastern and rebranded the stations Hess. Along the way Olixir, an Albany, New York area chain was purchased and rebranded Hess. In 1964, the first of the Christmas trucks were sold, branded Billups or Hess. Since that time their has been a promotional truck sold most every year at Christmas and these have become very desirable collectibles. In 1968 Hess Oil and Chemical merged with Amerada Petroleum, a production/exploration company with operations around the world. Hess currently markets through over 500 branded outlets in 15 states as well as suppling several strong southern private brands.

HESS - PERTH AMBOY, NJ

No collectible size Hess signs known at this time.

OLIXIR - GREEN ISLAND, NY

New York independent that sold out to Hess in the early 1960s.

No collectible size Olixir signs known at this time.

SOUTHERN OIL CO. - FT. LAUDERDALE, FL

Florida private brand that had been purchased by Delhi-Taylor in the late 1950s. When Hess purchased Delhi-Taylor in 1963, the Southern stations were rebranded Hess.

No collectible size Southern signs known at this time.

BILLUPS EASTERN PETROLEUM - JACKSONVILLE, FL

When Billups was broken up in 1959, control of Billups Eastern passed to independent refiner Delhi-Taylor. In 1963 Hess purchased Delhi-Taylor and in 1965 purchased the remaining portions of Billups Eastern. Billups stations from Delaware to Florida were then rebranded Hess.

See SIGNAL for collectible Billups signs.

HUDSON & AFFILIATES

HUDSON OIL COMPANY - KANSAS CITY, KS

Hudson Oil Company's storied history began in 1930 when Mary Hudson married Wayne Driver and moved to Kansas City. Mr Driver went to work with Mary's brothers at a Kansas City oil marketer operating under the Silver King name. The family was involved in two gas stations in Kansas City when, in 1933, Mr. Driver was killed in an accident with a fuel oil truck. Mary quickly sold off the two stations but within a year had purchased another station, at 25th and Broadway in Kansas City and, at age 21, was selling gasoline under the Hudson name. By 1937 the chain consisted of five stations, the original Kansas City location and stations at Leavanworth, Kansas, Norfolk and Omaha, Nebraska, and Moberly, Missouri. In the late 1930s Mary formed a partnership with her brothers and Hudson Oil was formed. The chain expanded coast to coast with hundreds of locations by the time family members went their separate ways in 1968, with the Hudson brothers operating competing service stations under the Highway (Workingman's Friend) and FISCA (First In Service Company of America) brands. The 1977 purchase of a Cushing, Oklahoma refinery led to a long downward spiral for Hudson and in 1985 the Mary Hudson operation was bankrupt. More than 300 Hudson brand stations closed, leaving only a dozen or so in operation today. Hudson Affiliates Highway and Fisca continue to market across the country.

SILVER KING - KANSAS CITY, KS

Silver King of Kansas City was a predasessor to Hudson.

No collectible size Silver King signs known at this time.

HUDSON OIL CO. - KANSAS CITY, KS

Hudson brand introduced in 1933.

IDENTIFICATION SIGNS

HUDSON OVAL 18x36

(5) 1946-1960 $150

Red porcelain oval with white "HUDSON" across center.

HUDSON PRICE SIGN PANEL 12x60

(5) 1960-1985 $50

White porcelain panel that bolted to top of price signs. Red "HUDSON" across panel.

BUILDING SIGNS AND LETTERS

HUDSON PORCELAIN LETTERS 18"

(3) 1946-1985 $100

Red porcelian channel letters for mounting on station roof.

HORNET OIL COMPANY - TULSA, OK

Hornet is a Tulsa based affiliate that re-branded Hudson in 1958.

No collectible size Hornet signs known at this time.

MAJOR'S

Major's signage identical to Hudson globes - company relation unknown but it is assumed that Major's was operated by one of the Hudson family members.

No collectible size Major's signs known at this time.

HIGHWAY OIL COMPANY

Hudson affiliate marketing under the Highway, Hi-Lo, and Workingman's Friend brands.

LUBRICATION/MOTOR OIL SIGNS

HIGHWAY OIL CANS 18x30

(5) 1970's $45

White embossed tin litho signs with detailed Highway "Workingman's Friend motor oil cans and blue arrows with prices.

FISCA OIL - KANSAS CITY, KS

Hudson affiliate with stations coast-to-coast today. Fisca is an acronym for "First In Service Company of America.

BUILDING SIGNS AND LETTERS

FISCA PORCELAIN LETTERS 18"
(3) 1963-1985 $100
Red porcelian channel letters for mounting on station roof.

PUMP SIGNS

FISCA
(5) 1963-1968 $350
White porcelain sign with red, white, and blue Fisc shield in center.

LUBRICATION/MOTOR OIL SIGNS

FISCA OIL CANS 18x30
(5) 1970's $45
White embossed tin litho signs with detailed Fisca Premium and Heavy Duty motor oil cans and blue arrows with prices.

HUSKY OIL COMPANY - CODY, WY

Husky was originally a brandname of Western Oil and Fuel, Minneapolis, Minnesota. About 1954 Western acquired H. Earl Clack, an independent marketer in Cody, Wyoming. Western assigned the Husky brand to the H. Earl Clack subsidary and expanded the Husky Hi-Power brand throughout the Rocky Mountain region. Western continued marketing under the Mileage name in the upper midwest, but some dealers in that area were using the Husky name as late as 1960. When Conoco purchased Western in the late 1950s, H. Earl Clack, renamed Husky Oil, was reoragnized as a separate company. Husky Oil of Cody, Wyoming merged with Frontier Oil in 1968. Frontier had earlier (about 1960) bought Western States Petroleum of Salt Lake City, Utah. The Husky name replaced Frontier and Beeline about 1970 and Husky expanded into Canada. In 1984 Husky Oil sold off the U.S. marketing to concentrate on Canadian marketing. In 1990 a Denver based marketer revived the Frontier brandname and "Rarin' to Go" trademark and currently operates a chain of staions under the Frontier trademark.

HUSKY/WESTERN OIL & FUEL - MINNEAPOLIS, MN:

IDENTIFICATION SIGNS

HUSKY GASOLINE 30"RD
(5) 1930-1935 $1800
White outlined blue porcelain sign with detailed multi-color orange and yellow Northern Lights scene at top. Extremely detailed blue and white dog's head at center of lights. Yellow "HUSKY/GASOLINE" in blue area below, with orange "GETS YOU THRU" arched around bottom.
HUSKY GASOLINE 28x30
(5) 1930-1935 $1500
Blue outlined keyhole shaped white porcelain sign with detailed multi-color orange and yellow Northern Lights scene at top. Extremely detailed blue and white dog's head at center of lights. Yellow "HUSKY" over white "GASOLINE" in blue area below circle.
HUSKY SERVICE 48x60
(5) 1935-1946 $1500
Blue outlined porcelain die-cut shield. Lower area is blue with red outlined yellow "HUSKY" logotype over white "SERVICE." Upper area is red/yellow/white sunrise scene with detailed blue and white husky dog jumping left to right.
HUSKY 72"RD
(5) 1946-1960 $1200
Red and blue round porcelain sign with white stripe separating upper blue area from lower red area.Yellow "HUSKY" in upper blue area, with detailed black and white jumping husky in center.

H. EARL CLACK, INC. - CODY, WY

Operated stations and truckstops under the HECCO brand until 1954 when Western purchased Clack and rebranded the stations Husky.
No collectible size Hecco signs known at this time.

HUSKY - HI POWER, INC. - CODY, WY

When Western purchased H. Earl Clack, Western's Husky brand was combined with Clack's Hi-Power brand to form Husky Hi-Power.

IDENTIFICATION SIGNS

HUSKY DIE CUT DOG 60x84
(5) 1954-1960 $1800
Multi-piece die cut porcelain sign. White outlined blue rectangle with white "HUSKY" across center. Brown and white detailed die-cut husky dog at top.
HUSKY ROUNDED RECTANGLE 24x84
(5) 1960-1980 $300
Blue outlined white rounded rectangle with red "HUSKY" across center.

PUMP SIGNS

HUSKY HI-POWER INC HUSKY HI-POWER 14x15
(5) 1954-1960 $450
HUSKY HI-POWER INC HUSKY HI-POWER 10x10
(5) 1954-1960 $450
Orange porcelain sign with blue rectagle across bottom and white arc above. Blue "HUSKY HI-POWER INC" arched around white arc, with blue and white detailed husky dog in center. White "HUSKY" on blue rectangle. Smaller blue rectangle below with white "HI-POWER" across.
HUSKY HI-POWER 12x15
(4) 1960-1970 $425
White porcelain sign with detailed blue and white husky dog at top. Two blue rectangles below, with orange "HUSKY" in top, "HI-POWER" below.
HUSKY TRI-POWER 12x15
(4) 1960-1970 $425
Orange porcelain sign with detailed blue and white husky dog at top. Two white rectangles belwo, with orange "HUSKY" in top, "TRI-POWER" below.
HUSKY DIESEL 12x15
(5) 1960-1970 $350
White porcelain sign with detailed blue and white husky dog at top. Two rectangles below, with blue "HUSKY" on white in top, white "DIESEL" in blue below.
HUSKY DIESEL 1 12x15
(5) 1960-1970 $350
White porcelain sign with detailed blue and white husky dog

at top. Two rectangles below, with blue "HUSKY" on white in top, white "DIESEL-1" in blue below.

HUSKY DIESEL 2 12x15
(5) 1960-1970 $350
White porcelain sign with detailed blue and white husky dog at top. Two rectangles below, with blue "HUSKY" on white in top, white "DIESEL-2" in blue below.

HUSKY HI-POWER 12x15
(5) 1960-1970 $200
Same as above, tin litho sign.

HUSKY TRI-POWER 12x15
(5) 1960-1970 $200
Same as above, tin litho sign.

HUSKY DIESEL 2 12x15
(5) 1960-1970 $200
Same as above, tin litho sign with "2" added after "DIESEL."

SECONDARY SIGNS

MEN 3x12
(5) 1960-1970 $50
Blue outlined rounded rectangle with red "MEN" across center.

LADIES 3x12
(5) 1960-1970 $50
Blue outlined rounded rectangle with red "LADIES" across center.

TRUCKERS SHOWER 3x12
(5) 1960-1970 $75
Blue outlined red rounded rectangle with white "TRUCKERS/SHOWER" across center.

HUSKY LARGE PETS 12x24
(5) 1960-1970 $125
White porcelain sign with red and blue Husky rounded rectangle at top. Blue "LARGE PETS" below.

HUSKY SMALL PETS 12x24
(5) 1960-1970 $125
White porcelain sign with red and blue Husky rounded rectangle at top. Blue "SMALL PETS" below.

HUSKY KITTIES 12x24
(5) 1960-1970 $125
White porcelain sign with red and blue Husky rounded rectangle at top. Blue "KITTIES" below.

HUSKY ? MILES 18x18
(5) 1954-1960 $250
Orange outlined blue die-cut tin litho sign with brown and white dog at top. White reflective "HUSKY" across top, with painted in mileage number below to the left of orange "MILES."

FRONTIER OIL - DENVER, CO

Denver based refiner/marketer founded in the mid-1930s, Frontier operated a chain of stations in the Rocky Mountain region. About 1960 they purchased Western States Petroleum and operated Beeline stations on the western slope of the Rockies. In 1968 Husky Oil purchased Frontier Oil and replaced both the Frontier and Beeline brands with Husky about 1970. In 1990 a Denver marketer reintroduced the Frontier brandname on a chain of retail stations in Colorado.

IDENTIFICATION SIGNS

FRONTIER 60x60
(4) 1950-1956 $650
White outlined red keyhole shaped die-cut porcelain sign. White "FRONTIER" across bottom, with detailed black and white bucking horse and rider above. White outlined black "RARIN'/TO GO" to right of horse.

FRONTIER 72x72
(4) 1950-1956 $600
White outlined red keyhole shaped die-cut porcelain sign. White "FRONTIER" across bottom, with detailed black and white bucking horse and rider above. White outlined black "RARIN'/TO GO" to right of horse.

FRONTIER RARIN TO GO 72"RD
(2) 1956-1970 $650
Round porcelain sign with white upper half and red lower half. Black sillouette of bucking horse and rider in upper white area, with black outlined white "FRONTIER" across lower red area. White "RARIN'-TO-GO" arched around bottom.

WESTERN STATES PETROLEUM - SALT LAKE CITY, UT

Dating from after WW2, Western States operated the Beeline stations on the western slope of the Rockies. They merged with Frontier Oil about 1960 but the Beeline name remained in use until replaced by Husky about 1970.
No collectible size Beeline signs are known at this time.

IOMA
INDEPENDENT OIL MEN OF AMERICA

(IOMA) - CHICAGO, IL

The "Independent Oil Men of America" was formed in 1925 in Chicago, Illinois to offer independent jobbers and gas/oil marketers a marketing organization and nationwide trademark to assure recognizable quality. The trademark selected was "Red Hat," registered in June 1925. The logo consisted of a large red hat with a black band and white stars. Standard Oil of Indiana, having long used the "Red Crown" trademark sued for trademark infringement and in 1929 the IOMA lost the right to use this valuable trademark. After that time their trademark consisted of the organization logo, the red "thunderbird" on a black triangle, the corporate logo that had been in use since 1925 but only as a secondary trademark.

IDENTIFICATION SIGNS

MOTOR OIL RED HAT GASOLINE 42"RD
(5) 1926-1932 $2500
Black porcelain sign with white center circle. Red and black hat with stars in center. Black "RED" to left, "HAT" to right, with small IOMA logo at bottom. White "MOTOR OIL" archeda round top, "GASOLINE" around bottom.

FLANGE SIGNS

APPROVED RED HAT MOTOR OIL 14"
(5) 1926-1932 $2200
White outlined black circle with white center circle. Black flange area. White "MOTOR OIL" arched around bottom, with red and black hat with stars in center. Black "RED" to left, "HAT" to right, with small IOMA logo at bottom. Black outlined white box above hat with black script "Approved" in box.

INDEPENDENT OILS

Not all IOMA members chose to use the RED HAT brand. The brands listed below identified their products with the IOMA black triangle/red thunderbird logo. This does not indicate that a globe with the IOMA logo as opposed to the Red Hat brand is from after the 1932 discontinuance of the Red Hat. No collectible size signs known using just the IOMA logo.

ROYAL AND ROYAL 400 - FT. DODGE, IOWA

IDENTIFICATION SIGNS

ROYAL 400 RED HAT GASOLINE 42"RD
(5) 1926-1932 $2800
White/green outlined orange porcelain sign with white center circle. Green outlined white "ROYAL 400" arched around top, "GASOLINE" around bottom. Red and green hat with stars in center. Green "RED" to left, "HAT" to right, with small IOMA logo at bottom.

LUBRICATION/MOTOR OIL SIGNS

LOYAL PENN MOTOR OIL 100% PURE PENNSYLVANIA A ROYAL 400 PRODUCT $2800

KENDALL OIL - BRADFORD, PA

Kendall Refining was founded as a lubricants manufacturer and refiner in Bradford, Pennsylvania in 1881. The company had been well established in motor oils when the first venture into gasoline came in 1922. Kendall became a unit of Witco Chemical in 1966. The company supplied stations in Pennsylvania and New York until after the first gas shortage in 1973 when the decision was made to concentrate on lubricants. The last Kendall branded station closed in 1976.

IDENTIFICATION SIGNS

GASOLINE KENDALL MOTOR OILS 42"RD
(5) 1930's $400
Black/red outlined white porcelain sign with black outlined red "KENDALL" across center. Black outlined red "GASOLINE" arched around top, "MOTOR OILS" around bottom with red map of Pennsylvania on lowe rarea of sign. White ribbon across map with black "BRADFORD GRADE" with script white "Guaranteed 100%' above and white "PENZBEST OIL" below.
KENDALL 24x84
(5) 1950's No Listing
Red outlined white die-cut trapezoid with red "KENDALL" across center.

PUMP SIGNS

KENDALL THE 2000 MILE OIL 12"RD
(3) 1946-1955 $75
White outlined red tinlitho sign with black outlined white "KENDALL" across center. Black outliend white hand at top, holding up two fingers to symbolize 200 miles. White script "The" above white "2000 MILE/OIL" below "KENDALL."
KENDALL DELUXE 12"RD
(3) 1946-1970 $75
Black outlined white tin litho sign with black outlined red "KENDALL" across to above black script "DeLuxe." Red "ARISTOCRAT OF MOTOR FUELS" below, over small Ethyl (EC) logo.

CURB SIGNS

KENDALL PENZBEST MOTOR OIL 24"RD
(2) 1930's $225
KENDALL PENZBEST MOTOR OIL 36"RD
(2) 1930's $225
Black/red outlined white porcelain sign with black outlined red "PENZBEST" across center. Black outlined red "KENDALL" arched around top, "MOTOR OILS" around bottom with red and black PGCOA seal on lower area of sign.
KENDALL THE 2000 MILE OIL 36"RD
(1) 1930's $200
Black outlined white porcelain sign with black outlined red "KENDALL" across center above "The 2000 MILE/OIL." Red and white hand logo at top.
SOLD HERE IN SEALED CANS 36"ARC
(5) 1930's $75
Die cut porcelain arc attachment for top of above sign. White "SOLD HERE IN SEALED CANS" on red.
KENDALL THE 2000 MILE OIL 24"RD
(1) 1940's $75
Same as above in tin litho.
KENDALL THE 2000 MILE OIL 24"RD
(1) 1950's $75
Same as above with colors reversed. Tin litho.
KENDALL MOTOR OILS 24" TIN WT/RED
Red tin litho sign with white "KENDALL" across center above "MOTOR OILS." White hand logo at top.

LUBRICATION/MOTOR OIL SIGNS

KENDALL THE 2000 MILE OIL CAN SIGN 20x34
(4) 1934-1940 No Listing
Red, white and black die cut porcelain sign in design of Kendall oil can.
KENDALL THE 2000 MILE OIL CAN SIGN 12x20
(4) 1934-1940 No Listing
Same as above, tin litho.
THE DEALER SIGN OF QUALITY KENDALL 12x72
(3) 1960's $150
Black/white outlined red self framed horizontal rectangle tin litho sign. White "KENDALL" on red area, with red Kendall logo at left and black "The Dealer/Sign of/Quality."
THE DEALER SIGN OF QUALITY KENDALL 12x72
(3) 1960's $150
Similar to above, in vertical format.

LUBESTER SIGNS

KENNDALL PENZBEST MOTOR OIL 7" No Listing

L.SONNEBORN SONS, INC - NEW YORK, NY:

Lubricants manufacturer founded in 1896. Original lubricants unit of Witco Chemical Company.
AMALIE

CURB SIGNS

AMALIE PENNSYLVANIA MOTOR OIL 18x32
(2) 1930-1960 $400
Red porcelain oval with black outlined white "AMALIE" arched across top over "PENNSYLVANIA/MOTOR OIL" Small PGCOA seal at bottom.

LUBRICATION/MOTOR OIL SIGNS

AMALIE PENNSYLVANIA MOTOR OIL 11.5x35.5
(2) 1946-1960 $75
AMALIE PENNSYLVANIA MOTOR OIL 38x60
(2) 1946-1960 $75

Powerlube porcelain sign. *Leo Mathieu.*

Standard Gasoline 36" porcelain sign. *Don Christensen.*

Humble porcelain No Smoking sign *Mark McKeown.*

Standard No Smoking Sign. *Mark McKeown.*

Humble porcelain in-laid with lead building sign. *Bruce Bednar.*

Jenney Aero 12" porcelain pump sign. *Leo Mathieu.*

Conoco Motor Oil porcelain flange. *Dennis Leith.*

Jenney Solvenized 12" porcelain pump sign. *Tom Kleeberg*

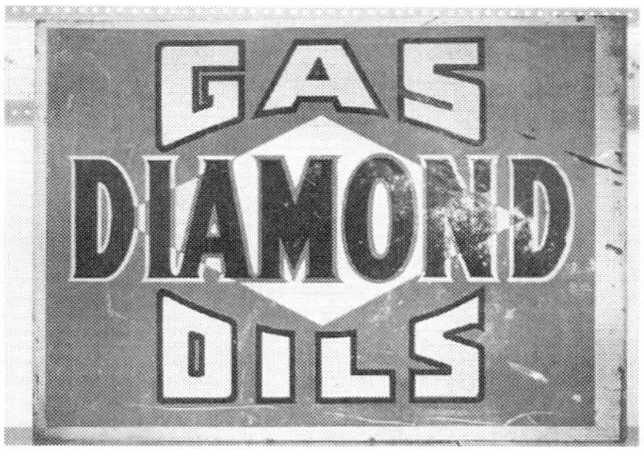

Diamond Gas Oil tin flange 12"x18."

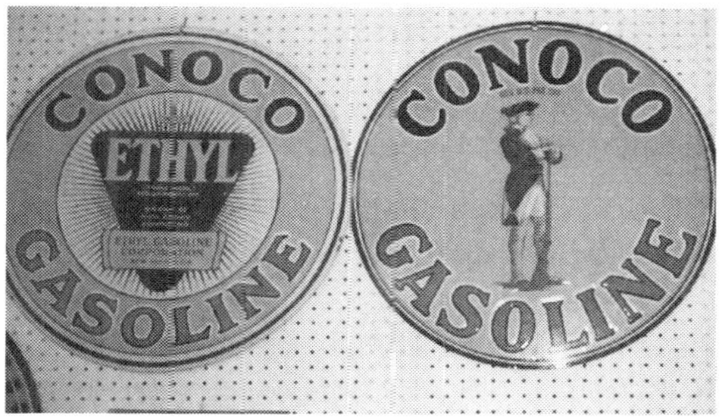

Pair of Conoco 30" porcelain signs.

Chevron 12" porcelain pump sign. *Mark McKeown.*

Red Crown 10" porcelain flange.

Standard of California Pearl Oil tin flange.

White Flash porcelain globe attachment for Atlantic Gasoline. *Tom Kleeberg.*

Red tin litho sign with black outlined white "AMALIE" arched across top over "PENNSYLVANIA/MOTOR OIL."
AMALIE PENNSYLVANIA MOTOR OIL 28x42
(3) 1946-1960 $125
Red embossed tin litho bubble sign. Oval with black outlined white "AMALIE" arched across top over "PENNSYLVANIA/ MOTOR OIL" Small PGCOA seal at bottom.

WAVERLY OIL WORKS CO. - PITTSBURGH, PA: Founded as a lubricants manufacturer in Pittsburgh, PA, Waverly briefly ventured into gasoline marketing in the 1920's. The gasoline operation was short lived and Waverly remained a lubricants manufacturer until their 1975 merger with Witco/Kendall.

CURB SIGNS
WAVERLY ETHYL GASOLINE 30"RD
(5) 1926-1930 $450
White porcelain sign with outline ring and outlined center circle. Ethyl (EGC) logo in center, with "WAVERLY" arched around top, "GASOLINE" around bottom.

LUBRICATION/MOTOR OIL SIGNS
LICENSED DEALER WAVERLY MOTOR OIL 24x24
(5) 1924-1928 No Listing
Red outlined white porcelain sign with white tab at top. Red outlined blue script "Waverly" diagonal across sign with white "ALL PENNSYLVANIA" on underline. Red "since 1880" above, "MOTOR OIL" below, with PGCOA seal to upper left. Black "LICENSED/DEALER" on upper white tab.

SECONDARY SIGNS
WAVERLY OILS AND GASOLINE DRUM SIGN 10x20
(5) 1915-1920 No Listing
White porcelain flange sign with red oil drum with blue highlights covering most of sign. White "WAVERLY/OILS AND/GASOLINE" on side of barrell. Detailed text on barrell head.

KERR MCGEE OIL INDUSTRIES - OKLAHOMA CITY, OK

Kerr McGee was founded in the Oklahoma City oilfields as Anderson-Kerr Drilling Co. in 1929. Through the early years they operated a single Oklahoma City service station, branded Phillips 66, which closed during the gas rationing days of World War 2. In 1946 the corporate name was changed to Kerr-McGee, shortly after the company had entered retail gasoline marketing having purchased Oklahoma City based Harrell Bros. The Harrell Bros. "Yellow Cab" gasoline trademark was replaced by the "Kermac" brand. This was the extent of KM's gas station involvement until 1955 when they purchased Deep Rock Oil Co. Deep Rock had been founded in the Cushing Oklahoma oil fields in 1913 and had been through a stormy existance, having sold out to Shaffer Oil and Refining of Chicago in the late 1920s and again been in receivership by the late 1930s. At the time of the KM merger, the Deep Rock brand appeared on over 1000 stations in 16 mid-continent states. In 1957 Kerr-McGee acquired Triangle Refineries, a primarily wholesale operation that had been founded in 1935. Knox Industries of Texas also sold out to KM about this time and the Knox brand survived into the 1960s. Triangle was active in purchasing Southern private brand operations which Kerr-McGee operated well into the eighties, including Peoples Oil of Nashville, 1959; Illinois based Mileage Mart in 1959; Trackside Gasoline Stations of Gulfport, Miss, 1962; Wallace Service Stations of Atlanta, 1962, along with Power Oil Co. of Macon, GA, Coast Oil Co. of Shreveport, Louisiana, and Tan-Kar Stations of Birmingham, Alabama.

DEEP ROCK OIL CO.

Founded in the Cushing field in 1913, Deep Rock had a series of financial problems and existed under several owners before being purchased by Kerr McGee in 1955. The Kerr McGee name was added to Deep Rock stations about 1960 and the "brand" Kerr McGee was introduced in 1965. Deep Rock stations were gradually converted over to the KM name, nearing completion by 1985. The Deep Rock brand survives in some areas, but this is thought to be jobbers who simply haven't updated signage at smaller locations.

IDENTIFICATION SIGNS
GASOLINE/DEEP ROCK/MOTOR OIL 16x30
(5) 1920's $250
White porcelain sign with blue and white Deep Rock logo in center. White "GASOLINE" in upper blue area, "MOTOR OIL" belwo with red "DEEP ROCK" across center.
DEEP ROCK 30"RD
(4) 1935-1948 $250
DEEP ROCK 48"RD
(3) 1935-1948 $250
Steel framed white porcelain sign with black band across top and bottom. Red "DEEP ROCK" across center, with white "GASOLINE" in upper band, "MOTOR OILS" below.
DEEP ROCK 60x96
(2) 1948-1985 $100
Blue outlined yellow porcelain vertical rectangle with blue band across center. Blue "DEEP" in upper yellow area, "ROCK" below.

PUMP SIGNS
BLUE DEEP ROCK 6.5x8.5
(2) 1948-1960 $75
Blue/yellow outlined blue porcelain sign with yellow band across center. Yellow "DEEP" in upper field, "ROCK" below.
RED DEEP ROCK 6.5x8.5
(2) 1948-1960 $75
Red/yellow outlined red porcelain sign with yellow band across center. Yellow "DEEP" in upper field, "ROCK" below.
REGULAR PANEL 2.5x12
(2) 1948-1960 $35
Blue strip with yellow "REGULAR"
PREMIUM PANEL 2.5x12
(2) 1948-1960 $35
Red strip with yellow "PREMIUM"

CURB SIGNS
KANT NOCK ETHYL GASOLINE 30"RD
(3) 1926-1935 $325
White porcelain sign with red "KANT-NOCK" across top above Ethyl
(EGC) logo with yellow rays. Red "GASOLINE" arched around bottom.

LUBRICATION/MOTOR OIL SIGNS
DEEP ROCK PRIZE MOTOR OIL 16x30

(5) 1920's $250

White tin litho oil rack sign with blue and white Deep Rock logo in center. White "PRIZE" in upper blue area, "MOTOR OIL" belwo with red "DEEP ROCK" across center. Blue script "Certified" at upper left of logo, with other red lettering below.

DEEP ROCK AIR RACE PREMIUM 18x30
(4) 1956-1960 $125

Red tin litho sign with yellow band across center. Yellow "DEEP" in upper field, "ROCK" in lower with blue "AIR RACE/ PREMIUM" and red airplane in center.

KERMAC - OKLAHOMA CITY

In 1945 when Kerr McGee purchased the former "Yellow Cab" gas stations of Harrell Bros. Oil Comapny the several stations were rebranded Kermac. Kerr McGee was more involved in production and refining than in marketing and the Kermac brand never appeared on more than a few stations.
No collectible Kermac signs known at this time.

KERR MCGEE

The Kerr McGee name was added to Deep Rock stations about 1960 and the Kerr McGee brand, standing alone, first appeared on a group of Oklahoma City stations in 1965. The KM brand and logo eventually replaced Deep Rock and continues in use today.

IDENTIFICATION SIGNS

KM SHIELD 48x96
(1) 1966-PRESENT $100

Die cut porcelain Kerr McGee shield with blue upper outline and red lower. Red and blue "K," blue "M" in center.

TRIANGLE REFINERIES

Founded in 1935, Triangle Refineries operated stations under the "HIOTANE" brand, as well as "TRIANGLE" and "CLOVERLEAF" and supplied various unbranded marketers. Kerr McGee purchased Triangle in 1957 and all of the private brand KM affiliates operated through Triangle.
No collectible size Triangle signs known at this time.

MILEAGE MART

Illinois based, early gasoline c-store operation bought out in 1959 and rebranded Deep Rock.
No collectible size Mileage Mart signs known at this time.

TRACKSIDE SERVICE STATIONS - GULFPORT, MS

Trackside was just that-an early trackside discount operation, founded in Gulfport in the early 1930s. Kerr McGee, through Triangle, purchased Trackside in 1962 and continued to operate the chain under the Trackside name until the late 1980s.
No collectible size Trackside signs known at this time.

PEOPLES OIL CO. - NASHVILLE, TN

Another mid-south discount operation, Peoples dates from the 1930s and was purchased by Triangle in 1959. Many of the stations were rebranded Kerr McGee by the 1980s.

IDENTIFICATION SIGNS

POWER-PLUS 42x72
(5) 1940S $100

Outlined white panel with red band across top. White lettering "POWER-PLUS" in band, with place to attach pricing numbers below.

POWER PEOPLES PLUS 60x60
(5) 1950s $100

Red porcelain can type diamond shaped with with large "PEOPLES" across center. White "POWER" at top, "PLUS" at bottom.

MUTUAL OIL CO. - BIRMINGHAM, AL

Mutual, a private brand founed in 1936, was purchased by Kerr McGee in 1985 and stations were rebranded Kerr McGee.

IDENTIFICATION SIGNS

MUTUAL 42"
(5) 1936-1970 $450

Blue outlined white porcelain sign with blue rabbit jumping in center. Red "MUTUAL" above rabbit, with red and blue ribbon below. White "GASOLINE" on ribbon.

PUMP SIGNS

MUTUAL KEROSENE
(5) 1940s $300

Blue outlined white porcelain sign with blue rabbit jumping in center. Red "MUTUAL" above rabbit, with red and blue ribbon below. White "KEROSENE" on ribbon.

No collectible size signs are known from other Kerr McGee/Triangle affiliates including Walco, Wallace, Coast, Knox, Power, Tan-Kar & Sparky.

LION OIL CO. - EL DORADO, ARKANSAS

Lion Oil Co was founded in 1921 in the south Arkansas oil fields. Primarily a exploration/drilling company and an early innovator in petrochemicals, Lion was operating a few service stations by the early 1920s. From the 1920s until the early 1950s, gasoline marketing was limited to Arkansas, Mississippi, Western Tennessee and Northern Alabama and Louisiana, although Lion Naturalube Motor Oils were distributed throughout the U.S. In 1954 the Monsanto Chemical Corporation purchased Lion and began to expand the marketing territory throughout the south central U.S., expanding as far west as central Kansas and as far north and east as Bristol, Virginia. In 1959 Lion introduced the jumping lion logo, replacing the "lion and mound" logo that had been in use since the 1920s. During the 1960s the Monsanto name appeared on gas pump signage along with the Lion brand and some St.Louis, Missouri area stations added Monsanto signage alongside the Lion signs, creating speculation that the Lion name would be replaced. The experiment was short lived as Monsanto decided to get out of retail gasoline and in 1971 Monsanto sold the petroleum divisions of Lion to "The Oil Shale Co." (TOSCO), retaining the petrochemical divisions. Lion branded marketing was discontinued in 1981, although TOSCO/Lion is still involved in unbranded marketing.

IDENTIFICATION SIGNS

LION MOTOR OILS/GAS 30"RD
(5) 1920's $500

Red outlined white porcelain sign with red outlined white center circle. Detailed lion in center above red "LION," with blue "GASOLINE" arched around top, "MOTOR OILS" around bottom.

LION 48"RD
(3) 1930-1959 $550

LION 72"RD
(3) 1940-1959 $300

White/black outlined orange porcelain sign with large black and white lion on black mound. White "LION" across mound.

LION 48"RD
(2) 1959-1981 $300

LION 72"RD
(2) 1959-1981 $200

Red and white porcelain sign with detailed black and white jumping lion in top red area. Black "LION" across lower white area.

LION SHIELD 36x60
(5) 1970-1981 $200

White background sign with red and white Lion shield logo in center. Black and white detailed jumping lion in upper red area. Black "LION" in lower white area.

PUMP SIGNS

ADVANCED LIONITE 12x20
(4) 1954-1956 $75

Black outlined white tin litho sign with black band across center. White "LIONITE" in band. Detailed Lion logo at top, over orange "Advanced." Orange "Gives You/TOP MILEAGE" below band.

NEW LION REGULAR CONTAINS LIONITE 12x20
(3) 1956-1959 $75

Black outlined white tin litho sign with black band across center. White "NEW LION/REGULAR" on center band. Orange "Supercharge" over black "WITH" above band, black "CONTAINS" over orange "Lionite" below band. Detailed Lion logo at bottom.

NEW LION PREMIUM CONTAINS LIONITE 12x20
(3) 1956-1959 $75

Orange outlined white tin litho sign with orange band across center. White "NEW LION/PREMIUM" on center band. Black "Supercharge" over orange "WITH" above band, orange "CONTAINS" over black "Lionite" below band. Detailed Lion logo at bottom.

LION/MONSANTO 12x20
(3) 1959-1965 $35

Black outlined tin litho sign. Colors split diagonally, with white upper area and red lower area. Red and black jumping lion Lion logo at top of white area, with white "M" with black dropshade and black "Monsanto" at bottom of red area.

CURB SIGNS

LION ETHYL GASOLINE 30"RD
(3) 1930's $375

Black outlined white porcelain sign with black outlined white center circle. Orange "LION" arched around top, "GASOLINE" around bottom with large Ethyl (EGC) logo in center.

LION NATURALUBE REINFORCED MOTOR OILS 32x36
(4) 1954-1959 $200

Keyhole shaped die-cut tin litho sign. Black outlined white circle, with red circle at top. Black "LION" across circle, with black script "Maturalube" across center above red "RE-INFORCED" and black "MOTOR OIL." White "SAVES WARE-ADDS POWER/REDUCES CARBON" on lower red area.

LUBRICATION/MOTOR OIL SIGNS

LION OILS 42"RD
(4) 1940's $500

LION OILS 48"RD
(4) 1940's $500

White/black outlined orange porcelain sign with large black and white lion on black mound. White "LION/OILS" across mound.

SECONDARY SIGNS

LION CREDIT CARDS ACCEPTED 10x18
(5) 1950'S $100

Outlined white tin litho flange sign with large "CREDIT" across center. Smaller "LION" across top, "CARDS ACCEPTED" across bottom.

LION SUPER TEST KEROSENE 10x18
(5) 1950's $200

White tin litho flange sign with red band across bottom. Black outlined red "LION SUPER TEST" above black "KEROSENE" on upper white area, white "BURNS CLEANER LASTS LONGER" on lower red area with Lion logo in center.

LION NATURALUBE REINFORCED MOTOR OIL 10x18
(4) 1950's $200

White tin litho flange sign with red band across bottom. Red circle at top. Black "LION" across circle, with black script "Maturalube" across center above red "RE-INFORCED" and black "MOTOR OIL." White "SAVES WARE-ADDS POWER/REDUCES CARBON" on lower red area.

LION MEN 5x8
(4) 1960's $50

White tin litho vertical rectangle with black "MEN" at top. Small Lion jumping lion logo below.

LION WOMEN 5x8
(4) 1960's $50

White tin litho vertical rectangle with black "WOMEN" at top. Small Lion jumping lion logo below

ADVANCE SIGNS

LION AHEAD 46x60
(4) 1959-1981 $200

Self framed tin litho sign with large red/white/black Lion logo at top over large black "AHEAD."

MARATHON AND AFFILIATES

MARATHON OIL CORPORATION - FINDLAY, OH

The Ohio Oil Company was founded in 1887 when several northern Ohio oil production firms joined together to market their crude oil, primarily to Standard Oil. So much so that in 1889 Standard purchased Ohio Oil and operated the company as a production subsidary until the breakup of Standard in 1911. After the breakup Ohio Oil continued as before with oil exploration and production, satisfied to sell their product as crude oil and leave the refining and marketing to others. Ohio Oil first ventured into marketing with the June 1924 purchase of Lincoln Oil Refining Co of Robinson, Illinois. Lincoln Oil operated a refinery and about 20 retail gasoline stations in the Robinson-Terre Haute area at the time of the purchase by Ohio Oil. With an assured source of crude oil supply for the small refinery, the Linco brand began to expand. By 1930 Ohio Oil had expanded the Linco brand throughout most of the current Marathon territory—Ohio, Indiana, Illinois, Michigan and Kentucky. In the meantime Ohio Oil had been involved in numerous successful oil production ventures and found they needed even more outlets for their products. In 1930 Ohio Oil purchased Transcontinental Oil, a refiner/marketer that had marketed gasoline under the trademark "MARATHON" since about 1920. They had acquired the brand from a Pittsburgh operation, Riverside Oil Co., that we believe to be a forerunner of Republic Oil Company, which will be mentioned later. Transcontinental sold gasoline under the Marathon trademark across the midwest and south, from North Carolina to New Mexico. Trasanscontinental can best be remembered for a significant first when in 1929 they opened several Marathon stations in Dallas, Texas in conjunction with Southland Ice Company's "Tote'm" stores (later 7-Eleven), creating the first Gasoline-Convienence Store tie-in. The Marathon brand proved so popular that by WW2 the name had replaced Linco at stations in the original five-state territory. During the war, the widespread Marathon territory couldn't be properly supplied and marketing operations outside the Great Lakes area were sold, primarily to Tydol. Concentrating their marketing in a smaller territory allowed Marathon to expand within the terrritory and the late 1950s brought the first of many acquisitions, Tower Oil Company of Cincinnati, Ohio, that would expand the company's marketing presence. Numerous other purchases are detailed in the listings below, with a group of marketing companies brought together by the 1980s under a subsidary, Emro Marketing. Ohio Oil, renamed Marathon Oil in 1962, was purchased by U.S. Steel (now USX) in 1981 and continues to market through dealer and jobber outlets under the Marathon name in the north central states and through Emro Marketing under several brandnames, primarily Speedway and Starvin Marvin, throughout the south and east.

LINCOLN OIL REFINING CO. - ROBINSON, IL

Purchased by Ohio Oil in 1924. Linco served as the company's primary brand until about 1939.

IDENTIFICATION SIGNS

LINCO GASOLINE/MOTOR OIL 48"RD
(4) 1924-1935 $450
Blue outlined white porcelain sign with blue circle in center. White cross on blue circle with red "LINCO" vertical and horizontal in cross. Blue "GASOLINE" arched around top, "MOTOR OIL" arched around bottom.

LINCO 24x48
(5) 1930-1939 $350
Red outlined white can type vertical rectangle porcelain sign with notched corners. Red "LINCO" vertical on sign.

CURB SIGNS

LINCO ETHYL 30"RD
(5) 1926-1935 $500
White porcelain sign with outline rings around border and white center circle. Large Ethyl (EGC) logo in center, with "LINCO" arched around top, "GASOLINE" around bottom.

LUBRICANTS SIGNAGE

LINCO MOTOR OIL 16x24
(5) 1924-1930 $250
Blue outlined white tin litho sign eith red "LINCO" above blue "MOTOR OIL" across sign. Small Linco cross logo at bottom with red "LINCO/QUALITY" to left and "NEVER/VARIES" to right.

RED FOX OIL COMPANY - FT. WAYNE, IN

Regional marketer purchased by Linco in 1928 and rebranded Linco at that time.
No collectible size Red Fox signs known at this time.

TRANSCONTINENTAL OIL CO. - TULSA, OK

Purchased by Ohio Oil in 1930, the Transcontinental/Marathon marketing was kept separate from Linco marketing until about 1935 and Marathon had completely replaced Linco by 1939.

IDENTIFICATION SIGNS

MARATHON PRODUCTS 42"RD
(3) 1925-1935 $1500
Orange outlined green porcelain sign with orange outlined black center circle. Detailed white/black runner in center circle with white banner at bottom. Black "BEST IN THE LONG RUN" on banner. Orange outlined white "MARATHON" arched around top, "PRODUCTS" around bottom.

BUILDING LETTERING & SIGNAGE

ROOFTOP RUNNER 48"
(5) 1925-1930 No Listing
Detailed cream, green and orange porcelain die-cut runner designed for rooftop mounting.

CURB SIGNS

MARATHON PRODUCTS 30"RD
(4) 1925-1935 $1800
Orange outlined green porcelain sign with orange outlined black center circle. Detailed white/black runner in center circle with white banner at bottom. Black "BEST IN THE LONG RUN" on banner. Orange outlined white "MARATHON" arched around top, "PRODUCTS" around bottom.

MARATHON ETHYL 30"RD
(3) 1926-1935 $1200
White porcelain sign with green "MARATHON" arched around top, orange outlined black "ETHYL" across center and large Ethyl (EGC) logo at bottom. Orange runner with black details at top positioned between and around "MARATHON" and "ETHYL"

LUBRICANTS SIGNAGE

MARATHON MOTOR OILS 9.5x13.5
(5) 1920-1925 $1100
Orange outlined green tin litho sign with cream runner with orange details to left of center. Orange outlined cream "MARATHON" across top above smaller "MOTOR OILS/BEST IN THE/LONG RUN." "TRANSCONTINENTAL OIL COMPANY/EXECUTIVE OFFICES: PITTSBURGH,PA" across bottom.

MARATHON MOTOR OILS 14x42
(4) 1920-1925 $1100
Orange/cream outlined two-tone green sign with orange race car and orange detailed cream runner at far left. Orange outlined cream "MARATHON/MOTOR OILS" across center of sign above orange "BEST IN THE LONG RUN." Cream "TRANSCONTINENTAL OIL COMPANY/NEW YORK CITY-

SOUIX CITY,-IOWA-DES MOINES,IA-PITTSBURGH,PA" across bottom.

GET MARATHON MOTOR OIL 14x42
(4) 1920-1925 $1200
Black tin litho sign with green bands with waves across bottom. White runner with black details at left with orange outlined white "GET/MARATHON" in upper blue area and "MOTOR OIL" in lower green area.

MARATHON MOTOR OILS 28x40
(4) 1920-1925 $1250
Black tin litho sign with white flange and white runner with black details covering most of sign. Green repeating pattern across upper sign area with orange outlined white "MARATHON/MOTOR OIL" across top. Green "waves" across bototm with black "TRANSCONTINENTAL OIL COMPANY" on waves. Orange "FOR GOOD/RUNNING" to left of runner at bottom.

SECONDARY SIGNAGE

MARATHON GASOLINE 36x72
(5) 1920-1925 $1000
Black tin litho wood framed sign with white runner with black details covering most of sign. Green repeating pattern across upper sign area with orange outlined white "MARATHON/GASOLINE" across top. Green "waves" across bototm with black "TRANSCONTINENTAL OIL COMPANY" on waves. Orange "BEST IN THE/LONG RUN" to left of runner at bottom.

THE OHIO OIL COMPANY & MARATHON OIL, FINDLAY, OH

See introduction for history.

IDENTIFICATION SIGNS

LINCO GLASS SIGN W/SPINNER 36x48
(5) 1935-1939 No Listing
Three dimension can type sign with upper area internally illuminated and covered with glass domes. Light motion action in glass area simulated runner logo running. Red outlined white lower panel with red "LINCO" across center.

MARATHON GLASS SIGN W/SPINNER 48x72
(5) 1935-1946 No Listing
Three dimension can type sign with upper area internally illuminated and covered with glass domes. Light motion action in glass area simulated runner logo running. Red outlined white lower panel with red "MARATHON/DEALER" across center.

LINCO RUNNER SIGN 30x48
(5) 1935-1939 $900
Red outlined white porcelain die-cut sign with circular area at top and elaborate octagonal area below. Outlined of upper circle is red "OHIO" repeating all the way around sign. Red runner in center. Black outlined red "LINCO/GASOLINE" on lower white area.

MARATHON RUNNER SIGN 30x48
(5) 1935-1946 $900
Red outlined white porcelain die-cut sign with circular area at top and elaborate octagonal area below. Outlined of upper circle is thick red line with elaborate design at top. Red runner in center, with black outliend white banner with red "BEST IN THE LONG RUN" in banner. Black outlined red "MARATHON/DEALER" on lower white area.

MARATHON RUNNER SIGN 48x60
(5) 1946-1962 $1200
Red and white outlined red die-cut keyhole shaped sign with blue lower area below upper circle. White runner with red and blue details in circle with blue banner below runner. White "BEST IN THE LONG RUN" on banner, with larger white "MARATHON" on blue area below circle.

MARATHON MULTI-PIECE RUNNER SIGN 72x90
(4) 1946-1962 $550
Complex pole framed round porcelain sign with red and white outlined around red center circle. White runner with red and blue details on circle above blue banner. White "BEST IN THE LONG RUN" on banner. Lower blue porcelain die cut pedestal piece bolts to pole structure below upper frame. White "MARATHON" across blue area. Very rare to find complete signs.

MARATHON "M" MINOR SIGN 42x48
(1) 1962-PRESENT $125
Blue outlined white elongated hexagon with large red "M" in upper sign area above blue "MARATHON."

MARATHON "M" MAJOR SIGN 72x80
(1) 1962-PRESENT $75
Blue outlined white elongated hexagon with large red "M" in upper sign area above blue "MARATHON."

BUILDING LETTERING & SIGNAGE

MARATHON 24"
(3) 1946-1962 $125
Red outlined white porcelain three dimensional channel lettering designed for rooftop mounting.

RUNNER CIRCLE 72"RD
(2) 1946-1962 $550
Round porcelain sign with red and white outlined around red center circle. White runner with red and blue details on circle above blue banner. White "BEST IN THE LONG RUN" on banner.

NEON SIGNS

Ordinary Marathon multi-piece runner signs listed above were retrofit with neon outline rings.

SECONDARY SIGNAGE

MARATHON CREDIT CARDS HONORED 12x28
(4) 1946-1962 $225
White tin litho sign with red band across center. Red "MARATHON" in upper white area, "HONORED HERE" in lower white area with white "CREDIT CARDS" on center red band.

BULK PLANT SIGNAGE

MARATHON VALVE MARKER 5x12
(5) 1939-1950 $100
White porcelain sign with red "MARATHON" across center.

MARATHON ETHYL VALVE MARKER 5x12
(5) 1939-1950 $100
White porcelain sign with red "MARATHON/ETHYL" across center.

THE OHIO OIL COMPANY PANEL 24x48
(2) 1946-1962 $200
Red porcelain sign with black outlined white "THE OHIO OIL COMPANY" across top and detailed red, white and blue Marathon logo at left. Remainder of sign is blank for "priviledge panel" addition of bulk plant name.

TOWER OIL CO. - CINCINNATI, OH

Tower was a Cincinnati based marketer that sold out to, and rebranded Mara-thon in 1958.

No collectible size Tower signs are known at this time.

AURORA GASOLINE CO. - DETROIT, MI

Aurora Gasoline Company was a Detroit refiner marketer that introduced the Speedway brand in the early 1930s. The company's ppremium product, Speedway 79 Stratofuel, was introduced before WW2 and the primary brandname used from the end of WW" until 1962 was "Speedway 79." In 1959 Marathon purchased the Aurora Gasoline Company and the Speedway 79 logo was redesigned. Gas pump globes were eliminated at this time. The Speedway 79 brand was replaced by Marathon in 1962, however,in 1975, the Speedway brandname was reintroduced at several company owned retail "gas only" stations. As Marathon began buying up regional private brands and consolidating them under Emro Marketing with the Speedway brandname. As of this writing, the Speedway name appears on petroleum products retailed at hundreds of Emro "Starvin Marvin" c-stores.

IDENTIFICATION SIGNS

SPEEDWAY/79 OVAL WITH FLAG 42x80
(4) 1946-1962 $300
Pole framed oval porcelain sign with lower edge die-cut flat. Blue upper half with red lower half and white line across center. White "SPEEDWAY" with blue and white flag in upper area, white "79" in lower area.

PUMP SIGNS

SPEEDWAY 79 ETHYL 11x13
(5) 1940-1950 $175
White porcelain sign with "SPEEDWAY" arched across top above "79." Small Ethyl
(EGC) logo below.

THIS PUMP CONTAINS ETHYL SPEEDWAY79 11x13
(3) 1950-1958 $175
White porcelain sign with blue outline. Red/white/blue Speedway flat oval on lower sign face with Ethyl
(EGC) logo to upper right. Red "This pump/contains" to left of Ethyl logo.

NEON SIGNS

SPEEDWAY GAS "79" 60x96
(5) 1935-1940 No Listing
Red porcealin die-cut can type sign punched for neon Square shape with the "steps" down in size. White "SPEEDWAY" arched across top above large "GAS." White "79" on smallest bottom step.

SPEEDWAY 79 60x96
(5) 1940-1950 No Listing
Red porcealin die-cut can type sign punched for neon. White "SPEEDWAY" arched across top above large "79."

OLD DUTCH REFINING CO. - MUSKEGON, MI

Founded in Muskegon about 1930, the Old Dutch brand appeared on a small number of stations, primarily jobber operated, in the Muskegon area. As best as can be determined, Aurora Gasoline Co. purchased Old Dutch Refining in the late 1950s, although jobbers kept the Old Dutch brand alive well into the 1970s.

IDENTIFICATION SIGNS

OLD DUTCH 72"RD
(4) 1946-1970 $450
White/red outlined white porcelain sign with fine red horizontal lines across center of sign above single thicker blue line. Blue windmill design in upper center with blue "Old" to left. Blue "Dutch" below centerline.

PUMP SIGNS

OLD DUTCH ETHYL 11x13
(4) 1946-1970 $225
White tin litho sign with red outline. Blue Old Dutch windmill logo in center, with blue rectangles above and below. White "ETHYL" in top rectangle, "REFINERY OUTLET" below.

REPUBLIC OIL CO. - PITTSBURGH, PA

Dating from 1923 and a successor to Riverside Oil Co (originator of the Marathon trademark), Republic supplied branded and unbranded jobbers throughout the southeast. Republic marketing was concentrated in small areas and no attempt was made at complete statewide coverage in the states where they marketed. By the 1950s the corporate name was Plymouth Oil Co and Marathon purchased the operation in 1962. Concentrating on the extensive unbranded market, the Republic brandname was eventually phased out, although some jobbers continue to use the brandname today.

IDENTIFICATION SIGNS

REPUBLIC GASOLINE & OILS 48"
(5) 1930-1946 $300
Blue/white outlined die-cut shield with blue area across top and vertical red and white stripes below. Three white stars above white "REPUBLIC" in upper blue area with blue "GASOLINE/AND/MOTOR OILS across striped area.

REPUBLIC SHIELD 45"
(4) 1946-1970 $350
Blue/white outlined die cut shield with blue area across top and vertical red and white stripes below. White band extending beyond shield edges across below blue area with blue "REPUBLIC" in band.

REPUBLIC SHIELD 66"
(4) 1946-1970 $300
Blue/white outlined die cut shield with blue area across top and vertical red and white stripes below. White band extending beyond shield edges across below blue area with blue "REPUBLIC" in band.

PUMP SIGNS

REPUBLIC 13"SQ
(3) 1946-1950 $150
White porcelain sign with blue/red outline. Red/white/blue Republic shield logo in center with blue "REPUBLIC" extending beyond edges of shield.

ETHYL REPUBLIC PREMIUM 13"SQ
(3) 1946-1960 $125
White porcelain sign with blue/red outline. Red/white/blue Republic shield logo in center with blue "REPUBLIC" extending beyond edges of shield. Blue "ETHYL" above shield, "PREMIUM" below.

ROYALE REPUBLIC REGULAR 13"SQ
(3) 1950-1960 $150
White porcelain sign with blue/red outline. Red/white/blue

Republic shield logo in center with blue "REPUBLIC" extending beyond edges of shield. Red and blue script "Royale" above shield, blue "REGULAR" below.

CURB SIGNS

ROCOLENE MOTOR OIL 24"RD
(5) 1930-1940 No Listing
Blue porcelain sign with small white circle in center and white rays extending out to form sunburst design. Blue "ROCOLENE" and PGCOA seal in center.

LUBRICANTS SIGNAGE

ROCOLENE LUBESTER SIGN 7"RD
(4) 1930's No Listing
Black outlined red circle with white "ROCOLENE" and PGCOA seal across center. Grade markings below.

SECONDARY SIGNAGE

REPUBLIC CREDIT CARDS HONORED 12x14
(5) 1941-1955 $225
White outlined blue porcelain sign with white "REPUBLIC" across top, "HONORED" across bottom and white script "Charges" in center. White "REPUBLIC OIL COMPANY" across bottom. Dated 1/41.

BONDED - SPRINGFIELD, OH

The Springfield, Ohio based affiliate of Gaseteria-Bonded of Indianapolis. Dating from after WW2, Bonded of Ohio remained independent after Gaseteria-Bonded sold out to Esso's Oklahoma division in 1957. Marathon purchased Bonded in January, 1975 and although many locations were converted to Speedway/Starvin-Marvin, the Bonded brand has remained in use at many locations.

NEON SIGNS

BONDED GAS 60x96
(5) 1946-1962 No Listing
Red outlined white porcelain can type sign with light sockets to either side forming "arrow" across center, punched for neon. Red "BONDED/GAS" across sign face.

CONSOLIDATED - OSHKOSH, WI

A Wisconsin discount operation from the postwar era, that operated about 30 stations throughout the state. Purchased by Marathon in July 1972 and rebranded Speedway about 1982. No collectible size Consolidated signs known at this time.

CHEKER OIL CO. - CHICAGO, IL

Cheker was a suburban Chicago collection of private brands, the Cheker name originally replacing "Chief" and "Road Chief" about 1968. No globes are known from these brands. Cheker purchased the Sun-Glo operation in Indiana about 1970 and converted those stations to the Cheker brand. In 1976 Cheker, then 50% owned by Exxon, purchased the Imperial Refineries operation. Imperial, founded in St. Louis in 1915, was in it's earliest days an intergrated oil company with production, refining and marketing. After WW2, Imperial resumed marketing as a discount chain operation, eventually operating stations in nearly 20 states. Cheker operated Imperial Refineries as a separate brand until the September, 1983 merger with Marathon/Emro and Emro replaced most of the Imperial brand with Ecol after the 1984 purchase of this Jackson, MS independent.
No collectible size Cheker signs known at this time.

SUN GLO

Purchased by Cheker about 1970.

IDENTIFICATION SIGNS

Sun-Glo 72"RD
(5) 1950-1970 $100
White porcelain sign with red center circle. Blue "SUN-GLO" arched around top, "GASOLINE" around bottom.

IMPERIAL REFINERIES - CLAYTON, MO

Founded in 1915 as a refiner/marketer, Imperial operated discount stations throughout the midwest and south until the early 1980s.

IDENTIFICATION SIGNS

Imperial Shield 96x120
(4) 1946-1960's $200
Black outlined red porcelain die cut shield assembled with two piece faces on can type structure. Black outlined yellow banner across shield separates upper and lower sections. Yellow "cross" logo on upper red area, place to attach pricing numbers on lower red area, with black "IMPERIAL" above very small "Refineries" on yellow banner.

PUMP SIGNS

Imperial Refineries Ethyl 13x13
(4) 1946-1965 $250
Yellow porcelain sign with yellow/red Imperial shield covering most of sign. Yellow/red split diagonal, with yellow banner across top. Black "IMPERIAL" in banner, with red "cross" logo above and small Ethyl
(EC) logo below.

JOHNSON OIL CO. - MT. PLEASANT, MI

Founded in 1926 in Lakeview, Michigan and currently Imperial Oil Co of Mt. Pleasant, Michigan—was apparently an Imperial jobber during the time Imperial Oil had refineries and a jobber marketing network. They continue to market in Michigan and Indiana under the Imperial name. This company is not related to Marathon marketing, and is listed here only because of the ties to Imperial.
No collectible size Imperial of Michigan signs known at this time.

ROCK ISLAND REFINING - INDIANAPOLIS, IN

Founded in Duncan, Oklahoma in the 1920s, Rock Island was a mid-continent area marketer until WW2. Having shifted to wartime refining and virtually abandoning retail gasoline, the company was purchased by the Koch family in 1946 and moved to Indianapolis, where it became a major source of unbranded gasoline for independent marketers. Over the years the company acquired several Indiana and Michigan marketers including Colonial/Progressive of Ft.Wayne, Tulsa of Detroit, United of Indianapolis, Naph-Sol of Muskegon and Golden Imperial(not listed below) of Indianapolis. By the mid 1980s "United" had become the primary brand, replacing the others where convenient. In 1989 Marathon's Emro Marketing purchased Rock Island and many United locations have converted to the Speedway brand.

IDENTIFICATION SIGNS

ROCK ISLAND GASOLINE 42"RD
(5) 1930-1946 No Listing

Standard Heating Oils porcelain building sign. *Jim Hollabaugh.*

White Rose of Cincinnati neon hanging sign.

Fort Pitt porcelain 18"x22" sign.

Porcelain standard flange sign attachment. *Mark McKeown.*

White Rose porcelain building sign.

Sunoco bottle rack with reflective letters. *Jim Hollabaugh.*

Red Indian Canadian sign. *Leo Mathieu.*

Firestone unusual shaped porcelain curb sign.

Early Bowser service station 34"x12" tin sign. *Dunbar's Gallery.*

Star Cars 36"x24" porcelain sign. *Dunbar's Gallery.*

Michelin Tires 60"x18" porcelain sign. *Dunbar's Gallery.*

Unusual Magnolia 27"x16" porcelain sign. *Dunbar's Gallery.*

Red/white outlined blue porcelain sign with white center circle. Red/white "Rock of Gibralter" design in center circle, with white "ROCK ISLAND" arched around top, "GASOLINE" around bottom.

CURB SIGNS

ROCK ISLAND ETHYL 30"RD
(5) 1930-1946 $550

Red/white outlined blue porcelain sign with white center circle. Ethyl (EGC) logo in center circle, with white "ROCK ISLAND" arched around top, "GASOASOLINE" arouund bottom.

LUBRICANTS SIGNAGE

ROCKILENE 18"RD
(5) 1930's $350

White porcelain sign with two parallel red lines across center. Red "ROCKILENE" between lines, with red rays radiating out from center.

COLONIAL/PROGRESSIVE - FT. WAYNE, IN

Independent marketer founded in the 1930s that operated Colonial stations in Indiana and Progressive stations in Michigan. Sold out to Rock Island in the 1960s.
No collectible size Colonial/Progressive signs known at this time.

TULSA OIL CO. - DETROIT, MI

Tulsa is an old Detroit independent brand that sold out to Crystal Refining in the early 1970s and was in turn purchased by Rock Island about 1978. The United brandname replaces Tulsa in 1985.
No collectible size Tulsa signs known at this time.

DURO OIL CO. - DETROIT, MI

Detroit based independent that merged with Tulsa in the 1970s.
No collectible size Duro signs known at this time.

UNITED - INDIANAPOLIS, IN

United was an Indianapolis based discount marketer with stations throughout Indiana and a few in Michigan. In 1985 United was purchased by their primary supplier, Rock Island, and became the brandname used to consolidate all of Rock Island's marketing. Most of the United stations were rebranded Speedway in 1990 but some locations remain United.
No collectible size United signs known at this time.

NAPH-SOL REFINING CO. - MUSKEGON, MI

Naph-Sol entered the gasoline market with the introduction of Zephyr gasoline in 1932. Marketing continued through jobbers and independent dealers in Michigan and Wisconsin until 1989 when Naph-Sol was bought out by Rock Island Refining. Several Michigan jobbers continue to use the Zephyr brand.

IDENTIFICATION SIGNS

ZEPHYR GASOLINE 48"RD
(4) 1946-1957 $150

Blue outlined white porcelain sign with blue tapered lines extending from left edge nearly all the way across sign. Red "Zephyr" above blue "GASOLINE" between lines.

ZEPHYR SMALL TIN 42x48
(3) 1957-1989 $125

White outlined red rounded triangle with spear extending from left side of lower sign face. White italic "Zephyr" across sign above spear.

ZEPHYR LARGE PORCELAIN 60x84
(3) 1957-1989 $100

White outlined red rounded triangle with spear extending from left side of lower sign face. Blue outlined white italic "Zephyr" across sign above spear.

OSCEOLA REFINING CO. - REED CITY, MI

Independent refiner/marketer purchased by Zephyr in the 1970s.
No collectible size Osceola signs known at this time.

WAKE UP/J.A.HOGSHIRE & SONS - INDIANAPOLIS, IN:

Founded in Lebannon,IN in 1923, J.A.Hogshire was one of the original IOMA members. They introduced the Wake Up brand in 1937, replacing the earlier Criteria brandname. They were operateing discount stations throughout central Indiana when purchased by Marathon's Emro Marketing Division in 1993. The Wake Up brand remains in use at a small number of stations.

IDENTIFICATION SIGNS

CRITERIA 42x84
(5) 1930-1937 No Listing

Red porcelain can type sign with white stylized "CRITERIA" with lines across center of sign above white "STANDARDS OF QUALITY."

WAKE UP GAS OIL 72"SQ
(5) 1937-1960 No Listing

Red outliend whit eporcelain can type sign with large red rooster at top, above red "WAKE UP." Smaller red "GAS OIL" across bottom.

MARTIN OIL SERVICE - ALSIP, IL: Refiner/marketer whose branded operation was purchased by Marathon's Emro Marketing in 1992. See Martin section in this book for more information.

MARTIN & AFFILIATES

MARTIN OIL COMPANY - CARBONDALE, IL

MARTIN OIL SERVICE - BLUE ISLAND, IL: Martin, an Illinois based trackside discounter dating from the early 1930's, was actually a collection of several family owned operations within the Martin chain. In the earliest days dual headquarters-Chicago and Carbondale,IL-were listed on products. Later the Chicago division was moved to Alsip and still later Blue Island,IL. About 1956 the chain separated into two distinct operations, the Chicago (Alsip) division becoming Martin Oil Service while the Carbondale division remained Martin Oil Company. Another family member established a base of operations in Dubuque, IA which continues to market under the Martin brand today.
Martin Oil Service expanded coast-to-coast, operating over 200 gas stations by the 1970's, while Martin Oil Company concentrated marketing through a small number of stations in the south central states. Cut off from supplies during the

Left column: Oilzum porcelain sign. *Leo Mathieu;* Southland porcelain pump plate sign. *Mark McKeown*; Musgo 48" porcelain sign. *C. Hook*; Right column: Fortune Gasoline porcelain pump plate sign. *Mark McKeown*; Plastic embossed pump plate. *Mark McKeown*; Auntie Chatter embossed early tin sign.

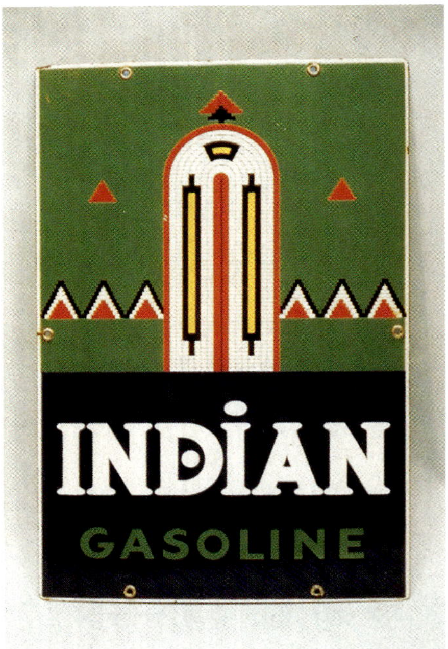

Left column: Quaker State porcelain curb sign. *Dunbar's Gallery*; Indian 12"x18" porcelain pump plate. *Dunbar's Gallery;* Tidewater Associated Oil Co. porcelain truck door sign 11"x7." *Dunbar's Gallery*; Right column: Texaco 15"x14" curved porcelain pump sign. *Dunbar's Gallery;* Sunoco pump plate 12"x8." *Dunbar's Gallery;* Phillips shield 48" porcelain sign. *Dunbar's Gallery.*

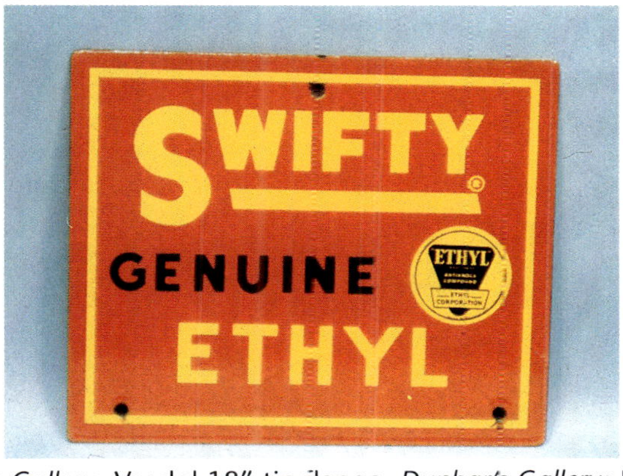

Left column: Atlantic Aviation Oil 17"x10" sign. *Dunbar's Gallery*; Veedol 18" tin flange. *Dunbar's Gallery*; Union 76 pump plate. *Dunbar's Gallery*; Champion Spark Plugs 29"x14" steel sign. *Dunbar's Gallery*; Right column: Cargray Gold porcelain pump plate. *Dunbar's Gallery*; Fisk Tires 38"x32" tin. *Dunbar's Gallery*; United Motors Service 48"x28" porcelain sign. *Dunbar's Gallery*; Swifty Ethyl - rare pump plate. *Dunbar's Gallery*.

Left column: Husky pair porcelain pump plates 12″x12.″ *Dunbar's Gallery*; Texaco 16″ porcelain sign. *Dunbar's Gallery*; Good Gulf 10″ porcelain pump plate. *Dunbar's Gallery*; Right column: Tydol curved porcelain pump sign. *Dunbar's Gallery*; Cushman 21″x18″ steel flange sign. *Dunbar's Gallery*; Magnolia 30″ porcelain sign. *Dunbar's Gallery*.

Left column: Authorized Nash Service 22″x36″ rare porcelain sign. *Dunbar's Gallery*; Veedol porcelain 20″x28″ sign. *Dunbar's Gallery*; Miscellaneous Flying A 10″ porcelain pump signs. *Dunbar's Gallery*; Right column: Texaco flange porcelain sign. *Dunbar's Gallery*; Kendall can shape sign. *Dunbar's Gallery*.

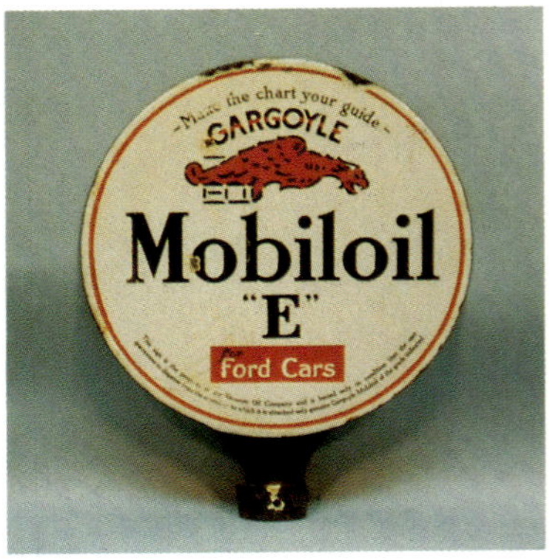

Left column: Pacer 200 HiTest porcelain pump plates. *Bruce Bednar*; T.P. Products 42" porcelain sign. *Dunbar's Gallery*; Super 100 Ethyl porcelain pump sign. *Mark McKeown*; Right column: Tydol Ethyl 9.75" pump sign. *Dunbar's Gallery*; Mobiloil "E" - 9" porcelain lubester sign. *Dunbar's Gallery*; Whitlock porcelain pump plate sign - scarce. *Mark McKeown*; Oilzum Motor Oil tin. *Dunbar's Gallery*.

Left column: Mars regular pump plate. *Mark McKeown;* Jenny Aero 12"x9" pump plate sign. *Dunbar's Gallery*; Pair of Crown and Fina porcelain pump plates. *Bruce Bednar;* Magnolia Oil Line 15"x8" porcelain sign. *Dunbar's Gallery*; Right column: Shamrock 16"x12" porcelain pump sign. *Dunbar's Gallery*; Blue Sunoco 20"x15" porcelain pump plate. *Dunbar's Gallery*.

Left column: Pair of 12" porcelain pump plates. *Mark McKeown*; Pan Am 13"x17" porcelain pump sign. *Dunbar's Gallery*; Sinclair Opaline truck door sign. *Mark McKeown*; Chevrolet 23"x12" tin sign. *Dunbar's Gallery*; Right column: Calso Gasoline porcelain pump sign. *Mark McKeown*; Wyeth Tires curved porcelain sign. *Leo Mathieu.*

1973-74 gas shortage, Martin Oil Service retreated to their home base, operating around 50 stations in Illinois and Indiana until the 1992 purchase by Marathon's Emro Marketing Division. Stations remain branded Martin as of this writing.

As for Martin Oil Company, they sold out to J.D.Street of Maryland Heights,MO in 1977 and the stations were rebranded with J.D.Street's "Zephyr" brand. In the 1980's Street became a branded jobber for several major oil companies, and today their Zephyr brand remains in use at only a small number of stations.

MARTIN OIL STATIONS - CARBONDALE, IL:

Operated stations in the south central states from the 1930's until purchased in 1977 by J.D.Street and Company.

IDENTIFICATION SIGNS

SMARTIN UP WITH MARTIN 36x72
(5) 1946-1977 $100
Blue porcelain sign with white band across center. White "SMARTIN' UP WITH" in upper blue area, red "MARTIN" across white center band.

ZEPHYR/J.D.STREET & COMPANY - MARYLAND HEIGHTS, MO:

Founded in 1884 as a lubricants distributor, J.D.Street entered the retail petroleum market with the introduction of the Zephyr brand in 1939. J.D.Street purchased Martin Oil Company in 1977 and converted their stations to the Zephyr brand. J.D.Street currently serves as a jobber for several major brands as well as operating a few remaining Zephyr branded stations.

IDENTIFICATION SIGNS

MOTOR OIL/ZEPHYR/GASOLINE 32x40
(5) 1954-1960 $175
Blue/white outlined oval blue embossed tin litho sign with white/red outlined white band diagonal across center. Red "ZEPHYR" on diagonal band, with white "MOTOR OIL" in upper blue area, "GASOLINE" below.

ZEPHYR 52x84
(3) 1954-PRESENT $150
Blue/white outlined oval blue porcelain sign with white/red outlined white band diagonal across center. Red "ZEPHYR" on diagonal band, with white "MOTOR OIL" in upper blue area, "GASOLINE" below.

BUILDING SIGNS AND LETTERS

ZEPHYR ROOFTOP LETTERS 24"
(3) 1946-PRESENT $100
Red porcelain channel letters for rooftop mounting.

MARTIN OIL SERVICE - BLUE ISLAND, IL:

Martin Oil Service is the division that expanded coast-to-coast beginning in the late 1950's. Sold out to Marathon's Emro Marketing division in 1992. Approximately 43 stations in Illinois and Indiana remain branded Martin as of this writing and for the forseeablve future.

PUMP SIGNS

SUPER REGULAR PLASTIC SIGN 10x12
(5) 1962-1970 $35
White vinyl embossed ridged plastic sign with red/blue Martin "M" logo on smooth area near top. Red and blue "SUPER/REGULAR" below.

xTRA SPECIAL ETHYL PLASTIC SIGN 10x12
(5) 1962-1970 $35
White vinyl embossed ridged plastic sign with red/blue Martin "M" logo on smooth area near top. Red and blue "xTRA SPECIAL/ETHYL" below.

PURPLE MARTIN PLASTIC SIGN 10x12
(5) 1962-1970 $35
White vinyl embossed ridged plastic sign with red/blue Martin "M" logo on smooth area near top. Purple "PURPLE MARTIN/ETHYL" below.

MIDWEST & ACE HIGH

MIDWEST OIL COMPANY - SOUIX FALLS,SD:

Founded in Minneapolis in 1919, Midwest was a somewhat loose federation of family owned companies that was also operating in Souix Falls by 1924 and in Fargo, ND by 1927. They originally marketed under the "Avio" brandname. In the 1930s they purchased Northwest Oil Company, believed to be based in Minneapolis, and in the early 1940s adopted Northwest's "ACE HIGH" brand. In 1958 they merged with Minneapolis based H.K.Stahl and all of the stations were rebranded Midwest at that time. In the mid 1977 Midwest retreated to a homebase operation in Souix Falls, where they remain today as a branded Sinclair jobber.

IDENTIFICATION SIGNS

AVIO GASOLINE 72x96
(5) 1930-1935 $350
Elaborate die-cut porcelain shield shaped sign with large red outlined white triangle forming lower shield. Blue outlined red "AVIO" above blue "GASOLINE" in upper triangle area with red outlined white oval on blue lower area of triangle. Blue "GUARANTEED/PRODUCT OF MIDWEST" with "M" in oval. Upper shield has blue outliend red circle in center with large white "A." Blue lined white wings to either side.

ACE HIGH GAS OIL "A NORTHWEST BRAND" 42x72
(5) 1930-1935 $1800
Blue porcelain oval sign with blue and white cloud scene at bottom. Large white "ACE HIGH" with red dropshading across center with red and white airplane above. White outlined red script "Gas-Oil" just above clouds, with red "A NORTHWEST BRAND" arched around top.

ACE HIGH 72"RD
(4) 1935-1958 $200
Black outlined red porcelain sign with whtie center circle. Large black "ACE" above "HIGH" covering most of sign. White "GASOLINE" in lowee red area below.

MIDWEST 48x96
(3) 1958-1980 $100
Red porcelain faces on can type sign assembled with aluminum band. Large white arrow across face with blue "MIDWEST" logotype on arrow. Sign is shaped to conform to arrow, and being that left and right sides are not identical, then left and right pointing arrows would be senn in two faces from same sign are compared. Most unusual in gasoline signage.

LUBRICATION/MOTOR OIL SIGNS

AVIO MOTOR OIL WITH AIRPLANE 12x20
(5) 1919-1925 $750
Blue outlined cream tin litho embossed with with blue area across bottom. Airplance and highway scene with race car in

center, with blue "AVIO" above and cream "MOTOR OIL" below. Cream "MIDWEST OIL COMPANY/SOUIX FALLS,S.D." across bottom.

ACE HIGH MOTOR OIL 12x20
(5) 1930's $900
White outlined black tin litho sign. White "ACE HIGH" arched around top above cloud/highway scene with race car and airplane. Red outlined oval logo with black "GUARANTEED/BY/MIDWEST/MIDWEST OIL COMPANY" just above scene. White "MOTOR OIL" in across bottom, with PGCOA seal to either side below.

ACE HIGH MOTOR OIL 48x96
(5) 1930s $900
Wood framed tin litho sign with white/blue race car and airplane scene in center. White "ACE HIGH" across top above orange and black PGCOA seal. White "MOTOR OIL" below scene, above small white "A PRODUCT OF MIDWEST OIL COMPANY."

CURB SIGNS

AVIO MOTOR OIL 17x23
(5) 1920's $900
Red outlined white oval porcelain sign with blue area across botom and to each side. Red "AVIO" in upper white area above red and blue race car/airplane scene. White "MOTOR OIL" across bottom, with small white "A PRODUCT OF MIDWEST OIL COMPANY" arched around botom edge.

ACE HIGH MOTOR OIL MASTER LUBRICATION 17x23
(5) 1930's $900
White outlined dark blue porcelain oval sign. White "ACE HIGH" across top above cloud scene with race car and airplane. Small PGCOA seal just above scene. White "MOTOR OIL" below scene, with red "MASTER LUBRICATION" arched around bottom edge.

ACE HIGH MOTOR OIL 17x23
(5) 1930's $900
White outlined dark blue porcelain oval sign. White "ACE HIGH" across top above cloud scene with race car and airplane. Small PGCOA seal just above scene. White "MOTOR OIL" below scene, with small white "MIDWEST OIL COMPANY/MINNEAPOLIS SOUIX FALLS" at bottom. Probably first identification of Ace High brand with Midwest Oil Company.

ACE HIGH MOTOR OIL 17x23
(5) 1930's $900
White outlined dark blue porcelain oval sign with red/white lines forming dark blue band across lower sign face. White "ACE HIGH" arched around top above cloud scene with race car and airplane. Red outlined oval logo with dark blue "GUARANTEED/BY/MIDWEST/MIDWEST OIL COMPANY" just above scene. White "MOTOR OIL" in lower band, with PGCOA seal to left, Bradford Crude seal to right.

H.K.STAHL OIL COMPANY - MINNEAPOLIS, MN:

H.K.Stahl, marketer of Trophy brand gasolines and Wil-Flo Motor Oils, merged with Midwest Oil Company in 1958. Stations rebranded Midwest at that time.

LUBRICATION/MOTOR OIL SIGNS

WIL-FLO MOTOR 18x24
(5) 1940's $650
Red outlined blue porcelain oval with light blue outlined white "WIL-FLO" arched around top. Light blue and white snow scene with red car below, with blue "at 30 below" across scene. Light blue outlined box with blue "THE/ALL WEATHER/OIL" to lower right.

MOBIL OIL CORPORATION - FAIRFAX, VA STANDARD OIL COMPANY OF NEW YORK/ SOCONY - NEW YORK, NY

Founded in 1882 as the administrative division of the Standard Oil Trust, Socony eventually became involved in refining and later marketing in New York and New England. With the breakup of Standard in 1911, Socony was assigned its refining and marketing properties but left without crude oil production. Exploration began, but most of the crude oil production came through acquisition of Magnolia Petroleum in 1918 and General Petroleum in 1926. Socony expanded it's marketing through Magnolia and General (and White Eagle) as well, marketing from coast-to-coast by 1930. In 1931 Socony merged with another Standard affiliate, Vacuum Oil, the new corporate entity to be named Socony-Vacuum Oil Company. Socony-Vacuum mated Socony's shield and flying horse with Vacuum's "MOBILGAS" trademark and expanded the Mobilgas brand to all of the affiliated companies by the end of WW2. In 1955 the company was renamed Socony-Mobil and in 1966 Mobil Oil Corporation. See SOCONY-VACUUM section below for more information.

NOTE: The Socony "SONY & Shield" logo, which will be referred to in the following descriptions is a blue circle with a large white/red outlined five-point shield positioned to lower edge of circle. A red line crosses the shield about 1/3 of the way down with red "SOCONY" above the line and red "MOTOR/GASOLINE" or "MOTOR/OIL" below the line. Small blue "STANDARD OIL CO. OF NEW YORK" around the bottom edge of the shield. Distorted white "S" and "O" in the blue area above the shield and elongated "N" to left and "Y" to right of shield.

IDENTIFICATION SIGNS

WE SELL SOCONY MOTOR GASOLINE 26x30
(5) 1915-1931 $450
Blue porcelain sign with SONY logo in center. Blue outlined white strip across top and white outlined blue strip across bottom. Red "WE SELL" across upper white area and white "STANDARD OIL CO of N.Y." across lower blue area.

SOCONY GASOLINE/MOTOR OIL SHIELD 42x48
(4) 1924-1931 $500
Blue outlined white porcelain shield red "SOCONY" across top of shield. Blue "GASOLINE/MOTOR OIL" across center with smaller red "STANDARD OIL CO." below. Red "OF NEW YORK" arched around bottom.

SOCONY SOCONY-VACUUM SHIELD 36"
(4) 1932-1946 $350

SOCONY SOCONY-VACUUM SHIELD 48"
(3) 1932-1946 $350
Blue outlined white porcelain five point shield. Blue "SOCONY" across bottom of shield with red "SOCONY-VACUUM" below. Blue outlined flying red horse covers top of shield area.

PUMP SIGNS

SOCONY GASOLINE 10x15

(5) 1915-1922 $225
Blue outlined white flat porcelain sign with blue "SOCONY/GASOLINE" across sign. Red "UNIFORM QUALITY" above, "BEST RESULTS" below.
SOCONY GASOLINE 10x15
(5) 1915-1922 $175
Blue outlined white curved porcelain sign with blue "SOCONY/GASOLINE" across sign. Red "UNIFORM QUALITY" above, "BEST RESULTS" below.
ASK FOR SOCONY MOTOR OIL FLAT 15x15
(4) 1915-1922 $225
Blue/red outlined white flat porcelain sign with blue "Ask for/SOCONY/MOTOR/OIL" on sign face.
ASK FOR SOCONY MOTOR OIL CURVED 15x15
(4) 1915-1922 $175
Blue/red outlined white curved porcelain sign with blue "Ask for/SOCONY/MOTOR/OIL" on sign face.
ASK FOR SOCONY PARABASE MOTOR OIL 10x15
(4) 1926-1930 $200
Blue outlined white flat porcelain signw ith blue "ASK FOR/SOCONY" across top. Red script "Parabase" above blue "MOTOR/OIL" below "SOCONY."
ASK FOR SOCONY PARABASE MOTOR OIL 10x15
(4) 1926-1930 $200
Blue outlined white curved porcelain signw ith blue "ASK FOR/SOCONY" across top. Red script "Parabase" above blue "MOTOR/OIL" below "SOCONY."
SOCONY WINTER ETHYL 10x16
(4) 1928-1930 $250
Die cut pennant shaped white tin litho sign. Black script "Socony" above black "WINTER/ETHYL" across pennant.

CURB SIGNS

STOP HERE SOCONY MOTOR OIL 24"RD
(2) $350
White porcelain sign with blue outlined red "SOCONY" across center. Blue "MOTOR/OIL" below "SOCONY," "Stop Here" above.

LUBRICATION/MOTOR OIL SIGNS

SOCONY MOTOR OILS 10x15
(5) 1915-1920 $250
White outlined blue porcelain sign with white "SOCONY/MOTOR/ OILS" across sign face.
SOCONY CHASSIS LUBRICANT 18x30
(3) 1915-1925 $175
Red/white outlined blue porcelain sign with white "SOCONY" across top. Red "CHASSIS" in white band in center. White "LUBRICANT/(PRESSURE GUN GREASE)" below band.
CRANK CASE SERVICE SOCONY MOTOR OIL 32x36
(3) 1920-1925 $250
Red/white outlined blue porcelain sign with white "Crank Case Service" across top. Red "Drain and Refill with" across center and white "SOCONY MOTOR OIL" across bottom.
SOCONY AIR-CRAFT OILS 12x20
(4) 1925-1932 $950
SOCONY AIR-CRAFT OILS 20x30
(2) 1925-1931 $950
Red/white outlined blue porcelain sign with white "SOCONY" across top and white "OILS" across bottom. Small white "STANDARD OIL COMPANY OF NEW YORK" across bottom edge. Blue and white detailed airplane in center with red "AIRCRAFT" across single white upper wing.

LUBESTER SIGNS

SONY SOCONY MOTOR OIL SHIELD 15' RD
(4) 1915-1931 $400
Blue/red outlined blue curved porcelain sign with red outlined white shield on lower sign face. Large white "SONY" wraps around shiled. Red "SOCONY" across upper shield with blue "MOTOR/OIL" below. Blue "Standard Oil Company of New York" arched around bottom.
FRONT SONY SOCONY MOTOR OIL SHIELD 15"RD
(3) 1915-1931 $400
Blue/red outlined blue flat porcelain sign with red outlined white shield on lower sign face. Large white "SONY" wraps around shiled. Red "SOCONY" across upper shield with blue "MOTOR/OIL" below. Blue "Standard Oil Company of New York" arched around bottom.
SONY OVAL SHIELD 10x16
(4) 1920-1925 $450
Blue porcelain rectangular sign with distorted SONY shield in white oval in center.
SONY SOCONY MOTOR OIL 8x9.5
(2) 1915-1931 $450
Blue/white outlined blue flat porcelain sign with red outlined white shield on lower sign face. Large white "SONY" wraps around shiled. Red "SOCONY" across upper shield with blue "MOTOR/OIL" below. Blue "Standard Oil Company of New York" arched around bottom. Blue outlined white base with grade listing in red across white area. The following grade listings are known
ALL $450
LIGHT MEDIUM
MEDIUM
990 MOTOR OIL FOR FORD CARS
SONY SOCONY AIRCRAFT OIL:
Blue/white outlined blue flat porcelain sign with red outlined white shield on lower sign face. Large white "SONY" wraps around shiled. Red "SOCONY" across upper shield with red "AIRCRAFT" above blue "MOTOR/OIL" below. Blue "Standard Oil Company of New York" arched around bottom. Blue outlined white base with grade listing in red across white area. The following grade listings are known
All $450
LIGHT MEDIUM
MEDIUM
SOCONY PARABASE MOTOR OIL 8x9.5
(4) 1926-1932
Blue outlined white porcelain keyhole shaped sign with blue "SOCONY" above red script "Parabase" on upper sign face, Blue "MOTOR OIL" across center with blue grade name below. Blue "STANDARD OIL COMPANY OF NEW YORK" arched around bottom. Red grade information on lower keyhole area. The following grades are known
All $450
990-A FOR FORD CARS

FLANGE SIGNS

WE SELL SOCONY MOTOR GASOLINE 22x24
(1) 1915-1920 $650
This sign features an unusual version of the SONY shield logo.

It is described as follows
White outlined blue circle with a large white/blue outlined five-point shield positioned to lower edge of circle. A pink band crosses the shield at the top with blue outlined red "SOCONY" across the band and blue outlined red "MOTOR/GASOLINE" below the band. Small blue "STANDARD OIL CO. OF NEW YORK" around the bottom edge of the shield. Distorted white "S" and "O" in the blue area above the shield and elongated "N" to left and "Y" to right of shield.
Blue porcelain flange sign with SONY logo in center. Blue outlined area arched around top and white outlined blue strip across bottom. Red "WE SELL" across upper white area and white "STANDARD OIL CO of N.Y." across lower blue area.
WE SELL SOCONY MOTOR GASOLINE 22x24
(1) 1920-1931 $650
Blue porcelain flange sign with SONY logo in center. Blue outlined white strip across top and white outlined blue strip across bottom. Red "WE SELL" across upper white area and white "STANDARD OIL CO of N.Y." across lower blue area.

MAGNOLIA PETROLEUM - DALLAS, TX

A Texas oil producer/refiner/marketer dating from 1911, Magnolia sold 45% interest to Socony in 1918. Socony completed the purchase in 1925. The Magnolia brandname was replaced by "MOBILGAS" about 1934, after the Socony-Vacuum merger. Magnolia remained a separate, named subsidiary, primarily involved in pipeline transportation, until 1960.

IDENTIFICATION SIGNS

MAGNOLIA GASOLINE MOTOR OIL 42"RD
(4) $750
White outlined red porcelain sign with white outlined blue center circle. White "MAGNOLIA" arched around top, "GASOLINE" around bottom. Small white magnolia flower with green leaves at top of center circle. Smaller flowers to either side of center on outer red band. White "MOTOR OIL" below flower in center.
MAGNOLIA MOBILGAS SHIELD$400
Blue outlined white porcelain five point shield with blue "Mobilgas"
(serif a) across lower sign face above red "MAGNOLIA." Flying red horse logo at top of sign above "Mobilgas."

CURB SIGNS

MAGNOLIA PETROLEUM COMPANY MAGNOLIA GASOLINE FOR SALE HERE 30"RD
(2) 1918-1925 $750
White outlined red porcelain sign with white outlined blue center circle. White "MAGNOLIA PETROLEUM" arched around top, "COMPANY" around bottom. Small white outlined red circle at top of center circle with white magnolia flower in center. White "MAGNOLIA/GASOLINE/FOR SALE HERE" below flower on center circle.
MAGNOLIA PETROLEUM COMPANY MAGNOLENE MOTOR OILS FOR SALE HERE 30"RD
(2) 1918-1925 $750
White outlined red porcelain sign with white outlined blue center circle. White "MAGNOLIA PETROLEUM" arched around top, "COMPANY" around bottom. Small white outlined red circle at top of center circle with white magnolia flower in center. White script "Magnolene" above white "MOTOR OILS/FOR SALE HERE" below flower on center circle.
MAGNOLIA GASOLINE MAGNOLIA SOCONY MOTOR OIL SHIELD 30"RD
(2) 1925-1932 $900
White outlined red porcelain sign with white outlined blue center circle. White "MAGNOLIA" arched around top, "GASOLINE" around bottom. Small white "MAGNOLIA" above white/red outlined white five point shield in center. Blue "SOCONY/MOTOR/OIL" on shield. White magnolia flower with green leaves to either side of shield in center.
MAGNOLIA ETHYL GASOLINE 30"RD
(4) 1926-1932 $550
White outlined red porcelain sign with white/blue outlined white center circle. White "MAGNOLIA" arched around top, "GASOLINE" around bottom. Large Ethyl (EGC) logo with yellow rays in center.

SECONDARY SIGNS

MAGNOLIA GASOLINE FOR SALE HERE 15x30
(5) 1918-1925 $400
Rectangular porcelain sign with red band across top, white band in center and blue band across bottom. White "MAGNOLIA" in upper red area with blue "GASOLINE" in center and white outlined red "FOR SALE HERE" on bottom band.
MAGNOLENE OILS AND GREASES FOR SALE HERE 15x30
(5) 1918-1925 $400
Rectangular porcelain sign with red band across top, white band in center and blue band across bottom. White "MAGNOLENE" in upper red area with blue "MOTOR OILS AND GREASES" in center and white outlined red "FOR SALE HERE" on bottom band.

FLANGE SIGNS

MAGNOLENE MOTOR OIL FOR FORDS 16x22
(5) 1918-1925 $400
White outlined blue porcelain flange sign with white "Magnolene" script logo across upper area of sign. White "MOTOR OIL/FOR FORDS/REDUCES VIBRATION/FOR SALE HERE" across lower sign face.

GENERAL PETROLEUM CORPORATION - VERNON, CA

A pioneer California oil producer and refiner, General was founded in 1910. In 1926 Socony purchased General Petroleum and maintained the General brands until the end of WWII, gradually replacing them with Mobilgas. As with Magnolia, General remained a separate, named subsidiary until 1960.

IDENTIFICATION SIGNS

GENERAL GASOLINE & LUBRICANTS 30"RD 1920-1930 $1200
Green/white/black outlined green porcelain sign with white outlined black band diagonal across center. White "GENERAL" on band. White "GENERAL/PETROLEUM/CORPORATION" in upper green area and "GASOLINE and LUBRICANTS" below.

BUILDING SIGNS & LETTERS

GENERAL GASOLINE W/ HORSE 54x60
(5) 1934-1940 $650
Die cut porcelain sign. Blue outlined flying red horse on white area with flat bottom for roofline mounting. Blue script "General" above "GASOLINE" on white area.

PUMP SIGNS

GENERAL GASOLINE & LUBRICANTS TAB $550

Green/white/black outlined green porcelain sign with white outlined black band diagonal across center. White "GENERAL" on band. White "GENERAL/PETROLEUM/CORPORATION" in upper green area and "GASOLINE and LUBRICANTS" below. Black tab for mounting to cylinder rods.

CURB SIGNS

GENERAL VIOLET RAY ANTI-KNOCK GASOLINE 30"RD
(5) 1925-1930 $1500

GENERAL VIOLET RAY ANTI-KNOCK GASOLINE 42"RD
(5) 1925-1930 $1400

Red outlined white porcelain sign with large green diamond covering most of sign face. White band across center of diamond with red outlined purple "VIOLET RAY" across center. White "GENERAL" and white outlined lightning bolt above. White outlined red "ANTI-KNOCK" below with red "GASOLINE" in white interruption of diamond near bottom.

GENERAL VIOLET RAY GASOLINE 30"RD
(5) 1930-1934 $1500

Red outlined white porcelain sign with large green diamond covering most of sign face. White "GENERAL" across top over large white "VIOLET/RAY" with red & purple block dropshading. Red"GASOLINE" in white interruption of diamond near bottom.

GENERAL ETHYL GAOSLINE EGC 30x30
(5) 1926-1934 $800

Red outlined white rounded triangular sign, point down, with red/white/green checkerboard pattern across flat top area. Green "GENERAL" above red "ETHYL/GASOLINE" across center of sign with Ethyl (EGC) logo at bottom.

LUBRICATION/MOTOR OIL SIGNS

SOCONY CERTIFIED LUBRICATION SERVICE 14x20
(5) 1928-1934 $450

Blue outlined white five point elongated white porcelain shield with blue lines around red "SOCONY/CERTIFIED/ LUBRICATION/SERVICE" CERTIFIED in blue. Blue script text below, with red outlined green General logo forming seal with blue ribbons at bottom left beside red "GENERAL PETROLEUM/CORPORATION/ OF CALIFORNIA."

MOBIL CERTIFIED LUBRICATION 14x20
(5) 1934-1946 $450

Blue outlined white five point elongated white porcelain shield with red shield outline around red "MobiLubrication" at top. Small blue "SOCONY VACUUM" (red horse in center) below "Mobil." Script blue "Certified" above "Lubrication." Blue script text below, with red outlined white & green General logo forming seal with blue ribbons at bottom left beside red "GENERAL PETROLEUM/CORPORATION/ OF CALIFORNIA." Small flying horse and gargoyle logos on revised seal.

WHITE EAGLE OIL AND REFINING - KANSAS CITY, MO

Mid-Continent producer/refiner/marketer with gas stations in eleven midwestern states. Chief White Eagle was a leader of the Ponca Indians in Oklahoma. White Eagle was first used as a corporate name in 1916 when L.L. Marcell founded the White Eagle Petroleum Company. That same year the new company built a refinery at Augusta, Kansas, having a capacity of 2000 barrels per day.

In 1919 the White Eagle Petroleum Company was reorganized and the name changed to the White Eagle Oil and Refining Company. This reorganization was in reality a merger of six independent oil companies and within a few years the new company enjoyed second position in the distribution of petroleum products in the Middle West. A second refinery was built at Casper, Wyoming, and one was purchased at Fort Worth, Texas.

The poised White Eagle was adopted as an emblem because of its dignity and general significance, particularly in their territory and because it was in keeping with the company name. As a service station identification, this emblem—the big iron eagle in front of the station and the illuminated glass eagles on the pumps—was very effective.

In January, 1930, the White Eagle Oil and Refining Company was purchased by the Standard Oil Company of New York (Socony) at which time the name was changed to the White Eagle Oil Corporation. When Socony and Vacuum merged in 1931, White Eagle Oil Corporation became a subsidiary company. In 1935 the White Eagle Oil Corporation was dissolved and the operation became the White Eagle division of Socony-Vacuum Oil Company, Inc.

IDENTIFICATION SIGNS

GASOLINE WHITE EAGLE AND KEYNOIL 42"
(5) 1925-1930 $750

White octagon shaped porcelain sign with black/white outlined red octagon in center. White outlined black band across center of octagon with white "WHITE EAGLE" on band. White "GASOLINE" arched around top, "Keynoil" around bottom.

CURB SIGNS

GASOLINE WHITE EAGLE AND KEYNOIL 30"RD
(4) 1925-1930 $750

White porcelain sign with black outlined white octagon around red circle in center. White outlined black band across center of circle with white "WHITE EAGLE" on band. White "GASOLINE" arched around top, "Keynoil" around bottom.

GASOLINE WHITE EAGLE AND KEYNOIL 30"OCT
(4) 1925-1930 $750

White octagon shaped porcelain sign with red/white outlined red circle in center. White outlined black band across center of octagon with white "WHITE EAGLE" on band. White "GASOLINE" arched around top, "Keynoil" around bottom.

100% PARAFFIN BASE KEYNOIL MOTOR OIL 30"RD
(5) 1920-1925 $500

White porcelain sign with red band across center. Black outlined white "Keynoil" on band with red and black diamond logo at top. White "PARAFFIN" on black band with white "100%" above and "BASE" below. Black "MOTOR OIL" across below red band.

BANNER WHITE EAGLE ETHYL GASOLINE 30"RD
(5) 1926-1930 $650

White porcelain sign with red/white outlined red circle with white center in center. Black band across center interrupted by large Ethyl (EGC) logo with yellow rays. White "WHITE" "EAGLE" on band. White "BANNER" arched around top, "GASOLINE" around bottom.

LUBRICATION/MOTOR OIL SIGNS

KEYNOIL STRIP 4x14

(5) 1920s $400

Red outlined white tin litho embossed sign with black outlined red "Keynoil" above red line to left. Black outlined white "LUBRICATES LASTS" on line. Black outlined white octagon with black circle with red center to right. Detailed standing white eagle in center of logo with white "WHITE EAGLE" in black band across center and white "MOTOR OIL" arched above.

VACUUM OIL COMPANY - ROCHESTER, NY

Founded as a lubricants manufacturer in 1866, Vacuum had developed an extensive line of specialty lubricants when Standard Oil purchased the company in 1879. During the Standard era, Vacuum's lubricants, under the Gargoyle and later Gargoyle Mobiloil brands, were sold by the various Standard marketing firms. With the breakup of Standard in 1911, Vacuum was left to market its oils through independent channels worldwide. They became involved with gasoline marketing, under the "MOBILGAS" brand, in the late 1920s and expanded the gasoline venture with the purchase of Lubrite Refining Co. of St. Louis, Missouri, in 1929. The following year Vacuum purchased Wadhams Oil Corporation of Milwaukee and the Vacuum "MOBILGAS" name was added to Wadhams extensive product line. Also in 1930 Vacuum purchased White Star Refining of Detroit and added over 1500 outlets for Mobilgas. In 1931 Socony and Vacuum merged to form Socony Vacuum Oil Company and after a two year reimaging program, the Mobilgas brandname became the company's primary brand, although several of the regional marketers retained their own identities until after WWII.

IDENTIFICATION SIGNS

GARGOYLE MOBILOIL/MOBILGAS 42x60

(4) 1925-1932 $400

One-sided porcelain sign faces assembled back to back can-style. White porcelain sign with series of progressively thicker red horizontal stripes progressing out from white band across center. Black "Mobilgas" across center band. Upper striped area is interrupted at center by 12" round red outlined white circle, projecting above horizontal edge of sign. Red and black Gargoyle logo above black "Mobiloil" in circle.

BUILDING SIGNS & LETTERS

GARGOYLE MOBILOILS GASOLINE STRIP 18x60

(5) 1920-1925 $400

Black porcelain sign with white rectangle at left beside white "Mobilgas." White "GASOLINE" below, with red and black gargoyle logo in white rectangle.

CURB SIGNS

MOBILOIL MAKE THE CHART YOUR GUIDE 24"RD

(5) 1920-1925 $375

Red outlined white tin litho sign with red and black Gargoyle logo at top. Black "Mobiloil" across center with black "Make the chart/your guide" below.

ASK HERE FOR MOBILOIL 28.5x32.5

(3) 1925-1930 $375

White rounded corner square porcelain sign with red band and stipes across lower sign face. Black "Ask here for" at upper left corner above red and black gargoyle logo above red bands. Black "Mobiloil" across red band, black "The World's Quality Oil" across bottom.

ASK HERE FOR GARGOYLE MOBILOIL 24"RD

(1) 1920-1931 $375

White porcelain sign with black "Ask here/for" across top of sign above red and black Gargoyle logo. Black "Mobiloil" below logo.

STOP HERE FOR GARGOYLE MOBILOIL 24"RD

(1) 1931-1935 $375

White porcelain sign with black "Stop here/for" across top of sign above red and black Gargoyle logo. Black "Mobiloil" below logo.

MOBILGAS ETHYL EGC 30"RD

(3) 1926-1931 $350

Red outlined white porcelain sign with Ethyl (EGC) logo with yellow rays at top. Black "Mobilgas/ETHYL" below.

LUBRICATION/MOTOR OIL SIGNS

GARGOYLE MOBILOIL VERTICAL STRIP 14x60

(4) 1920s $225

Black outlined red tin litho sign with white oval at top above black outlined white vertical "MOBILOIL." Red and black gargoyle logo in oval.

DRAIN AND REFILL WITH GARGOYLE MOBILOIL NOW 36x60

(4) 1920-1925 $225

Red/black outlined red porcelain sign with black outlined white "Mobiloil" diagonal across center with black outlined white oval at upper left. Red and black Gargoyle logo in oval with white outlined black "Drain and refill with" across top, "NOW!" across bottom of sign.

DRAIN AND REFILL WITH MOBILOIL 29x33

(3) $225

White porcelain sign with black "Drain and Refill with-" across above red and black Gargoyle logo. Black "Mobiloil" across center above black "Make the chart your guide."

GARGOYLE MOBILOIL CERTIFIED SERVICE 19.5SQ $300

White porcelain sign with irregular blue border around red outlined white center circle. Red and black Gargoyle logo at top with black "Mobiloil" across center above red "Certified Service."

LET US MOBILOIL YOUR CAR 42x60

(3) 1925-1930 $300

Black outlined self-framed red porcelain sign with white "Mobiloil" across center. White outlined black "Let us" above "Mobiloil," "your car" below. Black outlined white "ENGINE:CHASSSIS:GEARS" across bottom.

FLANGE SIGNS

GARGOYLE MOBILOIL FLANGE 15.5x24

(2) 1911-1931 $450

White porcelain flange sign with black "Mobiloil" across sign face above black "VACUUM OIL COMPANY." Black outlined and detailed gargoyle above "Mobiloil" with black "GARGOYLE" with red arched line through letter centers arched above gargoyle figure.

GARGOYLE MOBILOILS A GRADE... 15.5x24

(4) 1925-1930 $450

White porcelain flange sign with black "Mobiloil" across sign face above black "A grade for each type of motor." Black outlined and detailed gargoyle above "Mobiloil" with black "GARGOYLE" with red arched line through letter centers arched above gargoyle figure.

GARGOYLE MOBILOIL MAKE THE CHART... 15.5x24
(2) 1911-1931 $450
White porcelain flange sign with black "Mobiloil" across sign face above black "Make the chart your guide." Black outlined and detailed gargoyle above "Mobiloil" with black "GARGOYLE" with red arched line through letter centers arched above gargoyle figure.

LUBESTER SIGNS

MAKE THE CHART YOUR GUIDE GARGOYLE MOBILOIL $400
All of the following signs are a white/red outlined white porcelain sign with black "Mobiloil" across center. Black outlined and detailed gargoyle above "Mobiloil" with black "GARGOYLE" with red arched line through letter centers arched above gargoyle figure. Black Grade Name below "Mobiloil." Black "Make the chart your guide" arched above gargoyle figure. Black "This sign is the property of the Vacuum Oil Company and is loaned only on the condition that the user/guarantee to dispense from the container to which it is atached only genuine Gargoyle Mobiloil of the grade indicated. The following grade listings are known:
A,ARCTIC,B,BB,E/FOR FORD CARS,CW

OIL CABINET & RACK SIGNS

ASK FOR GARGOYLE MOBILOIL 19.5x24
(1) $600
White porcelain sign with black "Ask for" across top above red and black "Gargoyle" logo. Black "Mobiloil" below logo with black "Authorized Service/VACUUM OIL COMPANY" in lower right corner.

STOP HERE GARGOYLE MOBILOIL 19.5x24
(1) $600
White porcelain sign with black "Stop here" across top above red and black "Gargoyle" logo. Black "Mobiloil" below logo with black "Authorized Service/VACUUM OIL COMPANY" in lower right corner.

AUTHORIZED SERVICE GARGOYLE MOBILOIL
(5) 1920-1929 $600
Red outlined white porcelain rectangular sign with projecting area at top center. Small black "Property of/Vacuum Oil Company" in projecting area. Black "Authorized Service" across top of sign face with red script "Genuine" at left. Red and black "Gargoyle" logo in center with black "Mobiloil" below. Possibly part of an oil rack.

ASK FOR GARGOYLE MOBILOIL CHART 20x40
(3) 1920-1929 $350
The often referred to "Mobiloil Chart," the text changed annually with automotive introductions so exact text varies. White tin litho sign with red outline. Black "Ask for" above red and black gargoyle logo at top. Black "Mobiloil" above "Make this chart your guide" across below "Mobiloil." Black detailed text with automobile makes and oil reference below, with black "Authorized Service/VACUUM OIL COMPANY" across bottom.

GARGOYLE MOBIL 30C/35C 13.5x17
(4) 1920-1929 $300
White tin litho embossed with with red and black Gargoyle logo at top over black "Mobiloil." Black outlined red "30c" and "35c" below, with black text references pricing for various grades. Red outlined white band across bottom with black "VACUUM OIL COMPANY" in band.

LUBRITE OIL COMPANY - ST. LOUIS, MO

Lubrite was a St Louis based lubricants refiner that was purchased by Vacuum Oil in 1929. Lubrite stations were rebranded Vacuum-Lubrite and remained as such until after the Vacuum Oil merger with Socony.

BUILDING SIGNS & LETTERS

LUBRITE STATION PANEL 12x96
(5) 1926-1930 $200
Self framed white porcelain sign with blue "LUBRITE STATION" across panel.

CURB SIGNS

VACUUM LUBRITE 30"RD
(5) 1929-1932 $475
Blue outlined white porcelain sign with red "VACUUM across top above large red "V." Blue "Lubrite" interrupt "V" near bottom.

WADHAMS OIL CORPORATION - MILWAUKEE, WI

Founded in the 1870s as a lubricants manufacturer, Wadhams was a pioneer in gasoline marketing, operating Wadhams stations throughout Wisconsin prior to 1920. Wadhams was briefly affiliated with the Independent Oil Men of America about 1925 and expanded with the 1929 purchase of Milwaukee competitor Bartles-Maguire Oil Company. Noted for its Chinese pagoda style stations, Wadhams had developed a first-class gasoline marketing organization when the company was purchased by Vacuum in 1930. Vacuum's Mobilgas trademark was added to the Wadhams brands and finally replaced the Wadhams name about 1939.

IDENTIFICATION SIGNS

WADHAMS GASOLINE 48x72
(5) 1925-1936 $600
Red/yellow outlined black porcelain sign with white "Wadhams/Gasolines" across center.

MOBILGAS WADHAMS 48x52
(4) 1936-1942 $450
White/blue outlined white porcelain five point shield with blue "Mobilgas" (serif "a") across center of sign face. Flying red horse with white details above "Mobilgas," with blue "WADHAMS" below "Mobilgas."

BUILDING SIGNS & LETTERS

WADHAMS PANEL 18x96
(5) 1925-1939 $300
Yellow outlined black porcelain sign with white "Wadhams" logotype across center.

CURB SIGNS

WADHAMS ETHYL GASOLINE 42"RD
(5) 1926-1930 $400
Red/white outlined black porcelain sign with red outlined white center circle. White "Wadhams" arched around top, "Gasoline" around bottom with Ethyl (EGC) logo with yellow rays in center.

LUBRICATION/MOTOR OIL SIGNS

WADHAMS MOTOR OIL/EMPHATICALLY INDEPENDENT 16.5x20.5

(5) 1926-1930 $900
Black porcelain flange sign with white "Wadhams" across top above yellow "Tempered" and white "Motor Oil." Large red circle at left center with black and yellow oil cans on circle. Small IOMA logo at right center. Series of white vertical lines across sign below oil cans with yellow "Emphatically Independent" across bottom of sign face.
WADHAMS MOTOR OIL/IOMA 16.5x20.5
(5) 1926-1930 $900
Black porcelain flange sign with white "Wadhams" across top above yellow "Tempered" and white "Motor Oil." Large red circle at left center with black and yellow oil cans on circle. Series of white vertical lines across sign below oil cans with small IOMA logo in center at bottom of sign face.
WADHAMS TEMPERED MOTOR OIL 36x48
(4) $900
Black/white outlined black wood framed reflective tin sign with white "Wadhams" across top and "Motor Oil" in lower right corner. Yellow "Tempered" between, with yellow and white oil can and bottle to lower left.

SECONDARY SIGNS

WADHAMS TRUE GASOLINE FLANGE 18x22
(5) 1915-1920 $900
Black porcelain sign with white flange area. White script "Wadhams" across top of sign with detailed red and black gasoline can in center. White "True/Gasoline" below can.
WADHAMS TEMPERED MOTOR OIL 18x84
(5) 1920-1926 $375
Wood-framed black tin litho sign with white "Wadhams" across sign above yellow "Tempered" and white "Motor Oil." Red and white design at left with gold and white oil bottle on red circle at left.
WADHAMS TEMPERED MOTOR OIL 18x84
(5) 1920-1926 $375
Black wood framed reflective vertical tin sign with black outlined white "Wadhams" vertical down sign. Large red circle at top with yellow oil bottle over circle. Yellow "Tempered" over white "Motor/Oil" at bottom.
WADHAMS GAS 18x84
(5) 1920-1926 $350
Black wood framed reflective vertical tin sign with black outlined white "Wadhams" vertical down sign. Red square at top and bottom with black and white pump nozzle on upper square and white "Gas" diagonal on lower square.

BULK PLANT SIGNAGE

WADHAMS BRANCH OFFICE 48x144
(5) 1930-1939 $400
White outlined blue porcelain sign with large white "Wadhams" logotype across center. Smaller white "Branch Office" across bottom.

BARTLES-McGUIRE OIL CO. - MILWAUKEE, WI

Large Wisconsin marketer that merged with Wadhams in 1929, shortly before Vacuum purchased Wadhams.

CURB SIGNS

BARTLES ETHYL GASOLINE 30"RD
(5) 1926-1929 $425
Black outlined light blue porcelain sign with black outlined white center circle. Large Ethyl (EGC) logo in center with black outlined yellow "BARTLES" arched around top, "GASOLINE" around bottom.

SECONDARY SIGN

BARTLES BONDED OILS AND GREASES 12x20
(5) 1920-1925 $225
Red/yellow outlined black embossed tin litho sign with yellow and black station attendant in uniform at left end. Yellow "Bartles" (red arrow through "B") over red "Bonded." Smaller yellow "Oils & Greases" across bottom.

WHITE STAR REFINING CO. - DETROIT, MI

Also added to Vacuum's Mobilgas marketing was Detroit's White Star Refining. White Star operated over 1500 stations in Ohio, Indiana, and Michigan when purchased by Vacuum in 1930. Mobilgas replaced the White Star brand almost immediately, but former White Star outlets used a special Mobilgas shield sign with a small blue cirlce and white star at the bottom until after WWII.

IDENTIFICATION SIGNS

STAROLINE GASOLINE IS BETTER 36x72
(5) 1920s NO LISTING
Blue tin can type sign punched for opal glass insertion. White opal glass "Staroline" logotype over "GASOLINE/IS BETTER."
WHITE STAR GASOLINE/STAROLEUM 42"RD
(4) 1930-1934 $450
Blue porcelain sign with white outlined blue center circle. White "WHITE STAR" arched around top, "GASOLINE" around bottom. Large white star in center with blue "STAROLEUM/MOTOR/OILS" across star.
MOBILGAS WHITE STAR SHIELD 42x48
(4) 1934-1947 $450
White/red outlined white porcelain five point shield with blue "Mobilgas" (serif "a") across center of sign face. Flying red horse with white details above "Mobilgas," with blue circle with white star in center below "Mobilgas."

CURB SIGNS

STAROLEUM MOTOR OIL 30"RD
(5) $450
Blue/white outlined blue porcelain sign with large white star border to border. Blue "Staroleum" above "MOTOR/OIL" on star.
WHITE STAR GASOLINE "STAROLEUM" 30"RD
(3) $450
Blue porcelain sign with white outlined blue center circle. White "WHITE STAR" arched around top, "GASOLINE" around bottom. Large white star in center with smaller white stars in perimeter area to either side. Blue "STAROLEUM" across center star.
WHITE STAR GASOLINE BLUE CENTER 30"RD
(4) $450
Blue/white outlined blue porcelain sign with white outlined blue center circle. Large white star in circle, with white "WHITE STAR" arched around top in blue band, "GASOLINE" below.
WHITE STAR GASOLINE RED CENTER 30"RD
(4) $450
White outlined blue porcelain sign with white outlined red center circle. Large white star in circle, with white "WHITE STAR" arched around top in blue band, "GASOLINE" below.

WHITE STAR MOTOR OIL RED CENTER 30"RD
(5) $450
White outlined red porcelain sign with white outlined blue center circle. Large white star in circle, with white "WHITE STAR" arched around top, "MOTOR OIL" below.
VACUUM WHITE STAR LARGE "V" 30"RD
(5) 1930-1932 $450
Blue outlined white porcelain sign with red "VACUUM" across top above large red "V." Blue "WHITE STAR" interrupts "V" near bottom. Blue circle with white star in center between uprights of "V."
WHITE STAR ON BLUE NO LETTERING 30"RD
(5) 1932-1940 $400
Blue/white outlined blue porcelain sign with large white star border to border. No lettering.

INDEPENDENT OIL CO. OF PENNSYLVANIA - ALTOONA, PA

Purchased by Socony-Vacuum in 1934 and rebranded Mobilgas.
No signs specifically for Independent Oil are known at this time.

METRO - JAMESTOWN/OLEAN, NY

Small New York marketer purchased by Socony-Vacuum in 1934. The Mobilgas brand was added at Metro stations and Metro became a low-grade product sold at Mobil stations in the various marketing territories until after WWII.

IDENTIFICATION SIGNS

METRO GAS MORE POWER/MORE MILEAGE 30x72
(5) 1925-1930 $1000
Red porcelain sign with large black outlined white "Metro Gas" across top above black "MORE POWER MORE MILEAGE." Black outlined white oval in center above black outlined white "Mobiloil." Small black "Ask for" at upper left of "Mobiloil." White vertical rectangle at left of lettering with red and white visible gas pump with Metrol globe puctured.

GILMORE OIL CO. - LOS ANGELES, CA

Founded in 1900, Gilmore entered gasoline marketing in 1923. The company established an elaborate network of dealers in California, Oregon, and Washington, totaling more than 3500 outlets at it's peak in the late 1930s. In 1940 Socony-Vacuum purchased control of Gilmore and assigned Gilmore marketing to the General Petroleum subsidary. Transition from Gilmore to Mobilgas began in 1942 and was completed in 1945, just after WWII.

IDENTIFICATION SIGNS

GILMORE RED LION 48"RD
(5) 1932-1940 NO LISTING
Red outlined yellow porcelain sign with red outlined white center circle. Red and black charging lion in center circle with black outlined red "GILMORE" arched around top, "RED LION" around bottom.

BUILDING SIGNS

GILMORE GASOLINE 24x192
(5) 1926-1942 NO LISTING
Wood-framed red outlined yellow tin litho sign with black outlined "GILMORE GASOLINE" across sign. Circular "MONARCH OF ALL" signs listed below installed in the center between "GILMORE" and "GASOLINE."
BLU-GREEN MONARCH OF ALL 42"RD
(5) 1926-1933 $5000
Red outlined yellow porcelain sign with red outlined white center circle. Orange, red, and black lion's head in center circle with black outlined red "BLU-GREEN" arched around top, "MONARCH OF ALL" around bottom. Installed in center of Gilmore Gasoline panel listed above.
RED LION MONARCH OF ALL 42"RD
(5) 1933-1945 $5000
Red outlined yellow porcelain sign with red outlined white center circle. Orange, red, and black lion's head in center circle with black outlined red "RED LION" arched around top, "MONARCH OF ALL" around bottom. Installed in center of Gilmore Gasoline panel listed above.
ROOFTOP LION
(5) 1932-1940 NO LISTING
Cast full figure of standing lion for installaiton on station rooftops.
GILMORE DIE CUT FLAGMAN
(5) 1926-1935 $3000
Elaborate die cut tin litho sign featuring red and white suited racing flagman holding out a black and white checkered flag with the red Gilmore lion logo on flag. Black lower panel on sign with red outlined white "Try" at top and "GASOLINE" across bottom and white outlined red "GILMORE/The RECORD/BREAKER" in center.

NEON

GILMORE STRIP NEON 16x96
(5) 1935-1945 $2500
Red outlined yellow porcelain sign face with black outlined red "GILMORE" across center. Faces assembled back-to-back and pumched for neon "GILMORE."

CURB SIGNS

GILMORE GASOLINE MONARCH OF ALL 18x32
(5) 1923-1925 NO LISTING
White porcelain vertical rectagular sign with rounded top and bottom edges. Black outlined red "GILMORE" arched across top, "GASOLINE" across bottom with black "MONARCH OF ALL" across bottom of sign. Red outlined white circle in center with black detailed orange lion head in circle.
GILMORE RECORD BREAKER LANE 24"RD
(5) $2800
Yellow porcelain sign with red and black lion jumping across center. Red "GILMORE" logotype above lion with red "RECORD BREAKER/LANE" below. Used in conjunction with Gilmore racing functions.

LUBRICATION/MOTOR OIL SIGNS

GILMORE LION HEAD MOTOR OIL
$1800
Yellow outlined red embossed tin litho sign with white circle in lower left corner and white band diagonal across sign up from circle. Detailed orange and black lion head in circle with black "LION HEAD" on band. Black outlined yellow "GILMORE" and small "MONARCH OF ALL" and "2000 MILES OF LUBRICATION" above band,"Motor Oil" below with white outlined black "Purest Pennsylvania" & PGCOA seal just above motor oil. Small black outlined white "THE MOST HIGHLY FIL-

TERED/MOTOR OIL IN AMERICA" in lower right corner.
GILMORE CHEK/CHART LUBRICATION 24x30
(5) 1932-1940 $1200
White porcelain sign with black band across top and bottom. Black lines forming white band across center with black check mark and black "CHEK-CHART" with black outlined red "GILMORE" across top and "LUBRICATION" across bottom. Note:Chek Chart was a service of H.M.Gousha, the road map publisher. Each year they published a lubrication guidebook and series of lubrication point charts for the various major oil companies. This Gilmore sign acknowledges that they indeed used this guidebook system.

SECONDARY SIGNS

GILMORE BLU-GREEEN GASOLINE 24x54
(5) $3500
Black wood-framed green tin litho sign with black outlined white "GILMORE" in upper left corner, "Gasoline" in lower right with black "BLU-GREEN" with white block dropshading diagonal across center.

MOBILGAS/SOCONY-VACUUM

As is noted in the introduction, Socony and Vacuum merged in 1931. By 1934 it was decided to join the Socony shield and flying horse with Vacuum's Mobilgas to create a single brand-name for all the affiliates. Several, including Magnolia and White Eagle, readily converted to the new Mobilgas identity. Others, including General and Wadhams, combined the Mobilgas brand with their own brandnames, at least for the first few years. Conversion was nearly complete when WWII interrupted, and was completed as soon as possible after the war. Please note, however, that many manufacturers were involved in making the following signs and color variations, including some certain to exist but not listed, exist as well as typestyle and size variations. A complete listing would be nearly impossible, but two years of research has compiled the following list. The "Mobilgas" designations for products was changed to Mobil in 1962 and all of the older signage was replaced at that time.

IDENTIFICATION SIGNS

MOBILGAS W/ HORSE ON RECTANGLE 24x36
(4) 1932-1935 $200
White porcelain sign with flying red horse with white details in center at top. Blue "Mobilgas" across bottom of sign.
MOBILGAS W/ HORSE ON RECTANGLE 30x48
(4) 1932-1935 $425
White/red outlined white porcelain sign with flying red horse with white details in center at top. Blue "Mobilgas" across bottom of sign.
MOBILGAS PRODUCT OF A S-V COMPANY 48x52
(4) 1932-1946 $425
White/red outlined white porcelain five point shield with blue "Mobilgas" (serif "a") across center. Flying red horse with white details at top and red "PRODUCT OF A/SOCONY-VACUUM/COMPANY" below "Mobilgas."
MOBILGAS SOCONY VACUUM 48x52
(4) 1932-1946 $450
White/red outlined white porcelain five point shield with blue "Mobilgas" (serif "a") across center. Flying red horse with white details at top and red "SOCONY-VACUUM" below "Mobilgas."
MOBILGAS SOCONY VACUUM 48x52
(4) 1932-1946 $325
White/red outlined white porcelain five point shield with black "Mobilgas" (serif "a") across center. Flying red horse with white details at top and red "SOCONY-VACUUM" below "Mobilgas."
MOBILGAS SOCONY VACUUM SHIELD 42x48
(4) 1936-1947 $325
MOBILGAS SOCONY VACUUM SHIELD 58x62
(1) 1936-1947 $325
Thick blue outline around a white porcelain five point shield with blue "Mobilgas" (serif "a") across center. Blue outlined flying red horse with white details at top and red "SOCONY-VACUUM" below "Mobilgas."
MOBILGAS SOCONY VACUUM SHIELD 58x62
(1) 1936-1947 $350
Thick blue outline around a white porcelain five point shield with blue "Mobilgas" (round "a") across center. Blue outlined flying red horse with white details at top and red "SOCONY-VACUUM" below "Mobilgas."
MOBILGAS SHIELD 58x62
(1) 1947-1962 $350
Thick blue outline around a white porcelain five point shield with blue "Mobilgas" (round "a") across center. Blue outlined flying red horse with white details at top.
MOBILGAS SOCONY VACUUM SHIELD 58x62
(1) 1936-1947 $350
Thick black outline around a white porcelain five point shield with black "Mobilgas" (serif "a") across center. Black outlined flying red horse with white details at top and red "SOCONY-VACUUM" below "Mobilgas."
MOBILGAS SOCONY VACUUM shield 58x62
(1) 1936-1947 $350
Thick black outline around a white porcelain five point shield with black "Mobilgas" (round "a") across center. Black outlined flying red horse with white details at top and red "SOCONY-VACUUM" below "Mobilgas."
SOCONY VACUUM AVIATION PRODUCTS 72x72
(5) 1936-1946 $650
Red porcelain sign with white/black outlined white five point shield covering most of sign face. Black outlined flying red horse on upper sign face with black "SOCONY-VACUUM" above red "AVIATION PRODUCTS" on lower shield area.

BUILDING SIGNS & LETTERS

FLAT HORSES
Outlined as noted flat porcelain die-cut flying red horses with white details. Known in the following sizes:
29x37 White outline and details $1600
2' Blue outlined $1600
3' Blue outlined $1500
4' Blue outlined $1000
EMBOSSED HORSES
Embossed porcelain die cut flying red horses with white details. Known in the following sizes:
71x92 $1200
2' $1700
3' $1500
4' $1000
MULTI-PIECE HORSE FACING LEFT 48x60

Buick 42" porcelain sign. *Dunbar's Gallery.*

Cooper Tines 14"x60" tin sign. *Dunbar's Gallery.*

AAA porcelain sign. *Dennis Brilla.*

McColl-Frontenac Red Indian Canadian porcelain sign. *Dennis Leith.*

Oil Field Service 14" porcelain sign.

Lee Tires self-framed 4' metal sign. *Jim Hollabaugh.*

Hood Tires 72" porcelain die-cut sign. *Jim Hollabaugh*

United Service 2'x4' porcelain sign.

Albalube 11"x18" porcelain Australian sign.

Exide 14"x14" painted metal die-cut flange sign.

P ume Motor Spirit 14" porcelain foreign sign.

Gas Station 14"x3" porcelain sign.

Bendix 18" metal flange sign.

Castrol Motor Oil 12"x14" porcelain pump foreign sign.

(1) 1946-1966 $850

White outlined die-cut porcelain flying red horse with white details, facing left. Reverse channel condtruction with separate pieces for legs and tail.

MULTI PIECE HORSE FACING RIGHT 48x60

(1) 1946-1966 $850

White outlined die-cut porcelain flying red horse with white details, facing right. Reverse channel condtruction with separate pieces for legs and tail.

SOCONY VACUUM PRODUCTS WT/RED PANEL 12x96

(4) 1930s $250

Red outlined white self-framed embossed porcelain strip with red "SOCONY VACUUM PRODUCTS" across panel.

MOBILGAS WT/RED PANEL 10x60

(4) 1930s $250

Red self-framed porcelain strip with black outlined white "Mobilgas" across panel.

MOBILOIL WT/RED PANEL 10x60

(4) 1930s $250

Red self-framed porcelain strip with black outlined white "Mobiloil" across panel.

MOBILUBRICATION WT/RED PANEL 10x120

(4) 1930s $250

Red self-framed porcelain strip with black outlined white "Mobilubrication" across panel.

MOBILGAS BLUE PORCELAIN 12"SET

(3) 1946-1966 $200

Die-cut rounded embossed blue porcelain letter set spelling out "Mobilgas."

MOBILOIL BLUE PORCELAIN 12"SET

(3) 1946-1966 $200

Die-cut rounded embossed blue porcelain letter set spelling out "Mobiloil."

MOBILUBRICATION BLUE PORCELAIN 12"SET

(3) 1946-1966 $200

Die-cut rounded embossed blue porcelain letter set spelling out "Mobilubrication."

PUMP SIGNS

MOBILGAS ASK FOR MOBILOIL 10x15

(3) $300

White/blue outlined tin-litho die-cut sign. Upper area is rectangular with rounded corners, approximately 5"x10" and lower area is a 10" circle. Blue "Mobilgas" across upper rectangle with blue "Ask for" at top of circle above flying red horse. Blue "Mobiloil" across bottom of circle.

MOBILGAS SPECIAL ASK FOR MOBILOIL 10x15

(3) $300

White/blue outlined tin-litho die-cut sign. Upper area is rectangular with rounded corners, approximately 5"x10" and lower area is a 10" circle. Blue "Mobilgas" above red "SPECIAL" across upper rectangle with blue "Ask for" at top of circle above flying red horse. Blue "Mobiloil" across bottom of circle.

MOBILGAS 5x10

(4) $125

White tin litho sign with blue outline set in from edge. Red outline shield interrupts outline at upper right corner. Red flying horse in shield. Blue "Mobilgas" across center.

MOBILGAS SPECIAL 5x10

(4) $125

White tin litho sign with blue outline set in from edge. Red outline shield interrupts outline at upper right corner. Red flying horse in shield. Blue "Mobilgas" above red "SPECIAL" across center.

METRO 5x10

(4) $250

White tin litho sign with green outline set in from edge. Red outlined black "METRO" across center.

11"x12" BLUE BORDER SHIELDS

All of the following are die-cut five point shields with thick blue line border. Blue outlined red flying horse with white detailing at top of sign. Brandname, in sans-serif (Except Mobiloil) thick line lettering with round "a," as listed below appears across lower sign face below horse.

MOBILOIL 11x12

(1) 1947-1954 $250

Blue serif "Mobiloil" across bottom of shield.

BLANK WITH HORSE 11x12

(5) 1947-1954 $375

Flying red horse only, no writing across bottom of shield.

MOBILGAS 11x12

(1) 1947-1954 $125

Blue "Mobilgas" across bottom of shield.

MOBILGAS STUDDED 11x12

(5) 1947-1954 $200

Blue "Mobilgas" across bottom of shield. Studs welded to back of sign for mounting on pumps. No holes in sign.

MOBILGAS TWIN PUMP 7.5x7.5

(5) 1953-1962 $550

For use on two-product twin pumps. Blue "Mobilgas" across bottom of shield.

MOBILGAS SPECIAL 11x12

(1) 1947-1954 $125

Blue "Mobilgas" over large red "SPECIAL" across bottom of shield.

MOBILGAS SPECIAL STUDDED 11x12

(5) 1947-1954 $200

Blue "Mobilgas" over large red "SPECIAL" across bottom of shield. Studs welded to back of sign for mounting on pumps. No holes in sign.

MOBILGAS SPECIAL TWIN PUMP 7.5x7.5

(5) 1953-1962 $550

For use on two-product twin pumps. Blue "Mobilgas" above red "SPECIAL" across bottom of shield.

MOBILFUEL DIESEL 11x12

(1) 1947-1954 $250

Blue "Mobilfuel" over red "DIESEL" across bottom of shield.

MOBILGAS MARINE 11x12

(3) 1947-1954 $600

Blue "Mobilgas" over red "MARINE" across bottom of shield.

11"x12" WHITE/BLUE BORDER SHIELDS

All of the following are die-cut five point shields with a thin white outlined around a thick blue line border. Blue outlined red flying horse with white detailing at top of sign. Brandname, in sans-serif thick line lettering with round "a," as listed below appears across lower sign face below horse.

MOBILGAS 11x12

(2) 1954-1962 $125

Blue "Mobilgas" across bottom of shield.
MOBILGAS SPECIAL 11x12
(2) 1954-1962 $125
Blue "Mobilgas" over large red "SPECIAL" across bottom of sign.
MOBILFUEL DIESEL 11x12
(3) 1954-1962 $250
Blue "Mobilfuel" over red "DIESEL" across bottom of sign.
MOBILFUEL DIESEL SPECIAL 11x12
(4) 1954-1962 $350
Blue "Mobilfuel" over blue "DIESEL" over red "SPECIAL" across bottom of sign.
MOBILGAS MARINE 11x12
(5) 1954-1962 $600
Blue "Mobilgas" over red "MARINE" across bottom of shield.

12.5"x13" WHITE BORDER SHIELDS

All of the following are die-cut five point shields with white edging around a thin blue (except Metro) line border. Blue outlined red flying horse with white detailing at top of sign. Brandname, in sans-serif thin line lettering with round "a," as listed below appears across lower sign face below horse. Used primarily in the Magnolia Mobilgas territory.
MOBILGAS 12.5x13
(2) 1947-1962 $125
Blue "Mobilgas" across bottom of shield.
MOBILGAS SPECIAL 12.5x13
(3) 1947-1962 $125
Blue "Mobilgas" over red "SPECIAL" across bottom of shield.
METRO 12.5x13
(3) 1947-1962 $250
Green outline and green outlined horse. Green "Metro" across bottom of shield.
MOBILFUEL DIESEL 12.5x13
(3) 1947-1962 $250
Blue "Mobilfuel" over red "DIESEL" across bottom of shield.
MOBILGAS MARINE 12.5x13
(4) 1947-1962 $600
Blue "Mobilgas" over red "MARINE" across bottom of shield.
MOBILGAS AIRCRAFT 12.5x13
(4) 1947-1962 $1000
Blue "Mobilgas" over red "AIRCRAFT" across bottom of shield.
MOBILOIL AERO 12.5x13
(5) 1947-1962 $1100
Blue "Mobiloil" over red "AERO" across bottom of shield.

11"x12" BLACK BORDER SHIELDS

All of the following are die-cut five point shields with thick black border. The black border Mobilgas items were used primarily in the General Petroleum Company marketing territory on the west coast. Black outlined red flying horse with white detailing at top of sign. Brandname, in sans-serif thin line lettering with round "a," as listed below appears across lower sign face below horse. Used primarily in the General Mobilgas territory.
MOBILOIL 11x12
(2) 1947-1954 $225
Black outline/black serif letters "Mobiloil" across bottom of shield above red "SOCONY-VACUUM."
MOBILOIL SOCONY VACUUM SMALL 7.5x7.5
(5) $350
Black outline/black sans serif letters "Mobiloil" across bottom of shield above red "SOCONY-VACUUM."
MOBILOIL SOCONY VACUUM 11x12
(2) 1947-1954 $350
Black outline/black sans serif letters "Mobiloil" across bottom of shield above red "SOCONY-VACUUM."
MOBILGAS SOCONY VACUUM 11x12
(2) 1947-1954 $225
Black outline/black "Mobilgas" across bottom of shield above red "SOCONY-VACUUM."
MOBILGAS 11x12
(2) 1947-1954 $125
Black outline/black "Mobilgas" across bottom of shield above red "SOCONY-VACUUM."
MOBILGAS GASOLINE 11x12
(2) 1947-1954 $175
Black outline/black "Mobilgas" across bottom of shield above red "GASOLINE."
MOBILGAS SPECIAL 11x12
(2) 1947-1954 $125
Black outline/black "Mobilgas" across bottom of shield above large red "SPECIAL."
MOBILGAS SPECIAL GASOLINE 11x12
(2) 1947-1954 $175
Black outline/black "Mobilgas" across bottom of shield above large red "SPECIAL" over smaller red "GASOLINE."
MOBILFUEL DIESEL 11x12
(2) 1947-1954 $250
Black outline/black "Mobilfuel" across bottom of shield above red "DIESEL."
MOBILGAS AIRCRAFT 11x12
(5) 1947-1954 $1000
Black outline/black "Mobilgas" across bottom of shield above red "AIRCRAFT."
MOBILGAS MARINE 11x12
(5) 1947-1954 $600
Black outline/black "Mobilgas" across bottom of shield above red "MARINE."

11"x12" WHITE/BLUE BORDER SHIELDS

All of the following are die-cut five point shields with a thin white outline around a thick black line border. Black outlined red flying horse with white detailing at top of sign. Brandname, in sans-serif thick line lettering with round "a," as listed below appears across lower sign face below horse.
MOBILGAS 11x12
(2) 1954-1962 $125
Black outline/black "Mobilgas" across bottom of shield above red "SOCONY-VACUUM."
MOBILGAS GASOLINE 11x12
(2) 1954-1962 $175
Black outline/black "Mobilgas" across bottom of shield above red "GASOLINE."
MOBILGAS SPECIAL 11x12
(2) 1954-1962 $125
Black outline/black "Mobilgas" across bottom of shield above large red "SPECIAL."
MOBILGAS SPECIAL GASOLINE 11x12
(2) 1954-1962 $175
Black outline/black "Mobilgas" across bottom of shield above

large red "SPECIAL" over smaller red "GASOLINE."
MOBILFUEL DIESEL 11x12
(2) 1954-1958 $250
Black outline/black "Mobilfuel" across bottom of shield above red "DIESEL."
MOBILFUEL DIESEL SPECIAL 11x12
(4) 1954-1962 $350
Black outline/black "Mobilfuel" across bottom of shield above black "DIESEL" above red "SPECIAL."
MOBILGAS AIRCRAFT 11x12
(5) 1954-1962 $1000
Black outline/black "Mobilgas" across bottom of shield above red "AIRCRAFT."
MOBILGAS MARINE 11x12
(5) 1954-1962 $600
Black outline/black "Mobilgas" across bottom of shield above red "MARINE."
MOBIL DIESEL 11x12
(2) 1958-1962 $500
Black outline/black "Mobil" across bottom of shield above red "DIESEL."
MOBIL KEROSENE 11x12
(2) 1958-1962 $500
Black outline/black "Mobil" across bottom of shield above red "KEROSENE."

NEON

MOBILGAS WITH SMALL SHIELD 60"RD
(5) 1934-1940 $650
White outlined blue porcelain faces on can-type sign. White "Mobilgas" arched across top with red outlined white five point shield with red flying horse in shield.
MOBILGAS SOCONY VACUUM SHIELD 58x62
(4) 1938-1946 $650
Can type porcelain sign with neon outline, horse and "Mobilgas." Thick blue outline around a white porcelain five point shield with blue "Mobilgas" (serif "a") across center. Blue outlined flying red horse with white details at top and red "SOCONY-VACUUM" below "Mobilgas."
MOBILGAS SOCONY VACUUM SHIELD 58x62
(1) 1947-1962 $650
Can type porcelain sign with neon outline, horse and "Mobilgas." Thick blue outline around a white porcelain five point shield with blue "Mobilgas" (round "a") across center. Blue outlined flying red horse with white details at top and red "SOCONY-VACUUM" below "Mobilgas."
MOBILGAS/MOBILOIL HORSE 72x96
(5) 1935-1940 NO LISTING
Elaborate die-cut formed can type sign with wrap around two sided face. Upper section is flying red horse leaping to outside of sign structure. Base under horse is white with white outlined blue banner diagonal under white area. White "MOBILGAS" on one side, "MOBILOIL" on the other. Red outlined black panel forming horizontal base under banner.
PEGASUS VARIOUS SIZES
(3) 1935-PRESENT $1100
White outlined flying red horse with white details. These neon signs came in single-sided (fascia) and double-sided (rooftop) versions in many sizes by many different manufacturers. No attempt will be made to list the known varieties.

CURB SIGNS

MOBILGAS PRODUCT OF A SOCONY VACUUM COMPANY 30"RD
(4) 1932-1935 $375
White/red outlined white porcelain sign with black "Mobilgas" across center. Small flying red horse above, with red "PRODUCT OF A/SOCONY-VACUUM/COMPANY" below.
GARGOYLE MOBILOIL SOCONY VACUUM 24"RD $400
GARGOYLE MOBILOIL SOCONY VACUUM 30"RD $400
White porcelain sign with black "Mobiloil" in serif lettering across center. Small red "SOCONY-VACUUM" below with red and black gargoyle logo at top.
MOBILOIL SOCONY-VACUUM 24"RD $375
MOBILOIL SOCONY-VACUUM 30"RD $375
White porcelain sign with blue "Mobiloil" in serif lettering across center. Small red "SOCONY-VACUUM" below with flying red horse logo at top.
MOBILOIL MARINE 24"RD
(3) $650
White porcelain sign with blue "Mobiloil" in serif lettering across center. Small red "MARINE" below with flying red horse logo at top.
MOBILOIL 24"RD $450
MOBILOIL 30"RD $450
White porcelain sign with blue "Mobiloil" in serif lettering across center.
MOBILOIL DIE CUT WITH HORSE 30x36 $375
MOBILOIL DIE CUT WITH HORSE 42x48 $375
Blue outlined white porcelain sign with white circle at top above blue strip. Large flying red horse on circle with white "Mobiloil" across strip.

LUBRICATION/MOTOR OIL SIGNS

GARGOYLE MOBILOIL SHIELD 36x36
(5) 1932-1936 $450
Red outlined white porcelain five point shield with red and black Gargoyle logo at top and black "Mobiloil" across center.
GARGOYLE MOBILOIL SOCONY VACUUM 36x36
(5) 1936-1940 $375
Black outlined white porcelain five point shield with red and black Gargoyle logo at top and black "Mobiloil" across center.
MOBILOIL SERVICE SHIELD W/ HORSE 36x36
(5) 1954-1962 $375
Black outlined white porcelain five point shield with flying red horse with white details at top. Black sans serif "Mobiloil" across center with red script "Service" below.
GARGOYLE MARINE OILS SHIELD 42x42
(5) 1954-1962 $650
Black outlined white porcelain five point shield with flying red horse with white details at top. Black "Gargoyle/Marine Oils" below.
CERTIFIED MOBILUBRICATION BACKBAR 42x144
(5) 1940's $175
Large die-cut white tin litho sign with blue outline and blue flange trim panels. Small flying red horse at top over red script "Certified." Blue "MobiLubrication" logotype across center with red "SOCONY-VACUUM" in lower left corner.
MOBILUBRICATION BACKBAR 30x96
(4) 1950s $150
Blue outlined white self framed tin litho panel with blue

"MobiLubrication" across center. Small flying red horse at top.
MOBILOIL WORLD'S LARGEST SELLING MOTOR OIL 24x48
(3) 1950s $150
Red outlined white self-framed tin litho sign with blue "Mobiloil" across top above blue "WORLD'S LARGEST SELLING/MOTOR OIL" across bottom.
MOBILUBRICATION 10x26
(4) 1946-1950 $150
Red outlined white die-cut porcelain sign with blue "MobiLubrication" across center. Small flying red horse at top.

SECONDARY SIGNS

MOBIL TIRES VERTICAL PANEL 24x48
(3) 1946-1962 $175
Blue outlined white embossed tin litho sign. Blue "Mobil" across top, "Tires" across bottom with flying red horse logo in center.
SHIELD REST ROOM PLEDGE 7.5x7.5
(4) 1947-1962 $750
Die-cut five point shield with thick blue line border. Blue outlined red flying horse with white detailing at top of sign.
CLEAN REST ROOMS OCTAGON
(2) 1958-1966 $100
Three-piece sign. Each piece is white outlined blue porcelain flange sign designed to assemble one over the other. White "CLEAN," "REST," and "ROOMS" on individual signs.
SOCONY VACUUM CREDIT CARDS HONORED HERE W/B
$175
White porcelain sign with black "Socony Vacuum/CREDIT CARDS/Honored Here" across sign face.
SOCONY VACUUM CREDIT CARDS HONORED HERE R/W/B
$175
White porcelain sign with red band across top and blue band across bottom. Blue "CREDIT CARDS" in center with white "Socony-Vacuum" in upper red area and "Honored Here" in lower blue area.
FRIENDLY SERVICE TO MOTORISTS 48x72
(5) 1936-1940 $100
Black outlined red self-framed tin litho sign with white stripe across bottom. Red/white/black "Mobilgas" shield diagonal at left, with white banners with black "Mobilgas" and "Mobiloil" to right. Black "FRIENDLY SERVICE TO MOTORISTS" across white stripe.
MOBILOIL MARINE PANEL W/ WAVES 30x48
(4) 1954-1962 $450
White porcelain sign with blue "waves" that border the top and bottom. Small flying red horse at top with blue sans serif "Mobiloil" across center. Small red "MARINE" across bottom.

BULK PLANT SIGNAGE

MARINE PRODUCTS SOCONY VACUUM 11x11
(4) 1946-1950 $325
Die-cut five point white tin litho shield with blue line border. Blue outlined red flying horse with white detailing at top of sign. Blue "MARINE/PRODUCTS" above red "SOCONY-VACUUM" below horse.
MOBILHEAT SHIELD 13x14
(5) 1954-1962 $350
Die-cut five point white porcelain shield with blue line border. Blue outlined red flying horse with white detailing at top of sign. Blue "Mobilheat" below horse.
CORRECT FUEL OIL FOR HOME HEATING 6x14
(5) 1954-1962 $100
Attaches to bottom of Mobilheat shield. Red porcelain sign with white "CORRECT FUEL OIL/FOR HOME HEATING" across sign.
MOBIL-FLAME BOTTLE GAS SERVICE 12x26
(5) 1950-1962 $2000
White porcelain sign with blue "Mobil-flame" across sign above red "BOTTLED GAS SERVICE." Installed on hanging bracket with blue outlined red 24" horse above bracket.

TRUCK SIGNS

MOBILGAS ON RED PANEL 10x24
(4) 1947-1954 $650
Die-cut five point white porcelain shield with thick blue line border. Blue outlined red flying horse with white detailing at top of sign. Blue "Mobilgas," in sans-serif lettering with round "a" across lower sign face below horse. Red triangular extension to either side of shield for mounting on tank bodies at rear of truck.

ADVANCE SIGNS

JUST AHEAD MOBILGAS 48x60
(4) 1947-1958 $250
Blue self framed tin litho sign with white/blue outlined white five point shield in center. Flying red horse at top over blue "Mobilgas." Red "SOCONY-VACUUM" across bottom of shield. White "JUST AHEAD" above shield on blue background.

MOBIL

Although the Mobil name replaced Mobilgas as the primary station identification on a gradual basis between 1955 and 1958, gasoline brands remained Mobilgas and Mobilgas Special until 1962 when they were replaced with Mobil Regular and Mobil Premium. Mobil again reimaged in 1966, replacing the elongated Mobil shield with the modern "red O" logo.

IDENTIFICATION SIGNS

MOBIL
(1) 1958-1966 $150
Blue outlined white porcelain Mobil flat shield logo with red "V" forming lower outline. Small flying red horse at bottom below blue "Mobil." Blue porcelain upper frame/mounting bracket.
MOBIL 36x72
(4) 1958-1966 $150
MOBIL 48x96
(1) 1958-1966 $100
Blue outlined white porcelain Mobil flat shield logo with red "V" forming lower outline. Small flying red horse at bottom below blue "Mobil." Faces assembled back-to-back can-style over aluminum frame.
MOBIL FRP LARGE
(1) 1966-1980 $75
MOBIL FRP SMALL
(1) 1966-1980 $50
White fiberglass composite rectangular with rounded corners sign with white aluminum frame across top. Blue "M" and "bil" with red "o" across sign face.

MOBIL FRP FLAT SHIELD
(5) 1966-1980 $75
White fiberglass composite flat shield shaped sign with white aluminum frame across top. Blue "M" and "bil" with red "o" across sign face.

BUILDING SIGNS & LETTERS

PEGASUS ON WHITE CIRCLE PROTOTYPE 42"RD
(5) 1964 EXP NO LISTING
Formed white porcelain channel type sign with flying red horse, facing right, in center. This sign is the prototype for the plastic internally illuminated sign of this same design. It was used only in experimental locations on the "Pegasus" style station introduced in 1964.

PUMP SIGNS

MOBIL REGULAR 12x14
(4) 1962-1966 $50
Flat white porcelain sign with red and blue "Mobil" elongated shield across upper sign. Blue "Regular" across below logo.

MOBIL PREMIUM 12x14
(4) 1962-1966 $50
Flat white porcelain sign with red and blue "Mobil" elongated shield across upper sign. Red "Premium" across below logo.

MOBILFUEL DIESEL 12x14
(4) 1962-1966 $75
Flat white porcelain sign with red and blue "Mobil" elongated shield across upper sign. Blue "Mobilfuel/Diesel" across below logo.

MOBIL REGULAR STUDDED 12x14
(4) 1962-1966 $65
Embossed studded white porcelain sign with red and blue "Mobil" elongated shield across upper sign. Blue "Regular" across below logo.

MOBIL PREMIUM STUDDED 12x14
(4) 1962-1966 $65
Embossed studded white porcelain sign with red and blue "Mobil" elongated shield across upper sign. Red "Premium" across below logo.

MOBILFUEL DIESEL STUDDED 12x14
(4) 1962-1966 $85
Embossed studded white porcelain sign with red and blue "Mobil" elongated shield across upper sign. Blue "Mobilfuel/Diesel" across below logo.

MOBIL AVGAS STUDDED 12x14
(4) 1962-1966 $250
Embossed studded white porcelain sign with red and blue "Mobil" elongated shield across upper sign. Blue "Avgas" below logo with two holes drilled below for installation of grade strips.

MOBIL REGULAR TWIN PUMP STUDDED
(4) 1962-1966 $200
Embossed studded white porcelain sign with red and blue "Mobil" elongated shield across upper sign. Blue "Regular" across below logo.

MOBIL PREMIUM TWIN PUMP STUDDED
(4) 1962-1966 $200
Embossed studded white porcelain sign with red and blue "Mobil" elongated shield across upper sign. Red "Premium" across below logo.

MOBIL REGULAR 12x14
(1) 1966-1975 $50
Flat white porcelain sign with blue "Mobil" logotype (red "o") above blue "Regular" across sign.

MOBIL PREMIUM 12x14
(1) 1966-1975 $50
Flat white porcelain sign with blue "Mobil" logotype (red "o") above red "Premium" across sign.

MOBIL DIESEL FUEL 12x14
(2) 1966-1975 $75
Flat white porcelain sign with blue "Mobil" logotype (red "o") above black "Diesel Fuel" across sign.

MOBIL KEROSENE 12x14
(2) 1966-1975 $75
Flat white porcelain sign with blue "Mobil" logotype (red "o") above black "Kerosene" across sign.

MOBIL REGULAR STUDDED 12x14
(1) 1966-1975 $50
Embossed studded white porcelain sign with blue "Mobil" logotype (red "o") above blue "Regular" across sign.

MOBIL PREMIUM STUDDED 12x14
(1) 1966-1975 $50
Embossed studded white porcelain sign with blue "Mobil" logotype (red "o") above red "Premium" across sign.

MOBIL DIESEL FUEL STUDDED 12x14
(2) 1966-1975 $75
Embossed studded white porcelain sign with blue "Mobil" logotype (red "o") above black "Diesel Fuel" across sign.

MOBIL KEROSENE STUDDED 12x14
(2) 1966-1975 $75
Embossed studded white porcelain sign with blue "Mobil" logotype (red "o") above black "Kerosene" across sign.

MOBIL MARINE WHITE STUDDED 12x14
(2) 1966-1975 $250
Flat white porcelain sign with blue "Mobil" logotype (red "o") above small black "MARINE" over black outlined white "White" across sign.

REGULAR STUDDED 12x14
(5) 1971-1975 $100
Embossed studded white porcelain sign with red "Regular" across center. Used at some Mobil-affiliated secondary brand stations such as "Reelo" and "Sello."

PREMIUM STUDDED 12x14
(5) 1971-1975 $100
Embossed studded white porcelain sign with blue "Premium" across center. Used at some Mobil-affiliated secondary brand stations such as "Reelo" and "Sello."

CURB SIGNS

MOBILOIL
(2) 1958-1966 $200
White rectangular porcelain sign with red "V" design from identification sign across bottom of sign. Blue "Mobiloil" across sign face above small flying red horse.

MOBILUBRICATION
(4) 1958-1966 $200
White rectangular porcelain sign with red "V" design from identification sign across bottom of sign. Blue "Mobilubrication" across sign face above small flying red horse.

MOBIL SERVICE
(4) 1958-1966 $200

White rectangular porcelain sign with red "V" design from identification sign across bottom of sign. Blue "Mobil Service" across sign face above small flying red horse.

SECONDARY SIGNAGE

MOBIL CREDIT CARDS HONORED
(4) 1958-1966 $100
Red outlined white tin litho sign with blue "CREDIT CARDS" across center. Red "Mobil" across top and "HONORED" across bottom.

BULK PLANT SIGNAGE

MOBIL PRODUCTS 16x96
(4) 1958-1966 $100
White/red outlined white porcelain strip with red/white/blue Mobil elongated shield logo at left. Blue "products" across remainder of sign.

MOBIL DISTRIBUTOR STRIP 16x96 $100
White porcelain strip sign with red/white/blue "Mobil" flat shield logo beside blue "distributor" across strip.

MOBIL DISTRIBUTOR SQUARE 24x36
(4) 1958-1966 $100
White porcelain sign with red/white/blue "Mobil" flat shield logo above blue "distributor."

ADVANCE SIGNS

MOBIL ONE MILE 48x96
(5) 1958-1966 $100
White tin litho sign with blue strip across bottom. Red/white/blue "Mobil" flat shield logo on white area with white "1 MILE" on blue strip below.

MOBIL AHEAD 42x84
(5) 1966-1980 $509
Black tin litho sign with white rectangle across top. Blue "M" and "bil" with red "o" across rectangle with white "ahead" below logo.

PENNZOIL AND AFFILIATES

PENNZOIL COMPANY - HOUSTON TX/OIL CITY, PA

Pennzoil traces it's orgins to the founding of South Penn Oil Company, a production unit of the Standard Oil Trust. When South Penn became independent with the breakup of Standard in 1911 they remained primarily a production operation. In 1925 they purchased control of The Pennzoil Company, an Oil City refiner and lubricants marketer that had in 1921 opened the first of a small chain of branded Pennzoil staions. The gasoline marketing was rebranded "Pennzip" in 1936 and remained such until 1958 when the Pennzoil brand was reintroduced for gasoline. South Penn purchased the remaining interest in Pennzoil in 1955 and the company was renamed Pennzoil Company in 1963. Numerous purchases detailed below greatly expanded the scope of the company and today the company sells gasoline through 600+ stations in ten states as well as marketing Pennzoil lubricants around the world.

PENNZOIL LUBRICANTS MARKETING:

CURB SIGNS

100% SUPREME PENNSYLVANIA PENNZOIL 24"RD
(4) 1920s $550
Tan porcelain sign with gold bell in cetner. Black outlined red "PENNZOIL" diagonal across sign with black "100% Supreme Pennsylvania Quality" arched around top, "SAFE LUBRICATION" around bottom.

SUPREME PENNSYLVANIA PENNZOIL 24"RD
(4) 1920s $500
Tan porcelain sign with gold bell in cetner. Black outlined red "PENNZOIL" diagonal across sign with black "Supreme Pennsylvania Quality" arched around top, "SAFE LUBRICATION" around bottom. Modified Pennsylvania seal below bell.

SUPREME PENNSYLVANIA PENNZOIL OVAL 15x30
(4) 1928-1930 $450
Black pipe framed yellow porcelain oval sign with black outlined white bell in center. Red outlined black "PENNZOIL" across center, with small black "Supreme Pennsylvania Quality" arched across top, "Safe Lubrication" across bottom.

100% PURE PENNSYLVANIA PENNZOIL OVAL 15x30
(3) 1930-1936 $475
Black outlined yellow porcelain oval sign with red bell in center. Red outlined black "PENNZOIL" across center, with small black "100% Pure Pennsylvania" arched across top, "Safe Lubrication" across bottom.

BONDED PENNZOIL DEALER OVAL 16x31
(1) 1936-1950 $425
Yellow tin litho oval with yellow tab around top and red bell in center. Black "PENNZOIL" diagonal across center with black "100% Pure Pennsylvania" arched around top, "Safe Lubrication" around bottom. Black "BONDED PENNZOIL DEALER" in tab area.

SOUND YOUR Z PENNZOIL OVAL 16x31
(1) 1950-1970 $325
Yellow tin litho oval with yellow tab around top and red bell in center. Black "PENNZOIL" diagonal across center with black "100% Pure Pennsylvania" arched around top, "Safe Lubrication" around bottom. Black "SOUND YOUR Z" in tab area.

100% PURE PENNSYLVANIA PENNZOIL OVAL 14x31
(1) 1970-1975 $100
Yellow tin litho oval with red bell in center. Black "PENNZOIL" diagonal across center with black "100% Pure Pennsylvania" arched around top, "Safe Lubrication" around bottom.

SUPREME QUALITY PENNZOIL OVAL 14x31
(1) 1970-1975 $100
Yellow tin litho oval with red bell in center. Black "PENNZOIL" diagonal across center with black "Supreme Quality" arched around top, "Safe Lubrication" around bottom.

LUBRICATION/MOTOR OIL SIGNS

SUPREME PENNSYLVANIA QUALITY PENNZOIL 24"
(5) 1920s $850
Tan porcelain flange sign with gold bell in cetner. Black outlined red "PENNZOIL" diagonal across sign with black "Supreme Pennsylvania Quality" arched around top, "SAFE LUBRICATION" around bottom. Modified Pennsylvania seal below bell.

SOUND YOUR Z PENNZOIL FLANGE 14x24
(5) 1930s $275
Black outlined yellow porcelain flange oval with yellow tab around top and red bell in center. Black flange area. Black "PENNZOIL" diagonal across center with black "100% Pure Pennsylvania" arched around top, "Safe Lubrication" around bottom. Black "SOUND YOUR Z" in tab area.

PENNZOIL SUPREME PENNSYLVANIA QUALTIY SAFE LUBRICATION 12x36 $250
Black outlined yellow tin litho sign with brown bell in center. Black outlined red "PENNZOIL" diagonal across bell, with black "SUPREME/PENNSYLVANIA/QUALITY" in upper left corner, "SAFE/LUBRICATION" in lower right corner.

PENNZOIL SUPREME PENNSYLVANIA QUALTIY SAFE LUBRICATION 24x48 $250
Black outlined yellow porcelain sign with red bell in center. Black outlined red "PENNZOIL" diagonal across bell, with black "SUPREME/PENNSYLVANIA/QUALITY" in upper left corner, "SAFE/LUBRICATION" in lower right corner.

PENNZOIL SUPREME PENNSYLVANIA QUALTIY SAFE LUBRICATION 36x60 $225
Black outlined yellow tin litho sign with red bell in center. Black outlined red "PENNZOIL" diagonal across bell, with black "SUPREME/PENNSYLVANIA/QUALITY" in upper left corner, "SAFE/LUBRICATION" in lower right corner. Small PGCOA seal in upper right corner.

PENNZOIL 100% SUPREME PENNSYLVANIA QUALTIY SAFE LUBRICATION 24x48 $225
Black outlined yellow porcelain sign with red bell in center. Black outlined red "PENNZOIL" diagonal across bell, with black "100% SUPREME/PENNSYLVANIA/QUALITY" in upper left corner, "SAFE/LUBRICATION" in lower right corner.

BONDED PENNZOIL DEALER 12.25x16
(4) $325
Yellow outlined black tin litho sign with yellow oval logo with arched banner above logo. Yellow oval has red bell in center. Black "PENNZOIL" diagonal across center with black "100% Pure Pennsylvania" arched around top, "Safe Lubrication" around bottom. Black "BONDED PENNZOIL DEALER" in banner area.

WE ARE BONDED FOR $1000... TIN 10.5x13.5
(4) $300
Yellow outlined black tin litho sign with red "WE ARE BONDED/FOR $1000" across upper sign face. Small Pennzoil curb sign pictured at right, with other writing on sign.

PENNZOIL VERT STRIP W/LOGO AT TOP 12x60
$225
Black outlined yellow self framed tin litho sign with black band across top. Yellow Pennzoil oval on sunburst on black band across top with red outlined black "PENNZOIL" vertical down sign.

PENNZOIL SOUND YOUR Z VERT STRIP 11x59
$225
Yellow porcelain sign with black band across top and bottom. Yellow "PENNZOIL" on upper band, with yellow "BONDED PENNZOIL DEALER" logo on lower band. Small black "SOUND/YOUR" across upper yellow area above large red "Z" with black drop shading.

PENNZOIL STRIP WITH BONDED LOGO 12x70
$225
Red self framed tin litho sign with black oval at right. Yellow "BONDED PENNZOIL DEALER" logo in black area. Yellow "PENNZOIL" across sign.

PENNZOIL W/SOUND YOUR Z LOGO 12x70
$225
Red self framed tin litho sign with black oval at right. Yellow "SOUND YOUR Z" logo in black area. Yellow "PENNZOIL" across sign.

100% PURE PENNSYLVANIA PENNZOIL OVAL 35x62
(1) 1970-1975 $100
Yellow porcelain oval with red bell in center. Black "PENNZOIL" diagonal across center with black "100% Pure Pennsylvania" arched around top, "Safe Lubrication" around bottom.

PENNZOIL SAFE LUBRICATION 18x72
(4) 1950s $225
Red self framed tin litho sign with black square in lower left corner. Yellow "PENNZOIL/LUBRICATION" across sign with yellow script safe in black square.

OIL RACK SIGNS

BONDED PENNZOIL DEALER OVAL 12x16.5
(1) 1936-1950 $150
Yellow tin litho oval with yellow tab around top and red bell in center. Black "PENNZOIL" diagonal across center with black "100% Pure Pennsylvania" arched around top, "Safe Lubrication" around bottom. Black "BONDED PENNZOIL DEALER" in tab area.

SOUND YOUR Z PENNZOIL OVAL 12x16.5
(1) 1950-1970 $150
Yellow tin litho oval with yellow tab around top and red bell in center. Black "PENNZOIL" diagonal across center with black "100% Pure Pennsylvania" arched around top, "Safe Lubrication" around bottom. Black "SOUND YOUR Z" in tab area.

100% PURE PENNSYLVANIA PENNZOIL OVAL 12x16.5
(1) 1970-1975 $150
Yellow tin litho oval with red bell in center. Black "PENNZOIL" diagonal across center with black "100% Pure Pennsylvania" arched around top, "Safe Lubrication" around bottom.

SUPREME QUALITY PENNZOIL OVAL 12x16.5
(1) 1975-1980 $150
Yellow tin litho oval with red bell in center. Black "PENNZOIL" diagonal across center with black "Supreme Quality" arched around top, "Safe Lubrication" around bottom.

OUTBOARD MOTOR OIL PENNZOIL 12x16.5
(3) 1950-1970 $225
Yellow tin litho oval with yellow tab around top and red bell in center. Black "PENNZOIL" diagonal across center with black "100% Pure Pennsylvania" arched around top, "Safe Lubrication" around bottom. Black "OUTBOARD MOTOR OIL" in tab area.

LUBESTER SIGNS

PENNZOIL 2.5x8.5
(5) 1920s $275
Black outlined yellow porcelain sign with red bell in center. Black "PENNZOIL" diagonal across bell.

TRUCK SIGNS

PENNZOIL
$450
Series of black and red outline rings around a yellow porcelai oval sign. Red bell in center with black "PENNZOIL" diagonal across center.

Smith-Wolf Oil Co., Utica, Ohio metal sign.

U.S. Royal Tires embossed metal sign. *Jim Hollabaugh.*

Goodrich Silvertowns porcelain flange sign.

Goodyear 20"x60" porcelain sign. *Jim Hollabaugh.*

Chevrolet 9' neon sign.

Universal Batteries 60"x12" wood-framed metal sign.

Early Miller Cords 5'x14" wood-framed metal sign.

Unusual Gas-Oil-Lube and Coke 8'x15" sign. *Dunbar's Gallery.*

Bear Wheel Alignment die-cut metal sign.

Hudson large neon sign.

Zenith Carburetor porcelain sign.

Goodrich Truck Tires 24"x30" porcelain sign.

PENNZOIL/PENNZIP GASOLINE MARKETING:

Pennzoil entered gasoline marketing in 1921, operating stations in Pennsylvania and New York. Signage in this era did not usually refer to gasoline, specifically, but emphesized lubricants even though it was used for service station identification. In 1936 the marketing operation was rebranded Pennzip. Expansion of gasoline marketing was stagnent until a 1958 revitalization where the Pennzoil brandname was readopted. They continue to operate stations and convenience stores under the Pennzoil brand through jobber owned outlets in 10 states.

IDENTIFICATION SIGNS

PENNZOIL 23x24
(5) 1921-1926 $375
Brown porcelain die-cut bell with black details and black outlined cream "PENNZOIL" diagonal across center above black outlined red "SAFE/LUBRICATION." Black outlined cream "SUPREME PENNSYLVANIA QUALITY" arched around bottom.

PENNZOIL 48"RD
(4) 1926-1936 $375
Tan porcelain sign with gold bell in cetner. Black outlined red "PENNZOIL" diagonal across sign with black "Supreme Pennsylvania Quality" arched around top, "SAFE LUBRICATION" around bottom.

PENNZIP-A PENNZOIL PRODUCT 48"RD
(5) 1936-1957 $350
Tan porcelain sign with black outlined red "PENNZIP" (oversize "Z") diagonal across sign. Black "A Pennzoil Product" arched around top, "The (interrupted by "Z") Gasoline" arched around bottom.

PENNZOIL OVAL 42x96
(1) 1957-1990 $225
Black/white outlined yellow oval porcelain sign with black outlined white center oval. Black outlined red "PENNZOIL" across center.

NEON SIGNS

SAFE PENNZOIL LUBRICATION BELL SHAPED 72x96
(5) 1930-1936 $450
Elaborate die-cut porcelain can type bell shaped sign with band extending beyond edges across center. Gold bell with black outlined yellow band. Red "PENNZOIL" across band. Yellow "SAFE" across top, "LUBRICATION" across bottom.

CURB SIGNS

PENNZOIL ETHYL GASOLINE 30"RD
(5) 1926-1936 $450
Red porcelain sign with white center circle. Large Ethyl (EGC) logo in center with black "PENNZOIL" arched around top, "GASOLINE" around bottom.

SECONDARY SIGNS

PENNZIP A PENNZOIL PRODUCT 30x60
(5) 1940s $400
Black outlined yellow tin litho sign with red outlined black "PENNZOIL" across top over "GASOLINE/A PENNZOIL PRODUCT."

WOLVERINE-EMPIRE REFINING CO. - OIL CITY, PA

Founded in 1879 as Empire Oil Works, Wolverine-Empire was primarily a lubricants manufacturer. They entered the gasoline market under the Empire brand about 1928 and rebranded Wolf's Head, after their famous motor oils, about 1940. In 1963 Pennzoil purchased Wolf's Head and continued to operate the Wolf's Head stations as a secondary brand into the 1980s.

IDENTIFICATION SIGNS

WOLF'S HEAD OIL OVAL 24x42
(5) 1940s $275
Red outlined white oval with small red wolf's head at tp center of sign. Red "WOLF'S HEAD/OIL" across center with small "WOLVERINE-EMPIRE REFINING COMPANY" arched around bottom.

CURB SIGNS

PENN EMPIRE ETHYL GASOLINE 30"rd
(5) 1930S $375
Black outlined orange porcelainsign with black outlined white center circle. White outlined black "PENN-EMPIRE" arched around top, "GASOLIE" around bottom with Ethyl (EGC) logo in center.

WE SELL WOLF'S HEAD 23x30
(1) 1949-1979 $150
White tin litho sign with red area across top and green area across bottom forming white band across center. Black "WOLF'S/HEAD/Motor Oil" on center band with white "We Sell" in upper red area and white "100% Pure/Pennsylvania" in lower green area beside red outlined white Wolf's Head oval logo.

ASK FOR WOLF'S HEAD 23x30
(1) 1949-1979 $150
White tin litho sign with red area across top and green area across bottom forming white band across center. Black "WOLF'S/HEAD/Motor Oil" on center band with white "Ask for" in upper red area and white "100% Pure/Pennsylvania" in lower green area beside red outlined white Wolf's Head oval logo.

ASK FOR WOLF'S HEAD OIL 36"
(2) 1957-1960 $100
White tin litho sign with red area across top and green area across bottom forming white band across center. Black "WOLF'S/HEAD" on center band with white "Ask for" in upper red area and white "OIL" in lower green area.

100% PENNSYLVANIA MOTOR OIL WOLF'S HEAD 23x30
(1) 1972-1979 $50
White tin litho vertical oval sign with red wolf's head sillouette interrupting diagonal black band at top. Red "100% PENNSYLVANIA MOTOR OIL" across center above black "WOLF'S/HEAD." FINEST OF THE FINE SINCE 1879 WOLF'S HEAD 23x30
(1) 1979-PRESENT $50
White tin litho vertical oval sign with red wolf's head sillouette interrupting diagonal black band at top. Red "FINEST OF THE FINE SINCE 1879" across center above black "WOLF'S/HEAD." Small red "MOTOR OIL" across bottom.

LUBRICATION/MOTOR OIL SIGNS

WOLF'S HEAD OIL DETAILED WOLF 20SQ
(5) 1915-1920 $2500
Green porcelain sign with black outlined white circle around black outlined yellow circle in center. Grey, black, white, red

detailed wolf's head in center with red "WOLF'S HEAD" arched around top, "OIL" around bottom.

A REVELATION IN LUBRICATION 20x30
(3) 1920s $1200
Green porcelain sign with red outlined white oval covering most of center. White outlined red "A REVELATION IN LUBRICATION" across top with red "WOLF'S HEAD/OIL" in center of oval. Red wolf's head in center between "WOLF'S" and "HEAD." Small red "WOLVERINE LUBRICANTS COMPANY" arched around bottom.

ASK FOR WOLF'S HEAD 12x72
(2) 1949-1979 $125
White tin litho vertical strip sign with red area across top and green area across bottom forming white band across center. Black "WOLF'S/HEAD" vertical on stirp above "Motor Oil." White "Ask for" in upper red area and white "100% Pure/Pennsylvania" in lower green area beside red outlined white Wolf's Head oval logo.

FLANGE SIGNS

ASK FOR WOLF'S HEAD 17x22
(2) 1949-1979 $125
White tin litho sign with red area across top and green area across bottom forming white band across center. Black "WOLF'S/HEAD/Motor Oil" on center band with white "Ask for" in upper red area and white "100% Pure/Pennsylvania" in lower green area beside red outlined white Wolf's Head oval logo.

ASK FOR WOLF'S HEAD OIL & LUBES 17x22
(2) 1949-1979 $125
White tin litho sign with red area across top and green area across bottom forming white band across center. Black "WOLF'S/HEAD/Oils & Lubes" on center band with white "Ask for" in upper red area and white "100% Pure/Pennsylvania" in lower green area beside red outlined white Wolf's Head oval logo.

100% PENNSYLVANIA MOTOR OIL WOLF'S HEAD 17x22
1972-1979 $100
White tin litho vertical oval sign with red wolf's head sillouette interrupting diagonal black band at top. Red "100% PENNSYLVANIA MOTOR OIL" across center above black "WOLF'S/HEAD."

FINEST OF THE FINE SINCE 1879 WOLF'S HEAD 17x22
1979-PRESENT $75
White tin litho vertical oval sign with red wolf's head sillouette interrupting diagonal black band at top. Red "FINEST OF THE FINE SINCE 1879" across center above black "WOLF'S/HEAD." Small red "MOTOR OIL" across bottom.

ELK REFINING CO. - CHARLESTON, WV

Refiner/marketer that operated the Keystone stations in West Virginia and Pennsylvania. South Penn purchased control of Elk in 1952 and the remaining portion in 1963. South Penn/Pennzoil sold off the Keystone marketing to United Refining in the 1970s. United, an independent listed below, continues to operate a few stations under the Keystone brand.

IDENTIFICATION SIGNS

KEYSTONE DIE CUT 42x84
(5) 1946-1960 $175
Yellow/blue outlined red die cut porcelain keystone shaped sign with yellow band across center and extending to either side. Blue "KEYSTONE" on yellow band.

VITALIZED KEYSTONE GASOLINE 36x60
(5) 1930-1946 $175
Yellow porcelain sign with blue band and red line across top and bottom of sign. Large red keystone in center with blue "KEYSTONE" across center over keystone. Yellow "Vitalized" across top, "GASOLINE" across bottom.

PUMP SIGNS

KEYSTONE POWERFUEL MORE POWER TO YOU 12x14
(4) 1946-1965 $100
Yellow outlined blue porcelain sign with yellow outlined red and blue logo with yellow "KEYSTONE" in upper left corner. Yellow "POWERFUEL" with speed lines diagonal across center with yellow "MORE/POWER/TO YOU!" in lower right corner.

KEYSTONE POWERFUEL ETHYL 12x14
(3) 1946-1955 $100
Yellow outlined red porcelain sign with yellow outlined red and blue logo with yellow "KEYSTONE" in upper left corner. Yellow "POWERFUEL" with speed lines diagonal across center with Ethyl
(EC) logo with yellow rays in lower right corner.

UNITED REFINING CO. - WARREN, PA

Independent refiner/marketer that purchased the Keystone stations from Pennzoil in the 1970s. Many of the former Keystone locations were converted to United's "Kwik-Fill" brand, however, as of this writing a few Keystone branded stations remain.

IDENTIFICATION SIGNS

UNITED 42"
$275
Blue outlined white porcelain sign with series of red concentric circles around red center circle. White outlined blue band across center with white "UNITED" on band.

PUMP SIGNS

UNITED HIGH SPIRIT 10"RD
(5) 1935-1950 $250
Blue porcelain sign with series of red concentric circle around red center circle. White "UNITED" arched around top, "HI-SPIRIT" around bottom.

UNITED ETHYL 10"RD
(5) 1935-1950 $150
Blue porcelain sign with blue outlined white center circle. Large Ethyl
(EGC) logo in center with white "UNITED" arched around top, "ETHYL" around bottom.

CURB SIGNS

CHASES CARBON UNITED SOLVENIZED GASOLINE 30"RD
(5) 1930-1935 $550
Yellow porcelain sign with red band across center. United in white with grey dropshading across band. Black "CHASES CARBON" across top, with grey and black solvenized logo to

right with speed lines. Black "SOLVENIZED GASOLINE" arched around bottom.

UNITED ETHYL 30"RD
(5) 1926-1935 $325
Blue outlined cream color porcelain sign with blue "UNITED" arched across top over large Ethyl (EGC) logo at bottom.

FLEET WING CORPORATION - CLEVELAND, OH

Pennzoil purchased the Fleet Wing operation from Sohio in 1970 and as continues to operate a small number of Fleet-Wing branded stations today.

See Sohio section for Fleet-Wing sign listings.

PENNSYLVANIA REFINING COMPANY - BUTLER, PA

Founded in Butler in 1878; Pennsylvania Refining, later Penn-reco, marketed lubricants under the Penn-Drake brand. During the 1920s and 1930s the company branded a small number of Penn Drake stations in Pennsylvania. The company withdrew from gasoline marketing at the start of WW2 and later, in 1973, was purchased by Pennzoil Company. They continue to market a line of specialty oils today.

IDENTIFICATION SIGNS

PENN DRAKE 30x54
(5) 1940s $450
Oval porcelain sign with white on red comet & sunburst in center. "penn drake" across center."

CURB SIGNS

PENN DRAKE MOTOR OILS 20x28
(5) 1920s $450
Steel framed black outlined yellow tin litho sign with red and black PGCOA seal in upper left corner. Detailed black/tan/green Drake's well scene in cente with black outlined red "PENN-DRAKE" diagonal across center. Red outlined black "MOTOR OILS" across bottom.

MOTOR PENN-DRAKE OIL ORANGE/BLK 21x27
(3) 1930-1946 $450
Black outlined orange porcelain chevron shaped sign iwth white band across lower face. Black/white Drake's well scene on upper orange portion of sign with orange "MOTOR" above band, "OIL" below. Black PENN-DRAKE" across band. Small PGCOA seal in uppere left corner.

PENN DRAKE OIL 21x27
(2) 1946-197 $475
Black outlined white chevron shaped tin litho sign with black line above grey lines across top and red chevron line above grey chevron lines across bottom. Red/black/white Drake's well scene at top with black "penn/drake" below. Red "the double mileage" across center above black "oil." Black PG-COA seal in chevron point at bottom.

LUBRICATION/MOTOR OIL SIGNS

USE PENN DRAKE MOTOR OIL 10x28
(4) 1930s $250
Orange outlined black tin litho sign with white outlined orange and black oil well Penn Drake logo at left. White "USE" above white outlined orange "PENN-DRAKE/MOTOR OIL" across sign.

PENN DRAKE LUBRICANTS 10x14
(4) 1960-1970 $175
White tin flange sign with red band across top and bottom. Black and white Drake's well on black band across center. White "penn/drake" on black band with white "LUBRICANTS" across red band at bottom. White "the original drake well, 1859" in upper red band.

LUBESTER SIGNS

PENN DRAKE THE DOUBLE MILEAGE OIL 5x8
(4) 1930s $175
White outlined black chevron shaped tin litho sign with black line above grey lines across top and red chevron line above grey chevron lines across bottom. Red/black/white Drake's well scene at top with white "penn/drake" below. White outlined red "the double mileage" across center above white "oil." Black PGCOA seal in chevron point at bottom.

PHILLIPS 66

PHILLIPS PETROLEUM - BARTLESVILLE, OK Phillips Petroleum was founded in 1917 in Bartlesville, Oklahoma from various production and refining companies owned by Frank Phillips and others. The company entered gasoline marketing in November 1927 when they opened a small station in Wichita, Kansas. Other stations were quickly added and other marketing operations were purchased such that the company operated over 6000 stations in 12 states by 1930. Details on some of the companies purchased are listed in individual company histries, below. Phillips was primarily a mid-continent marketer until 1953 when they opened their first stations in Florida, setting off a marketing chain-reaction that saw them contracting with dissatisfied jobbers from other marketers and rebranding stations to Phillips 66 as far north as Virginia. Pleased with the market expansion, Phillip's "drive to Maine" expansion began, eventually pushing the brand all the way through the north Atlantic region and on to Maine. Never before had an oil company attempted such a systematic market expansion. With the purchase of Tidewater's west coast Flying A stations in 1966, Phillips found themselves operating stations in 49 states. A single unit in Alaska was opened in 1967, making Phillips the second marketer to brand stations in all 50 states at the same time (Texaco was first and Shell has been in all 50 states but not at the same time). The 1970s gas shortages forced Phillips to abandon many areas of the country, but currently the Phillips 66 brand appears on nearly 8000 stations in 29 mid-continent and southeastern states.

IDENTIFICATION SIGNS

PHILLIPS GASOLINE MOTOR OIL 42"RD
(5) 1927-1928 $750
White porcelain sign with bar and circle design across center. "PHILLIPS" across center bar and circle, with "GASOLINE" arched around top, "MOTOR OIL" arched around bottom.

PHILLIPS GASOLINE MOTOR OIL 30"RD
(3) 1928-1930 $350

PHILLIPS GASOLINE MOTOR OIL 48"RD
(4) 1928-1930 $350
Black/red outlined black porcelain sign with red "Phillips" logotype across center. White "GASOLINE" arched around top, "MOTOR OILS" around bottom.

PHILLIPS SHIELD 48"
(1) 1930-1959 $250
PHILLIPS SHIELD 72"
(1) 1946-1959 $200
Die cut red porcealin shield with black band across upper shield area. Red logotype "Phillips" on black band with black "66" slanted on lower red field below.
PHILLIPS SHIELD 48"
(1) 1959-1968 $125
PHILLIPS SHIELD 72"
(1) 1959-1968 $75
Black/white outlined die cut red porcelain shield with white band across upper shield area. Black "PHILLIPS" on white band with white "66" on lower red field below.
PHILLIPS SHIELD 48"
(2) 1968-1982 $100
PHILLIPS SHIELD 72"
(2) 1968-1982 $75
Black outlined die cut red porcelain shield with white band across upper shield area. Much wider white border around lower red area of sign. Black "PHILLIPS" on white band with white "66" on lower red field below.

BUILDING SIGNS & LETTERS

PHILLIPS PETROLEUM CO. STRIP 12x96
(4) 1929-1946 $325
Black self framed porcelain strip with gold "Phillips Petroleum Co." across center.
PHILLIPS CAST ALUMINUM LETTERS 18"
(2) 1946-1959 $300
Die cast aluminum channel lettering for facade or rooftop mounting. Painted maroon from 1946 through 1959, they were repainted red with the 1959 reimage. Continued in use until about 1990, although those installed after 1959 were plastic.

NEON

PHILLIPS SHIELD 48"
(2) 1936-1959 $750
Die cut embossed red porcealin shield with black band across upper shield area. Red logotype "Phillips" on black band with black "66" slanted on lower red field below. Outer neon border is green, with red neon "66."
PHILLIPS SHIELD 72"
(N/A) 1946-1959 Not Listed
Die cut red porcealin shield with black band across upper shield area. Red logotype "Phillips" on black band with black "66" slanted on lower red field below. Green neon tubing added around sign perimenter.

CURB SIGNS

PHILLIPS 66 SHIELD 30"RD
(4) 1930-1935 $400
White porcelain sign with red shield with black band across upper shield area covering most of sign. Red logotype "Phillips" on black band with black "66" slanted on lower red field below.
PHILLIPS 66 SHIELD 30"
(1) 1935-1940 $425
Die cut red porcealin shield with black band across upper shield area. Red logotype "Phillips" on black band with white outlined black "66" slanted on lower red field below. Possibly experimental.
PHILLIPS 66 SHIELD 30"
(1) 1935-1956 $425
Die cut red porcealin shield with black band across upper shield area. Red logotype "Phillips" on black band with black "66" slanted on lower red field below.
PHILLIPS 66 SHIELD WHITE BORDER 30"
(1) 1956-1959 $400
Die cut red porcealin shield with black band across upper shield area. Thick white outline around entire shield. Red logotype "Phillips" on black band with black "66" slanted on lower red field below.
PHILLIPS ETHYL 66 30"RD
(4) 1928-1930 $450
White porcelain sign with black "Phillips" logotype lettering arched around top, "66" with scrollwork at bottom and Ethyl (EGC) logo with yellow rays in center.
PHILLIPS 66 ETHYL SHIELD 30"RD
(4) 1930-1935 $475
White porcelain sign with red shield with black band across upper shield area covering most of sign. Red logotype "Phillips" on black band with black "66" slanted on lower red field below.
PHILLIPS 66 ETHYL 30"
(3) 1935-1946 $450
Die cut red porcealin shield with black band across upper shield area. Red logotype "Phillips" on black band with black "66" slanted on lower red field below. Ethyl (EGC) logo superimposed over "66" on lower sign face.
PHILLIPS MOTOR OIL SHIELD 30"
(1) 1935-1956 $400
Die cut red porcealin shield with black band across upper shield area. Red logotype "Phillips" on black band with black "MOTOR/OIL" slanted on lower red field below.
PHILLIPS MOTOR OILS SHIELD 30"
(1) 1935-1956 $400
Die cut red porcealin shield with black band across upper shield area. Red logotype "Phillips" on black band with black "MOTOR/OILS" slanted on lower red field below.

LUBRICATION/MOTOR OIL SIGNS

PHILLIPS MOTOR OILS 10x24
(5) 1928-1930 $250
Black tin litho sign with red "Phillips" logotype across upper sign area. Red "MOTOR OILS" below with red line across botom of sign.
PHILLIPS 66 MOTOR OIL LUBESTER SIGN 20"RD
(4) 1935-1940 $300
White porcelain sign with red shield with black band across upper shield area covering most of sign. Red logotype "Phillips" on black band with black "66" slanted on lower red field below. Black "MOTOR OILS" below shield.
PHILLIPS 66 MOTOR OILS 16x28
(4) 1946-1950 $200
Self framed tin tan litho sign with red/black Phillips 66 shield at top above black folded ribbon design. White "MOTOR/OILS" on ribbon.

OIL RACK SIGNS

THE WORLD'S FINEST OIL FOR YOUR MOTOR 12x24

(3) 1936-1940 $450
Green porcelain sign wht red stripe across top and bottom. Phillips 66 shield to left, with black script "The" over "WORLD'S/FINEST/OIL" over script "for your motor.

PHILLIPS 66 MOTOR OIL
(CAN) 12x24
(3) 1940-1946 $200
Red porcelain sign with white circle at left. Phillips 66 motor oil can in circle with black script "The" over black outlined white "SHIELD/OF/PROTECTION."

SECONDARY SIGNS

PHILLIPS 66 TIRES 12x60
(3) 1946-1959 $100
Self framed tin tan litho sign with maroon border and red/black Phillips 66 shield at top above vertical maroon thin lettering "TIRES." Maroon band across bottom with tan "SERVICE" in band.

PHILLIPS 66 TIRES 12x60
(3) 1946-1959 $100
Self framed tin tan litho sign with maroon border and red/black Phillips 66 shield at top above vertical maroon thick lettering "TIRES." Maroon band across bottom with tan "SERVICE" in band.

PHILLIPS 66 BATTERIES 12x60
(3) 1946-1959 $100
Self framed tin tan litho sign with maroon border and red/black Phillips 66 shield at top above vertical maroon thin lettering "BATTERIES." Maroon band across bottom with tan "SERVICE" in band.

PHILLIPS 66 BATTERIES 12x60
(3) 1946-1959 $100
Self framed tin tan litho sign with maroon border and red/black Phillips 66 shield at top above vertical maroon thick lettering "BATTERIES." Maroon band across bottom with tan "SERVICE" in band.

CERTIFIED CLEAN REST ROOMS 10x18
(4) 1935-1940 $125
Tan tin litho flange sign with Black outline. Maroon script "Certified" over black "CLEAN/REST ROOMS." Small Phillips 66 shield at bottom.

CERTIFIED CLEAN REST ROOMS NUMBER### 12x24
(3) 1959-1973 $35
Blue tin litho sign with white bands across top and bottom. White "CLEAN/REST ROOMS" between bands with red "Certified" in upper band and red "Number ????" on bottom band.

LADIES CAST ALUMINUM 4x14
(2) 1953-1959 $25
Die-cast aluminum lettering "LADIES" on projecting strip

MEN CAST ALUMINUM 4x14
(2) 1953-1959 $25
Die-cast aluminum lettering "MEN" on projecting strip

NO SMOKING PHILLIPS 66 4x6
(5) 1935-1940 $125
Red outlined black porcelain sign with small Phillips 66 shield at top over red "NO SMOKING."

BULK PLANT SIGNAGE

PHILGAS 36x72
(4) 1946-1959 $200
White self framed tin sign with small Phillips 66 shield at upper left above red "Philgas" logotype with black dropshade. Black "THE ALL PURPOSE FUEL" across bottom.

TRUCK SIGNS

PHILLIPS 66 12" SHIELD $250
Die cut red porcealin shield with black band across upper shield area. Red logotype "Phillips" on black band with black "66" slanted on lower red field below.

THE INDEPENDENT OIL AND GAS CO. - TULSA, OK

Sucessors to Waite Phillips Oil Co., merged into Phillips Petroleum in 1930.

IDENTIFICATION SIGNS

PHILLIPS HIGRADE PRODUCTS 30"RD
(5) 1925-1930 $450
Black/white outlined porcelain sign with red/black vertical stripes making up sign face. White band across top with black "PHILLIPS," across bottom with "PRODUCTS" and diagonal between with "HIGRADE"

INDEPENDENT "I" GASOLINE 42"RD
(4) 1926-1930 $375
White porcelain sign with red irregular border and white and red outline rings around white center circle. Large blue "I" in center, with red "INDEPENDENT" superimposed over "I."

CURB SIGNS

INDEPENDENT ETHYL GASOLINE 30"RD
(4) 1926-1930 $350
White porcelain sign with red irregular border and white and red outline rings around white center circle. Blue "INDEPENDENT" arched around top, "GASOLINE" around bottom with large Ethyl
(EGC) logo in center.

MANHATTAN GASOLINE CO. - ST. PAUL, MN

Manhattan originated the motor oil trademark "TROP ARTIC" in 1912 and marketed Trop-Artic oils through Manhattan stations in the north central states until Phillips purchased Manhattan in 1934. Stations were at that time rebranded Phillips 66, but Phillips adopted the Trop-Artic motor oil brand and continues to use it today.

IDENTIFICATION SIGNS

MANHATTAN GASOLINE 30"RD
(5) 1925-1934 $675
Black outlined white porcelain sign with black outline around white center circle. Logo in center with black concentric squares forming diamond in center. Black outleind box across center, with red "MANHATTAN" in box. "MANHATTAN" arched around top, "GASOLINE" around bottom.

CURB SIGNS

TROP ARTIC MOTOR OILS 30"RD
(5) 1920-1925 $2000
Red porcelain sign with yellow center circle. Yellow "TROP-ARTIC" arched around top, "MOTOR OILS" around bottom. Black world globe grid in cente with slogan "Satisfies From Pole to Pole" and small "MANHATTAN" diamond logo across globe.

THE QUAKER OIL CO. - ST. LOUIS, MO

Quaker Oil sold the Paraland gasoline brand from the early 1930s until after WW2. The brand shifted to Paraland Oil Co. of Omaha, Nebraska by the 1950s and it is not known if Phillips owned the company at that time. By the early 1960s it became known that Paraland was Phillips' "secret" secondary brand. In later years unsuccessful Phillips stations were converted to Paraland. The brand was last used about 1972.

IDENTIFICATION SIGNS

PARALAND MOTOR OIL/GASOLINE 40x66
(4) 1930's $250
Black outlined yellow oval with black band across sign above center and series of black and white lines below. Black and white lines interrupted at center with small black outlined white circle. Black tree and base with white "STABILITY" in circle. Yellow outlined white "PARALAND" on black band, with black "MOTOR OIL" above and "GASOLINE" below.

PARALAND 42x96
(3) 1953-1972 $100
Black outlined yellow oval with black outlined white band across center. Black "Paraland" on center band.

CURB SIGNS

PGCOA PENNFIELD MOTOR OIL 30"RD
(4) 1930's $1800
Black outlined yellow porcelain sign with blue lower field. Black detailed oil derrick scene in center, with black outlined red "PennField" across center. Yellow "MOTOR OIL" in lower blue area, with blue "PGCOA" seal in uper left area.

PENNFIELD MOTOR OIL 18x31
(5) 1940's $1600
Yellow porcelain oval sign with blue lower field. Black oil derrick scene in center, with black outlined red "PennField" across center. Yellow "MOTOR OIL" over small "THE QUAKER PETROLEUM CO., Inc. in lower blue area, with blue PGCOA seal at upper left and blue script "Finest from/Pennsylvania/Fields" in upper right.

LUBRICATION/MOTOR OIL SIGNS

PENNFIELD MOTOR OIL 12x24
(5) 1930's $450
Yellow tin litho square sign with blue lower field. Black oil derrick scene in center, with black outlined red "PennField" across center. Yellow "MOTOR OIL" over small blue script "Finest from/Pennsylvania/Fields," with blue PGCOA seal at upper left.

WASATCH OIL CO. - SALT LAKE CITY, UT

Wasatch was a Salt Lake City based collection of refining and marketing companies that branded stations under the Wasatch trademark in Utah, Idaho Chief in Idaho, Washington Chief in Washington and Golden Eagle in California. Phillips purchased Wasatch in 1947 and most stations were rebranded Phillips 66 at that time. The California "Golden Eagle" operation was sold off to become a leading independent marketer and early self-service operator.

IDENTIFICATION SIGNS

WASATCH 48"RD
(5) 1936-1947 $3500
Blue outlined white porcelain sign with red band across lower sign area. White "WASATCH" in red band, with detailed red/white/blue Indian head in full headdress above band.

IDAHO CHIEF 60"RD
(5) 1936-1947 $3000
Red/yellow outlined blue porcelain sign with yellow circle in center. Lower third of center circle is red. Upper yellow section has fully detailed multi-color Indian head. Yellow script "Chief" in lower red section. White "IDAHO" arched around top, "GASOLINE" around bottom.

WASHINGTON CHIEF 60"RD
(5) 1936-1947 $3000
Red/yellow outlined blue porcelain sign with yellow circle in center. Lower third of center circle is red. Upper yellow section has fully detailed multi-color Indian head. Yellow script "Chief" in lower red section. White "WASHINGTON" arched around top above small "IDAHO-OREGON-MONTANA," "GASOLINE" around bottom.

SILVER GAS - HOME OIL COMPANY - GREAT FALLS, MONTANA: Montana refiner/marketer purchased by the Wasatch companies in the late 1940's, shortly before Wasatch was purchased by Phillips.

ILLUMINATED SIGNS

SILVER GAS 48"
(5) 1940-1947 $1200
Octagonal shaped can type sign with neon outline ring. Blue painted 42" diameter glass face on each side with white "Silver/Gas" logotype in white below red "HOME/OIL." Internally illuminated.

SECONDARY SIGNS

SILVER GAS 18x36
(5) 1935-1940 $125
Blue tin litho sign with white reflective "SILVER GAS/A TREASURE STATE PRODUCT" across sign face.

STAR SERVICE AND PETROLEUM - CENTRAILIA, IL: Early independent discounter with marketing spread all around the country. Founded in the late 1930's, they were owned breifly by Phillips Petroleum in the early 1960's before becoming independent again in the 1970s. Closed in bankruptcy in 1991.

No collectible size Star Service signs are known at this time.

PURE

THE PURE OIL CO. USA - CHICAGO, IL

Pure Oil, founded in 1891 as the Producers Oil Co. in Oil City, Pennsylvania. The name was changed to The Pure Oil Company in 1895. Gasoline marketing began in 1914, and in 1917 ownership of the company passed to the Dawes family who operated public utility companies in Ohio and elsewhere. By the early 1920s the utilities had been sold off and the company was firmly in the gasoline business. In the 1920s Pure grew through a series of mergers with other regional marketers including Wofford Oil Co. of Atlanta, Georgia; Sherrill Oil Co of Pensacola, Florida; Seaboard Oil Company of Jacksonville, Florida; and the American Oil Co. of Hattiesburg, Mississippi. Also in the late 1920s the company introduced their famous "cottage" service station design that was to dominate Pure marketing forever and influence the competion until after WW2. In 1965 Pure Oil merged with Union Oil of California and began a five-year program to replace the Pure brand with Union 76. The Pure signs came down in 1970, thus for a time passing into history one of the most distinctive images in gasoline marketing. In 1992 Union Oil, by that time known as Unocal, announced intentions to withdraw from most marketing in the former Pure Oil territory. A joint venture operation was created between Unocal and the Venezuelan state-run oil company. As a result Uno-Ven continues to market under the Unocal 76 brand in the north central states. In the southeast Unocal jobbers formed a co-operative and licensed the Pure brand from Unocal. Supply arrangements were made and Pure signs were re-installed in 10 states from Virginia to Texas. Unocal truck stop operators also formed a co-operative and licensed the 76 brand from Unocal. Both of these operations have proven successful, and continue today. The last remnant of Unocal marketing in the southeast is the tie to Nascar Racing. Pure had early on formed an alliance with stock car racing and Unocal continues to supply racing fuel to the popular sport throughout the southeast.

IDENTIFICATION SIGNS

PUROL MOTOR GASOLINE 24x30
(5) 1919-1924 $700
Self framed tin-litho sign with black frame area. Red area to left forming point at right edge. Yellow above and below red point area. Black outlined white arrow pointing left to right with black outlined white "Purol" above arrow and "Motor Gasoline" below. Black "Pure Oil Company/(DIVISION THE OHIO CITIES GAS COMPANY)" at lower left.

PUROL GASOLINE 42x72
(3) 1924-1929 $400
PUROL GASOLINE 30x48
(3) 1924-1929 $450
White outlined blue porcelain sign with large white "Purol" across upper sign face. Smaller "GASOLINE" below with "Product of THE PURE OIL CO. U.S.A." across bottom.

PURE THE PURE OIL COMPANY, USA 42"RD
(1) 1929-1953 $250
White porcelain sign with blue seal border. Blue "PURE" in center, with blue "THE PURE OIL" arched around top, "COMPANY, U.S.A." around the bottom.

PURE PRODUCTS OF THE... 42"RD
(3) 1935-1938 $300
White porcelain sign with blue seal border. Blue "PURE" in center, with blue "PRODUCTS OF THE PURE OIL" arched around top, "COMPANY SOLD HERE." around the bottom.

PURE BE SURE WITH PURE 42"RD
(4) 1938-1953 $300
White porcelain sign with blue seal border. Blue "PURE" in center, with blue "BE SURE" arched around top, "WITH PURE" around the bottom.

PURE 42"RD
(4) 1953-1969 $250
White porcelain sign with blue seal border. Blue "PURE" in center.

PURE PRODUCTS OF THE... 72"RD
(3) 1935-1938 $150
White porcelain sign with blue seal border. Blue "PURE" in center, with blue "PRODUCTS OF THE PURE OIL" arched around top, "COMPANY SOLD HERE." around the bottom.

PURE BE SURE WITH PURE 72"RD
(3) 1938-1953 $150
White porcelain sign with blue seal border. Blue "PURE" in center, with blue "BE SURE" arched around top, "WITH PURE" around the bottom.

PURE 72"RD
(3) 1953-1969 $100
White porcelain sign with blue seal border. Blue "PURE" in center.

BUILDING SIGNS AND LETTERS

PUROL GASOLINE
(two logos) 18x70
(4) 1924-1926 $250
Blue/white outlined blue porcelain sign with white "Purol/GASOLINE" in center of sign face. White arrow between "Purol" and "GASOLINE." White "THE/PURE/OIL/CO" in white circle to either side.

PUROL GASOLINE
(one logo) 18x70
(4) 1926-1929 $250
Blue/white outlined blue porcelain sign with white "Purol/GASOLINE" to left side of sign. White arrow between "Purol" and "GASOLINE." White "THE/PURE/OIL/CO" logo to right with "Product of" above and "U.S.A." below.

TIOLENE 18x70
(4) 1926-1929 $250
Blue/white outlined blue porcelain sign with white "Tiolene/100% Super Pennsylvania MOTOR OIL" to left of sign face. White "THE/PURE/OIL/CO" logo to right with "Product of" above and "U.S.A." below.

BE SURE WITH PURE ALUMINUM LETTERS 12"
(4) 1938-1950 $125
Cast aluminum lettering pained blue mounted on single rail.

PUMP SIGNS

PUROL GASOLINE 15"RD
(5)1924-1930 $275
Blue/white outlined blue curved porcelain sign with white "Purol" logotype above "Gasoline" across center of sign. White "THE PURE OIL CO." arched around bottom.

ENERGEE TRUE GASOLINE 15"RD
(5)1924-1927 $250
Blue/white outlined blue curved porcelain sign with white "Energee!" logotype above "TRUE/GASOLINE" across center of sign. White "THE PURE OIL CO." arched around bottom.

DETONOX GASOLINE 15"RD
(5)1927-1930 $250
Blue/white outlined blue curved porcelain sign with red oval across center. White "DETONOX" across oval. White "THE PURE OIL CO." arched around top, "GASOLINE" around bottom.

PURE PEP 10x12
(1) 1947-1957 $90
Red outlined white porcelain sign with small Pure seal in center. Seal has "THE PURE OIL" arched around top, "COMPANY, USA" around bottom. Blue "PURE PEP" across top with blue "Be Sure With Pure"
("Sure" is red) in serif lettering across bottom.

PURE PEP REGULAR 10x12
(1) 1947-1957 $100

Delco Batteries metal sign. *Leelan Glanzer.*

Chevrolet Service neon sign. *Leelan Glanzer.*

Unique Boyce Motometer die-cut metal flange. *Leo Mathieu.*

Kelly Tires 42" porcelain sign. *Leo Mathieu.*

Early Firestone Tires Garage die-cut flange sign. *Leo Mathieu.*

United Service die-cut porcelain sign. *Leo Mathieu.*

Hood Tires 36" die-cut porcelain sign. *Leo Mathieu.*

Colorful 5' tall metal muffler sign. *Jim Hollabaugh.*

Goodyear 24" porcelain hanging sign.

Early 18.5"x13.5" metal flange tire sign.

Early 19.5"x13.5" embossed tin garage sign.

Early Goodrich 11.5"x36" tin sign.

Diamond Tires 18" x27" die-cut metal flange sign.

Dodge 8' neon sign with attached wings above.

Red outlined white porcelain sign with small Pure seal in center. Blue "PURE PEP" across top with blue "Be Sure With Pure" ("Sure" is red) in serif lettering across bottom. For use in North Carolina and Florida where gasoline grade was required to be posted on the pump.

PURE PEP 10x12
(2) 1957-1961 $100
Red outlined white porcelain sign with small Pure seal in center. No other wording in seal. Blue "PURE PEP" across top with blue "Be Sure With Pure"
("Sure" is red) in sans serif lettering across bottom.

PURE PREMIUM 10x12
(1) 1947-1957 $90
White outlined blue porcelain sign with white band across bottom. Small Pure seal in center. Seal has "THE PURE OIL" arched around top, "COMPANY, USA" around bottom. White "PURE/PREMIUM" across top with blue "Be Sure With Pure" ("Sure" is red) in serif lettering across bottom.

PURE PREMIUM 10x12
(3) 1957-1961 $75
White outlined blue porcelain sign with white band across bottom. Small Pure seal in center. No other wording in seal. White "PURE/PREMIUM" across top with blue "Be Sure With Pure"
("Sure" is red) in sans serif lettering across bottom.

PURE ENERGEE DIESEL FUEL #1 10x12
(4) 1957-1961 $250
Blue outlined white porcelain sign with blue band across lower sign face. Small Pure seal in center of lower blue band with white "Be sure" to left and "with Pure" to right. Blue "ENERGEE/DIESEL/FUEL" on upper sign face with red diagonal strip just above blue band at far right. White "NUMBER #1" on strip.

PURE ENERGEE DIESEL FUEL #2 10x12
(4) 1957-1961 $250
Blue outlined white porcelain sign with blue band across lower sign face. Small Pure seal in center of lower blue band with white "Be sure" to left and "with Pure" to right. Blue "ENERGEE/DIESEL/FUEL" on upper sign face with red diagonal strip just above blue band at far right. White "NUMBER #2" on strip.

PURE FIREBIRD REGULAR 10x12
(2) 1961-1965 $50
Embossed vinyl plastic pump sign mounted on rolled edge tin frame. Ridged lower half is silver, with white upper half. Large red and white Firebird logo embossed in center with embossed blue "PURE" over embossed red "Firebird" in upper white area. Blue "REGULAR" just above Firebird logo.

PURE FIREBIRD SUPER 10x12
(2) 1961-1965 $50
Embossed vinyl plastic pump sign mounted on rolled edge tin frame. Ridged lower half is gold, with white upper half. Large red and white Firebird logo embossed in center with embossed blue "PURE" over embossed red "Firebird" in upper white area. Blue "SUPER" just above Firebird logo.

PURE ENERGEE DIESEL FUEL #1 10x12
(4) 1961-1970 $200
Blue outlined white porcelain sign with blue band across lower sign face. Small Pure seal in center of lower blue band with white "Be sure" to left and "with Pure" to right. Blue "ENERGEE" above red "DIESEL" on upper sign face with red diagonal strip just above blue band. White "FUEL #1" on strip.

PURE ENERGEE DIESEL FUEL #2 10x12
(4) 1961-1970 $100
Blue outlined white porcelain sign with blue band across lower sign face. Small Pure seal in center of lower blue band with white "Be sure" to left and "with Pure" to right. Blue "ENERGEE" above red "DIESEL" on upper sign face with red diagonal strip just above blue band. White "FUEL #2" on strip.

NEON SIGNS

PURE THE PURE OIL COMPANY USA NEON 48"RD
(5) 1930's $1100
White porcelain can-type sign with blue seal border. Blue "PURE" in center, with blue "THE PURE OIL" arched around top, "COMPANY, U.S.A." around the bottom. Blue neon outline and "PURE."

PURE PRODUCTS OF THE..GLASS FACES 48"RD
(4) 1935-1938 $1500
White domed glass sign faces with aluminum frame. Neon ring inside frame for illumination. Sign has blue seal border with blue "PURE" in center. Blue "PRODUCTS OF THE PURE OIL" arched around top, "COMPANY SOLD HERE" around the bottom.

PURE THE PURE OIL CO GLASS FACES 48"RD
(4) 1935-1953 $1500
White domed glass sign faces with aluminum frame. Neon ring inside frame for illumination. Sign has blue seal border with blue "PURE" in center. Blue "THE PURE OIL" arched around top, "COMPANY, U.S.A." around the bottom.

CURB SIGNS

PUROL GASOLINE THE PURE OIL CO. 26"RD
(3) 1924-1929 $475
Blue/white outlined blue porcelain sign with white "Purol" above "GASOLINE" across center of sign. White arrow between "Purol" and "GASOLINE." White "THE PURE OIL CO." arched around bottom.

PUROL PEP THE PURE OIL COMPANY USA 26"RD
(2) 1929-1935 $450
Blue porcelain sign with white seal border. White "Purol/Pep" across center. Red semi-circles above and below, with white "THE PURE OIL" arched around top, "COMPANY U.S.A." around bottom.

PUROL ETHYL 30"RD
(5) 1929-1935 $450
White porcelain sign with blue seal border. Blue "Purol" across top, above Ethyl
(EGC) logo with yellow rays. Blue "THE PURE OIL CO. U.S.A." arched around bottom.

PURE THE PURE OIL CO USA 26"RD
(5) 1929-1938 $600
White porcelain sign with blue seal border. Blue "PURE" in center, with blue "THE PURE OIL" arched around top, "COMPANY, U.S.A." around the bottom.

100% SUPER-PENNA TIOLENE MOTOR OIL 26"RD
(3) 1920-1925 $425
Blue/white outlined blue porcelain sign with white "Tiolene" above "MOTOR/OIL" across center of sign. White arrow through bullseye between "Tiolene" and "MOTOR/OIL."

White "100% SUPER-PENNA." arched around top, "THE PURE OIL CO." arched around bottom.
100% SUPER PENNA TIOLENE PGCOA 26"RD
(3) 1925-1929 $425
Blue/white outlined blue porcelain sign with white "Tiolene" above "MOTOR OIL" across center of sign with white PGCOA seal below.. White arrow through bullseye between "Tiolene" and "MOTOR OIL." White "100% SUPER-PENNA." arched around top, "THE PURE OIL CO." arched around bottom.
TIOLENE PGCOA 26"RD
(3) 1929-1935 $425
Blue porcelain sign with white seal border. White "Tiolene" across center, with white PGCOA seal just below. White "THE PURE OIL" arched around top, "COMPANY U.S.A." around bottom.
TIOLENE AVIATION QUALITY W/ PLANE 26"RD
(5) 1935-1940 $675
Blue porcelain sign with white seal border. White "Tiolene" across center, with small white "AVIATION QUALITY" and airplane just below. White "THE PURE OIL" arched around top, "COMPANY U.S.A." around bottom.

LUBRICATION/MOTOR OIL SIGNS

PURITAN MOTOR OILS 18x30
(5) 1915-1920 $650
Blue porcelain sign with blue/white policeman in white area to right. White outlined red "GASOLINE" over white "AND" above white script "Puritan." Blue "MOTOR OILS" in underline, with red "THE SIGN OF THE COP" in white band at bottom.
PURITAN MOTOR AND TRACTOR OILS 12x35
(5) 1920-1924 $650
Black/yellow outlined black tin litho sign with large yellow "Puritan/The Pure Pennsylvania/Motor and Tractor Oils to right side of sign. "STOP/YOUR/ENGINE/TROUBLES" on policeman's hand.
PURITAN PURE PENNSYLVANIA MOTOR OIL 12x35
(5) 1920-1924 $600
Black/yellow outlined black tin litho sign with large yellow "Puritan/Pure/Pennsylvania/Motor Oil to right side of sign. Yellow "THE PURE OIL CO." across bottom with policeman in oval at left. "STOP/YOUR/ENGINE/TROUBLES" on policeman's hand.
DRAIN & REFILL WITH TIOLENE 30x42
(3) 1920-1929 $225
White porcelain sign with blue "Drain and/refill with" across top. Below is blue/white outlined circle with white "Tiolene" above "MOTOR/OIL" across center of circle. White arrow through bullseye between "Tiolene" and "MOTOR/OIL." White "100% SUPER-PENNA." arched around top, "THE PURE OIL CO." arched around bottom.
DRAIN & REFILL WITH TIOLENE 30x42
(3) 1929-1940 $225
White porcelain sign with blue "Drain and/refill with" across top. Below is blue circle with white seal border. White "Tiolene" across center. White "THE PURE OIL" arched around top, "COMPANY U.S.A." around bottom.

FLANGE SIGNS

PUROL GASOLINE "THE PURE OIL CO." 18x20
(5) 1924-1929 $375
Blue/white outlined blue tin litho sign with white "Purol" above "GASOLINE" across center of sign. White arrow between "Purol" and "GASOLINE." White "THE PURE OIL CO." arched around bottom.
100% SUPER PENNA. TIOLENE MOTOR OIL 18x20
(5) 1920-1925 $350
Blue/white outlined blue tin litho sign with white "Tiolene" above "MOTOR/OIL" across center of sign. White arrow through bullseye between "Tiolene" and "MOTOR/OIL." White "100% SUPER-PENNA." arched around top, "THE PURE OIL CO." arched around bottom.

SECONDARY SIGNS

PUROL PEP AUTHORIZED DEALER OVAL 8x18
(5) 1929-1939 $250
White tin litho oval sign with blue outline around edge and around center oval. Blue "Purol-Pep" in center oval with "COURTESY CARDS" arched around top, "ACCEPTED HERE" around bottom.
THE PURE OIL COMPANY CREDIT CARDS 10x16
(4) 1939-1946 $200
Blue porcelain sign with white band across center. White "THE PURE OIL COMPANY" in upper blue area, "HONORED HERE" below with blue "CREDIT CARDS" in center white band.
THE PURE OIL COMPANY CREDIT CARDS 12x16
(2) 1946-1960 $150
Blue porcelain sign with white band across center. White "THE PURE OIL COMPANY" in upper blue area, "HONORED HERE" below with blue "CREDIT CARDS" in center white band.
PURE OK'D REST ROOMS 12x24
(2) 1946-1960 $100
Blue porcelain sign with white panel on left side. Blue/white Pure seal in white area, with white "OK'd/REST/ROOMS" on blue area.
CUSTOMER OK'D REST ROOMS 12x28
(2) 1960-1970 $100
White porcelain sign with red band at top, adjacent to Pure seal. White "CUSTOMER OK'D" in red band. Blue "clean/rest/rooms" with red underlines below band and seal.
REST ROOMS LOCKED FOR YOUR PROTECTION
(4) 1940-1970 $50
Blue/white outlined white porcelain sign with blue band across bottom. Small Pure seal in center with white lettering in blue area below.
LADIES 2.5x10
(2) 1927-1950 $35
White porcelain rectangle with rounded ends. Blue "LADIES" across center.
MEN 2.5x10
(2) 1927-1950 $35
White porcelain rectangle with rounded ends. Blue "MEN" across center.
LADIES CAST ALUMINUM 3x14
(2) 1950-1970 $40
Cast aluminum "LADIES" on rail, painted blue.
MEN CAST ALUMINUM 3x14
(2) 1950-1970 $40
Cast aluminum "MEN" on rail, painted blue.
PURE TIRES BATTERIES ACCESSORIES 28x48
(3) 1930's $100

White outlined blue tin litho sign with Pure seal logo at top. White "PURE/TIRES/BATTERIES/ACCESSORIES" below logo.

TRUCK SIGNS

TIOLENE PGCOA 15"RD
(3) 1929-1935 $400
White porcelain sign with blue seal border. Blue "Tiolene" across center, with blue PGCOA seal just below. Blue "THE PURE OIL" arched around top, "COMPANY U.S.A." around bottom.

TIOLENE AVIATION QUALITY W/ PLANE 15"RD
(5) 1935-1940 $475
White porcelain sign with blue seal border. Blue "Tiolene" across center, with blue white "AVIATION QUALITY" and airplane just below. Blue "THE PURE OIL" arched around top, "COMPANY U.S.A." around bottom.

PUROL PEP THE PURE OIL/COMPANY USA 15"RD
(2) 1929-1935 $400
White porcelain sign with blue seal border. Blue "Purol/Pep" across center. Red semi-circles above and below, with blue "THE PURE OIL" arched around top, "COMPANY U.S.A." around bottom.

PURE PEP THE PURE OIL/COMPANY USA 15"RD
(2) 1929-1935 $350
White porcelain sign with blue seal border. Blue "PURE/PEP" across center. Red semi-circles above and below, with blue "THE PURE OIL" arched around top, "COMPANY U.S.A." around bottom.

WOFFORD OIL CO. - ATLANTA, GA

In 1925 Pure Oil merged with Wofford Oil Company of Georgia and Wofford Oil Co of Alabama. Wofford had marketed WOCO-PEP, a benzol blend, for a number of years and had used the seal type logo that Pure Oil adopted in 1930. Purol branded stations throughout the south continued to market "WOCO-PEP," a benzol base premium grade fuel, until WW2.

IDENTIFICATION SIGNS

WOCO PEP KING OF MOTOR FUEL 42x96
(5) 1920-1925 $500
Yellow porcelain sign with black border. Large black outlined red "WocoPep" across center with red "KING OF MOTOR FUELS" below. Black "More Miles-Less Carbon" above, with black "Does What Gasoline CAN'T DO" below.

WOCO PEP KING OF MOTOR FUEL 36x60
(3) 1930-1939 $450
Yellow porcelain sign with blue seal border. Large blue outlined red "WocoPep" across center with red "KING OF MOTOR FUELS" below. Blue "More Miles" across top, "Less Carbon" below.

WOCO PEP KING OF MOTOR FUEL 30x72
(3) 1930-1939 $400
Yellow porcelain sign with blue seal border. Large blue outlined red "WocoPep" across center with red "KING OF MOTOR FUELS" below. Blue "More Miles" across top, "Less Carbon" below.

CURB SIGNS

MORE MILES/WOCO PEP/LESS CARBON 16"RD
(5) 1925-1930 $650
Yellow porcelain sign with black border. Large black outlined red "WocoPep" across center with red "KING OF MOTOR FUELS" below. Black "More Miles" arched across top, "Less Carbon" below.
May also have been used as a pump sign.

MORE MILES/WOCO PEP/LESS CARBON 30"RD
(5) 1925-1930 $500
Yellow porcelain sign with black border. Large black outlined red "WocoPep" across center with red "KING OF MOTOR FUELS" below. Black "More Miles" arched across top, "Less Carbon" below.

HICKOCK OIL CO. - TOLEDO, OH

Founded in 1917, Hickock Oil Co. marketed gasoline under the HI-SPEED brandname. Pure purchased 51% of Hickock Oil Co in 1928. Hickock operated the "HI SPEED" stations throughout Ohio and Michigan, and were the originators of the successful Truck Stop concept that was greatly expanded on by Pure after the purchase of Hickock was completed in 1951 and their stations and truck stops were rebranded Pure.

IDENTIFICATION SIGNS

HI SPEED GASOLINE 30x48
(3) 1930-1946 $325
Green porcelain sign with red/white colonial style border area. White dots on red border area. White "HI-SPEED" above large white "GAS" inside border.

HI SPEED 18x48
(4) 1946-1951 $350
Green porcelain sign with red/white colonial style border area. White dots on red border area. White "HI-SPEED" across center inside border.

CURB SIGNS

SPEEDOLEUM MOTOR OIL SUPER QUALITY 30"RD
(5) 1930's $375
Green/white outlined green porcelain sign with white "SPEEDOLEUM" across upper sign face. Green "SUPER QUALITY" in underline of "SPEEDOLEUM." White "MOTOR/OIL" below, with white "HICKOCK OIL CORP" arched around bottom.

SECONDARY SIGNS

PURE OIL AND/HI SPEED/COURTESY CARDS 12x18
(5) 1935-1946 $225
White porcelain sign with green "PURE OIL AND/HI-SPEED" in large letters across top. Smaller green "COURTESY CARDS/HONORED HERE" below.

W.H. BARBER AND CO. - CHICAGO, IL

Founded in Minneapolis in the 1910s, they introduced the Fyre Drop brand on stations about 1915 and later operated the Meteeor stations from about 1925 on. They introduced the SAVE MORE brand in 1940. W.H. Barber had purc hased control of the Central West/Super Par operation by the time they were bought out by Pure in 1945. Pure continued to operate the Save More, Super Par and Pacer operations wint the 1960's. W.H. Barber was spun off as an independent after the Union/Pure merger in 1965 and markets lubricants today.

IDENTIFICATION SIGNS

FYRE DROP GASOLINE 30x52
(5) 1920's $650
Yellow outlined black porcelain sign with yellow outlined red "FYRE DROP/GASOLINE" across center of sign. Yellow letter-

ing "Starts Quickly - Explodes Completely" below.
IT'S FYRE DROP 42"RD
(5) 1930's $700
Yellow outlined black porcelain sign with red "IT'S" at top, "GASOLINE" across bottom and yellow "FYRE DROP" in bowtie lettering across center.

SECONDARY SIGNS

PENNO MOTOR OIL 30"RD
(5) 1920's $450
Orange outliend black porcelain sign with orange outlined white "PENNO" logotype across center. White "Unexcelled" above orange outlined white "MOTOR OIL' below "PENNO," white PGCOA seal at top.

SAVE MORE: W.H. Barber discount brand introduced in 1940 for company operated locations.

PUMP SIGNS

SAVE MORE SERVICE SYSTEM NO MIDDLEMAN
REFINERY DIRECT $150
White porcelain pump sign with large circle in center. White lettering on bands arched above and below circle "NO MIDDLEMAN" around top, "REFINERY DIRECT" around bottom. Large "SAVE-MORE" across "$" in center, with "SERVICE" above and "SYSTEM" below.

CENTRAL WEST OIL CO. - SOUTH BEND, IN

A subsidary of W.H. Barber. Operated the Super Par and Wisco 99 stations.

PUMP SIGNS

SUPER PAR REGULAR 100 10.5x12
(4) 1950-1965 $250
Porcelain pump sign with red upper half and black lower half. White "Super/Par" with black dropshading in upper red area, with white outlined yellow circle on lower black area. White band across center of circle with black "REGULAR," small black airplane above with black "100" below."

SUPER PAR ETHYL EC 10.5x12
(4) 1950-1965 $100
Porcelain pump sign with red upper half and black lower half. White "Super/Par" with black dropshading in upper red area, with white circle on lower black area. Ethyl
(EC) logo in circle.

SUPER PAR DIESEL FUEL 10.5x12
(4) 1950-1965 $100
Porcelain pump sign with red upper half and black lower half. White "Super/Par" with black dropshading in upper red area, with white outlined red circle on lower black area. Black "DIESEL/FUEL" in circle.

SUPER PAR KEROSENE 10.5x12
(4) 1950-1965 $100
Porcelain pump sign with red upper half and black lower half. White "Super/Par" with black dropshading in upper red area, with white outlined yellow circle on lower black area. White band across center of circle with black "KEROSENE" across band.

PACER OIL CO. - SOUTH BEND, IN

Super Par affiliate.

PUMP SIGNS

200 PACER HI-TEST 8.5x14
(4) 1950-1965 $125
Porcelain pump sign with red upper half and black lower half. White "Pacer" with black dropshading in upper red area with yellow "200" above. Yellow "HI-TEST" in black area below.

400 PACER ETHYL 8.5x14
(4) 1950-1965 $125
Porcelain pump sign with red upper half and black lower half. White "Pacer" with black dropshading in upper red area with yellow "400" above. Ethyl
(EC) logo in white circle on black area below.

WISCO 99: Another Central West affiliate brand, operating in Wisconsin.
No collectible size Wisco 99 signs are known at this time.

DICKEY, INC. - PARKWOOD, IA

Independent jobber who became a branded Pure jobber in the 1950s and continues to market Union 76 products.
No collectible size Dickey signs known at this time.

QUAKER STATE OIL REFINING CO. - OIL CITY, PA

In 1931 Sterling Oil Company of St. Mary's, West Virginia, Quaker State Oil Refining Company of Oil City, Pennsylvania and 16 other independent oil producers, refiners, lubricant compounders and marketers joined together to form Quaker State Oil Refining Corporation. It was decided that Sterling would remain the gasoline brand as Sterling was the only company involved with any significant gasoline marketing. The primary motor oil brand would be Quaker State, originally introduced in 1914 and distributed nationwide through lubricant jobbers, often jobbers of competing gasoline brands. The merger was the first time the famous motor oil brandname had been tied to any particular gasoline brand and it was thought that marketing gasoline under a different name would lessen the problems associated with gasoline jobbers selling a "competitors" motor oil. Sterling would remain the gasoline brandname until after 1960 when stations were rebranded Quaker State. Quaker State gasoline is currently sold through a small number of stations in Pennsylvania and Ohio.

PUMP SIGNS

STERLING QUALITY 12.25x14
(3) 1960-1974 $100
White porcelain sign with large black outlined green "Q" in center. Red "Sterling" at upper left, with black outlined red pound sign in center of "Q." Black "QUALITY" below pound sign.

STERLING SUPER BLEND 12.25x14
(3) 1960-1974 $100
White porcelain sign with large black outlined green "Q" in center. Red "Sterling" at upper left, with black outlined gold "SUPER/BLEND" in center of "Q."

TOMBSTONE SIGNS

QUAKER STATE CERTIFIED/GUARANTEED 26.5x29
(2) 1931-1940 $325
White outlined green tombstone shaped porcelain sign with white "QUAKER STATE" arched across top above "QSMO"

monogram. White "MOTOR OIL" below monogram with smaller white "CERTIFIED /GUARANTEED" across bottom. Large PGCOA seal at top of sign.
QUAKER STATE MOTOR OIL W/PGCOA 26.5x29
(2) 1940s $325
White outlined gren tombstone shaped sign with white "QUAKER/STATE above smaller "MOTOR OIL" across center of sign. Small PGCOA seal at top, with white "CERTIFIED/ GUARANTEED" across bottom of sign.
QUAKER STATE MOTOR OIL 26.5x29
(1) 1950s $250
White outlined gren tombstone shaped sign with white "QUAKER/STATE above smaller "MOTOR OIL" across center of sign. Small white oval at top, with oil derrick and detailed wording at top.
QUAKER STATE MOTOR OIL 26.5x29
(1) 1960-1980 $200
White outlined gren tombstone shaped sign with white "QUAKER/STATE above smaller "MOTOR OIL" across center of sign. Small white "Q" at top, with oil derrick scene in "Q."
QUAKER STATE MOTOR OIL LT GREEN 26x42
(1) 1980s $100
Bright green tombstone shaped sign with large white "Q" above double white lines. White "Quaker/State" inside "Q," with white "MOTOR OIL" below lines.
QUAKER STATE FAST LUBE LT GREEN 26x42
(2) 1980s $100
Bright green tombstone shaped sign with large white "Q" above double white lines. White "Quaker/State" inside "Q," with white "FAST LUBE" below lines.

LUBESTER SIGNS

QUAKER STATE MOTOR OIL 4.5x10
(3) 1920's $250
Green rectangular tin litho sign with white notched outline. White "QUAKER STATE" arched across top over "MOTOR OIL"
USE QUAKER STATE COLD TEST OIL 6x26.5
(4) 1920's $250
Green rectangular tin litho sign with white notched outline. White "QUAKER STATE COLD TEST OIL" across center. White "USE" at top, with white "FOR WINTER DRIVING" across bottom.
QUAKER STATE MEDIUM W/ MONOGRAM 13.5x18.5
(4) 1920's $300
Green tin litho sign with white notched border and white "QUAKER STATE" arched across top. "QSM" monogram in center, with white "MEDIUM" below. Black "AUTOMOBILE OIL" across bottom.

LUBRICATION/MOTOR OIL SIGNS

ASK FOR QUAKER STATE MOTOR OIL BUBBLE 24"RD
(2) 1960s $175
White outlined green with litho embossed bubble sign with white "ASK FOR" across top, "MOTOR OIL" across bottom and white "QUAKER/STATE" in center.
QUAKER STATE MODERN BUBBLE 24"RD
(1) 1980s $100
White outlined bright green tin litho embossed bubble sign with white "Q" above double lines in center. White "Quaker/State" inside "Q" with white "MOTOR OIL" below lines.
QUAKER STATE MOTOR OIL 12x72
(2) DATED 1932 $200
Black outlined green self-framed tin litho vertical sign with white "QUAKER STATE" vertical on sign. Black band at bottom with white "MOTOR OIL" on band.
QUAKER STATE OIL 12x60
(3) 1930's $200
Green self-framed tin litho sign with two diagonal white lines across face. White "QUAKER STATE OIL" with black dropshade between lines.
QUAKER STATE RACING OIL 5x26
(3) DATED 1967 $100
White tin litho sign with black and white checkered flag with green and white Quaker State "Q" logo at left, black "RACING OIL" across sign.

SECONDARY SIGNS

QUAKER STATE MOTOR OIL 42x96
(2) 1940s-1960 $150
Green self framed tin litho sign with two black lines diagonal across sign face. White "QUAKER/STATE" with black dropshading between lines with white "MOTOR OIL" across bottom.
BUY QUALITY QUAKER STATE 34.5x94.5
(3) 1960-1980 $100
Yellow embossed tin litho sign with green and white Quaker State can at right and black "Buy/Quality" at left.

STERLING OIL COMPANY - ST. MARY'S, WV

Pennsylvania producer/refiner/marketer that originated the Pennsylvania Grade Crude Oil Association. Sterling was involved in gasoline marketing by the early 1920s and remained the brandname for gasoline marketed by Quaker State when that company entered the gasoline market after merging with Sterling and others in 1931.

IDENTIFICATION SIGNS

STERLING GASOLINE OVAL 30x54
(3)1936-1960 $350
Steel framed red outlined yellow oval with red outlined white oval in center. Black "STERLING" arched around top of yellow area, "GASOLINE" below. Black outlined red pound sign in center, with black "QUALITY" below, "TRADE MARK" to left, "REG.U.S.PAT.OFF" to right.

PUMP SIGNS

STERLING GASOLINE 8.5x11
(3) 1946-1950 $150
Red outlined yellow oval with red outlined white oval in center. Black "STERLING" arched around top of yellow area, "GASOLINE" below. Black outlined red pound sign in center, with black "QUALITY" below, "TRADE MARK" to left, "REG.U.S.PAT.OFF" to right.
STERLING ETHYL GASOLINE 8.5x11
(3)1946-1950 $125
Red outlined yellow oval with red outlined white oval in center. Black "STERLING" arched around top of yellow area, "GASOLINE" below. Ethyl
(EC) logo with yellow rays in center.
STERLING GASOLINE A QUAKER STATE PRODUCT 10x11.5
(3)1950-1960 $200
Red outlined yellow oval with red outlined white oval in center. Black "STERLING" arched around top of yellow area,

"GASOLINE" below. Black outlined red pound sign in center, with black "QUALITY" below. Green outlined white banner across bottom with green "A QUAKER STATE PRODUCT" in banner.

STERLING ETHYL A QUAKER STATE PRODUCT 10x11.5
(3)1950-1960 $200
Red outlined yellow oval with red outlined white oval in center. Black "STERLING" arched around top of yellow area, "GASOLINE" below. Ethyl
(EC) logo with yellow rays in center. Green outlined white banner across bottom with green "A QUAKER STATE PRODUCT" in banner.

GLOBE TOPPER

A QUAKER STATE PRODUCT 6x12
(4) 1946-1950 $150
White outlined green tin litho arched banner with white "A QUAKER STATE PRODUCT" on banner. Designed to attach to top of globe.

CURB SIGNS

STERLING ETHYL GASOLINE 30"RD
(3) 1926-1940 $375
Black/red outlined yellow porcelain sign with black/red outlined white center circle. Black "STERLING" arched around top, "GASOLINE" around bottom with Ethyl
(EGC) logo in center.

LUBRICATION/MOTOR OIL SIGNS

WE SELL STERLING MOTOR OIL 15x30
(4) 1926-1930 $225
White porcelain sign with red outline and red band acrossc enter. Black outlined white "STERLING" on red band. Sterling logo and black "We Sell" above band, "MOTOR OIL" and PGCOA seal below.

STERLING OIL STERLING MOTOR OIL 12x72
(4) 1930's $225
Black outlined yellow self-framed tin litho sign with white area in center. Red "STERLING" vertical on white area, with Sterling logo at botto.

ASK FOR STERLING OILS 13.5x20
(4) 1924-1926 $250
White tin litho sign with large yellow/black/red Sterling Oils logo in center. Red "ASK FOR" at top, with yellow and black predassessor to PGCOA seal below logo.

STERLING OILS FLANGE
(5) 1924-1930 $225
Black/red outlined yellow porcelain sign with black/red outlined white center circle. Black "STERLING" arched around top, "OILS" around bottom with black outlined red pound sign in center. Black flange area.

SECONDARY SIGNS

STERLING CARDS HONORED 10x18
(4) 1940s $150
Green/white outlined green tin litho sign with black outlined white "STERLING/CARDS/HONORED" across sign.

LUBESTER SIGNS

PGCOA SEAL OIL PERMIT NO 1 10x12
(4) 1926-1930 $325
Yellow porcelain die-cut sign with black PGCOA seal above black "PERMIT NO.1."

ENTERPRISE OIL COMPANY, INC - BUFFALO, NY:

One of the many lubricants manufacturers that came together in 1931 to form Quaker State Oil Company.

LUBRICATION/MOTOR OIL SIGNS

DUPLEX MARINE ENGINE OIL 24x48
(4) 1926-1931 $225
White tin litho sign with blue outline and blue "DUPLEX MARINE/ENGINE OIL" above blue PGCOA seal in center. Red "AUTHORIZED DEALER" across top, with red and blue "MANUFACTURED BY/ENTERPRISE OIL CO.,INC., BUFFALO, N.Y.ESTABLISHED 1884/MAKERS ALSO OF DUPLEX MOTOR OIL SERVING PIERCE-ARROW SINCE 1901"

DUPLEX MARINE ENGINE OIL 10x20
(3) 1931-1935 $200
Mint green tin litho sign with blue "DUPLEX MARINE/ENGINE OIL across center. Red "AUTHORIZED DEALER" across top, "USE KASSON WATERPROOF GREASE"

RICHFIELD
RICHFIELD OIL CORPORATION - LOS ANGELES, CA

RICHFIELD OIL COMPANY OF NEW YORK - NEW YORK, NY
Richfield, a west coast producer/refiner/marketer was founded in California about 1901. By 1915 Richfield was marketing on the west coast. The company entered east coast markets in the mid-1920s, and created a subsidary, Richfield Oil Co. of New York with the 1929 purchase of Walburn Petroleum Company. During the 1930s, as detailed below, Richfield was reorganized and the eastern marketing was sold to Sinclair. The remaining west coast operations merged with Atlantic Refining in 1966 to form Atlantic Richfield. See Atlantic section for more information.

RICHFIELD OIL COMPANY

Sign listings for the entire company prior to 1933 breakup. Includes both east and west coast marketing.

IDENTIFICATION SIGNS

RICHFIELD GASOLINE OF POWER 60x60
(4) 1919-1937 $1000
Steel framed porcealin die-cut nine point shield. Blue/white outlined with with yellow outlined blue shield in center. Yellow "RICHFIELD" across top with blue/white outlined panel across lower sign face. White eagle with blue details and spread wings on top of panel. Blue "The GASOLINE/OF POWER" on panel.

CURB SIGNS

RICHFIELD GASOLINE OF POWER 24x24
(5) 1919-1937 $2800
Die-cut nine point shield. Blue/white outlined with with yellow outlined blue shield in center. Yellow "RICHFIELD" across top with blue/white outlined panel across lower sign face. White eagle with blue details and spread wings on top of panel. Blue "The GASOLINE/OF POWER" on panel.

RICHFIELD ETHYL 24x24
(5) 1926-1934 $2800
Die-cut blue/yellow outlined blue porcelain shield with yellow "RICHFIELD" across top over script yellow "Ethyl." White eagle with spread wings above "RICHFIELD," white circle with Ethyl

(EGC) logo below "Ethyl."

RICHFIELD OIL COMPANY OF NEW YORK

Eastern marketing division of Richfield Oil Corporation that was formed in 1929 with the purchase of New York marketers Walburn Petroleum Company and Acewood Petroleum Company. Richfield Oil Company of New York was created as a "jobber marketer" and former company owned outlets were turned over to the new company. Richfield Oil Company of New York was purchased by Sinclair in 1933 during the Richfield bankruptcy and operated as a secondary brand until 1964 when all Richfield jobbers and dealers were converted to the Sinclair brand.

IDENTIFICATION SIGNS

FLAT SHIELD 36x48
(5) 1936-1939 $500
Blue outlined yellow die cut shield shaped porcelain sign, flat across the top, with detailed blue and white eagle at top center with spread wings above blue "RICHFIELD" across lower sign.

RICHFIELD 48"RD
(4) 1939-1964 $350
RICHFIELD 60"RD
(4) 1939-1964 $200
RICHFIELD 72"RD
(3) 1939-1964 $150
Blue porcelain sign with white outlined yellow five point shield covering most of sign face. Blue and white eagle with blue/white upright wing facing right at top of sign above blue "RICHFIELD."

BUILDING SIGNS AND LETTERS

RICHFIELD PANEL 18x120
(5) 1935-1947 $250
Self framed two-piece blue porcelain panel with yellow "RICHFIELD" across center.

NEON SIGNS

RICHFIELD FLAT SHIELD 60x96
(5) 1935-1939 $1000
Yellow die cut shield shaped porcelain sign faces assembled back-to-back over can-type frame. Detailed blue and white eagle at top center witrh spread wings above blue "RICHFIELD" across lower sign.

RICHFIELD 72x72
(5) 1939-1964 $1000
Die cut can type shield shaped porcelain sign with two piece faces. White outlined around yellow center shield. Blue/white eagle, facing right, at top with blue "RICHFIELD" across center.

CURB SIGNS

RICHFIELD ETHYL EGC 24"
(5) 1934-1940 $2800
White porcelain sign with yellow outlined blue nine point shield covering most of sign face. Yellow "RICHFIELD" across top above yellow script "Ethyl" in center. White eagle with spread wings above "RICHFIELD," white circle with Ethyl (EGC) logo below "Ethyl."

RICHFIELD MOTOR OIL 24"RD
(5)1930s $3200
Yellow porcelain sign with two blue bands across center. White "RICHLUBE" in upper band, "MOTOR OIL" in lower. Small blue race car at top of sign with white PGCOA seal in blue block at bottom.

RICHLUBE MOTOR OIL 24"RD
(5)1930s $2800
Blue porcelain sign with yellow "RICHFIELD" with white dropshading arched arcross center. Yellow "1/4 MORE/HEAT RESISTANCE" above with white "MOTOR OIL/100% PURE PENNSYLVANIA" below.

SAFETY RICHLUBE MOTOR OIL 24"RD
(4)1935-1940 $3200
Yellow porcelain sign with blue/white outline area around yellow outlined blue nine point shield. Three color blue target in lower left corner with yellow race car heading into target from upper right. Yellow "RICHLUBE" with light blue dropshading arched across above target with white script "Safety" above "Richlube" and grey "MOTOR/OIL" below.

SECONDARY SIGNS

RICHFIELD CREDIT CARDS 12x20
(5) 1929-1946 $150
Yellow/blue outlined yellow tin litho sign with blue "RICHFIELD/CREDIT CARDS/HONORED HERE" across sign.

RICHFIELD CREDIT CARDS 12x20
(5) 1946-1964 $150
Yellow tin litho sign with blue band across center. Blue "RICHFIELD" across top, "HONORED" across bottom with yellow "CREDIT CARDS" in center.

ADVANCE SIGNS

JUST AHEAD RICHFIELD 54x72
(5) 1936-1939 $400
Yellow outlined blue tin litho sign with blue outlined yellow shield shaped logo, flat across the top, with detailed blue and white eagle at top center with spread wings above blue "RICHFIELD/GASOLINE" across lower sign. White JUST AHEAD" above shield.

RICHFIELD OIL CORPORATION

Richfield of California emerged from the bankruptcy in 1937 with marketing throughout the Western United States. Ownership of Richfield had transferred to Sinclair and Cities Service and the "new" Richfield consolidated with Sinclair's Rio Grande marketing subsidiary. Rio Grande continued to be used as a secondary brand to the new Richfield, as well as "Rocket" being used at some stations. In 1966 Richfield Oil of California merged with Atlantic Refining to form Atlantic Richfield. In 1970 the brandname Richfield was replaced by the Arco brand and Arco continues to market throughout the old Richfield territory today.

BUILDING SIGNS AND LETTERS

RICHFIELD PANEL 18x192
(5) 1946-1970 $350
Blue outlined yellow porcelain strip sign with blue "RICHFIELD" across center.

ROOFTOP EAGLE 96x144
(5) 1937-1970 $3500
Elaborate die cut porcelain can type sign. Upper section is highly detailed eagle with upright two tone blue wing, facing left. White body and tail feathers, yellow and red beak. Eagle is perched on blue area.

Early 8′ porcelain garage sign with light bulbs forming the letters.

Vacuum Cup 16″x22″ porcelain tire sign.

BP Motor Spirit porcelain flange sign. *Al Alford.*

Gas-Oils early 12″x24″ wood-framed tin sign. *Don Christensen.*

Kelly Tires porcelain sign. *Leo Mathieu.*

Amalie Motor Oil oval sign. *Jim Hollabaugh.*

Shell 12″ porcelain pump sign. *Tom Kleeberg.*

Texaco Marine 12″x18″ white pump sign.

Socony porcelain die-cut flange sign. *Pete Clark.*

Sinclair Dino 5′x44″ porcelain sign. *Dunbar's Gallery.*

Cities Service 5″ clover-shaped hanging sign. *Dunbar's Gallery.*

Veltex 12.5″x15″ porcelain pump sign.

Pencoil 7″ porcelain lubester sign.

Rare Havoline 18″x9″ tin No Smoking sign.

Miscellaneous pump plate signs. *Bruce Miller.*

NEON SIGNS

NEON EAGLE 96x144
(5) 1937-1970 $2500
Elaborate die cut porcelain can type sign. Upper section is highly detailed eagle with upright two tone blue wing, facing left. White body and tail feathers, yellow and red beak. Eagle is perched on blue area above blue outlined yellow panel. Blue "RICHFIELD" across yellow panel. Neon "RICHFIELD" and eagle outlines.

LUBRICATION/MOTOR OIL SIGNS

QUALITY/PROTECTED/RICHFIELD/MOTOR OIL 12x24
(4) 1937-1940 $200
Blue tin litho sign with red outlined yellow "RICHFIELD" diagonal across center. White outlined yellow "QUALITY/PROTECTED" in upper left corner, "25c/PER QUART" in lower right with white "MOTOR OIL" under "RICHFIELD"

SECONDARY SIGNS

RICHFIELD CREDIT CARDS ACCEPTED HERE 12x42
(5) 1950's $150
Yellow porcelain sgin with blue "RICHFIELD" above red "CREDIT CARDS/ACCCEPTED HERE" across sign face.

BORON ARROW 20x36
(3) 1955-1965 $100
Red porcelain die-cut arrow with white "BORON" across center. Auxillary sign attached to perimeter poles when the element Boron was added to gasolines in 1955.

BULK PLANT SIGNAGE

RICHFIELD RICH-HEAT HEATING OIL 20x20
(5) 1958-1970 $150
White outlined red tin litho sign with white outlined yellow Richfield "flat shield" logo in center. White eagle with blue vertical wing, facing left, at top of logo above blue "RICHFIELD." Smaller red "RICH-HEAT/HEATING OIL" below.

WALBURN PETROLEUM COMPANY: First Ethyl marketer in the New England region. In 1929 Walburn Petroleum was merged with existing Richfield company marketing and Acewod Petroleum Co. to form the Richfield Oil Company of New York.
No collectible size Walburn signs known at this time.

SHERWOOD BROS. - BALTIMORE, MD

Large Richfield/Sinclair jobber operation which had marketed under the Betholine trademark prior to branding Richfield and continued to dual brand "Betholine/Richfield" and "Betholine/Sinclair" until the late 1950s.

LUBRICATION/MOTOR OIL SIGNS

SHERWOOD RICHLUBE/REXOLINE MOTOR OIL 18x30
(5) 1935-1940 $350
White porcelain oil rack panel with block projecting from top center. Blue rectangle to one side with yellow "RICHLUBE/MOTOR OIL" over white "GUARANTEED 100% PENNSYLVANIA" across rectangle and red rectangle opposite with blue outlined yellow "REXOLINE/MOTOR OIL" above white 100% PURE PARAFFINE" across rectangle. Red and blue areas reversible for opposite sides of rack. Blue "SHERWOOD/PRODUCTS" in projecting area at top.

LAMSON OIL CO. - PROVIDENCE, RI

Jobber brand that rebranded Richfield about 1929.
NO COLLECTIBLE SIZE LAMSON SIGNS KNOWN AT THIS TIME.

RIO GRANDE OIL COMPANY

Independent California refiner/marketer purchased by Sinclair in 1932. When Richfield was reorganized in 1937, Sinclair merged their interest in Rio Grande with the new Richfield Oil Corporation. Rio Grande continued to be a Richfield/Arco secondary brand into the 1970s.

NEON SIGNS

RIO GRANDE NEON 36x96
(5) 1950's $350
Red porcelain can type sign with white "RIO/GRANDE" across sign face. Neon "RIO/GRANDE."

SECONDARY SIGNS

RIO GRANDE PENNSYLVANIA FLANGE 14x16
(5) 1926-1930 $350
White/blue outlined white porcelain die cut flange sign with blue upper area and red band across center. White "RIO GRANDE" in upper blue band and "PENNSYLVANIA" in red band across center. Similar version of logo is repeated in lower white area with white "100% PURE" and "MOTOR OIL" separated by red "PENNSYLVANIA."

ROCKET

Rocket was a Richfield/Arco secondary brand.

NEON SIGNS

ROCKET 36x96
(5) 1950's $250
Red porcelain can type sign wtih white "ROCKET" across center. Neon "ROCKET."

DIAMOND SHAMROCK, INC. - SAN ANTONIO, TX

Diamond Shamrock, founded in the 1930s as Shamrock Oil and Gas Corp. in Amarillo, Texas, is a regional marketer in the south central and southwestern U.S. Little is known about the company's early history, but evidence of marketing from the mid-1930s up exists. In 1967 Shamrock Oil and Gas merged with Diamond Alkali Comapny, a chemical manufacturer to form Diamond Shamrock. The company has gradualy expanded it's market in recent years and now brands over 1900 locations. Shamrock has historically marketed through jobbers, several, including Sigmor (purchased by Shamrock in 1983) and Cliff Brice, have became large marketing organizations in themselves.

IDENTIFICATION SIGNS

SHAMROCK 72"
(3) 1946-1962 $150
Green outlined white porcelain with large green clover in center. White "SHAMROCK" across clover.

PUMP SIGNS

SHAMROCK POLYMERINE GASOLINE 10.5x12.5
(5) 1940-1950 $350
White porcelain with green "POLYMERINE" logo in center. "SHAMROCK" arched around top, "GASOLINE" around bottom of logo.

SHAMROCK TRAIL MASTER REGULAR 10.5x12.5
(1) 1950-1972 $175
White porcelain with green Shamrock logo at top. Yellow, brown & green "TRAIL/MASTER" logo below. Green "REGULAR" across bottom.

SHAMROCK CLOUD MASTER PREMIUM 10.5x12.5
(1) 1950-1972 $175
White porcelain with green Shamrock logo at top. Light green, white, & black "CLOUD/MASTER" logo below. Green "PREMIUM" across bottom.

SHAMROCK NO.1 DIESEL 10.5x12.5
(4) 1950-1972 $100
White porcelain with green Shamrock logo at top. Black outlined white "NO.1/DIESEL" below.

SHAMROCK NO.2 DIESEL 10.5x12.5
(4) 1950-1972 $100
White porcelain with green Shamrock logo at top. Black "NO.2 DIESEL" below.

SHAMROCK KEROSENE 10.5x12.5
(4) 1950-1972 $100
White porcelain with green Shamrock logo at top. Red KEROSENE diagonal below.

CLIFF BRICE - PUEBLO, CO

Cliff Brice is a Pueblo, Colorado based Shamrock regional jobber operating dual branded Cliff Brice/Shamrock stations.

IDENTIFICATION SIGNS

BRICE GASOLINE 30"RD
(5) 1940s $300
Green/white outlined green porcealin sign with green outlined white center circle. White "BRICE" arched around top, "GASOLINE" around bottom with large green outlined red "B" in center circle.
CLIFF BRICE QUALITY GAS FOR LESS OVAL 42x72
(3) 1960s $175
Red porcelain oval with white and wavy green outline. White "Cliff/Brice" with green dropshade at top. Yellow "Quality" to left below, "for Less" to right with white "GAS" in center.

SHELL
SHELL OIL COMPANY - HOUSTON, TX

Shell Oil Company, a unit of the British/Dutch "Royal Dutch Shell" entered U.S. markets with the purchase of a terminal in Washington state in 1912. In 1917 Shell purchased St. Louis based Roxana Petroleum and by 1929 was marketing in all 48 states. Since that time Shell has proved to be most innovative marketers, often leading the way in new gasoline marketing concepts. Several Shell "firsts" include the 1947 introduction of internally lit plastic signage, the 1957 introduction of landscaped "rancher" style stations, and one of the first franchised Auto Care programs. Shell continues to market through company direct and jobber operated locations in 41 states and around the world today.

LOGO DESCRIPTION: In an effort to simplify the description of Shell signage, the following basic descriptions are given.
1. Early Shell Logo white with two arcs 1912-1917
2. Rounded Shell logo with arc and serif lettering 1917-1922
3. Rounded Shell logo with arc and sans serif lettering 1917-1922
4. Non-detailed Shell logo with pointed scallops and thick letters 1922-1929
5. Detailed Shell logo without arc and pointed scallops 1922-1929
6. Minimum detailed rounded scallop USED 1929-1946
7. Maximum detailed rounded scallop USED 1946-1960

IDENTIFICATION SIGNS

SHELL GASOLINE WITH WHITE SHELL 30"
(4) 1912-1917 $3000
SHELL GASOLINE WITH WHITE SHELL 45"
(5) 1912-1917 $3000
Black/red outlined yellow porcelain sign with red center circle. White shell scallop with black details in center circle. Black outlined red "SHELL" arched aorund top, "GASOLINE" around bottom. Small red diamond to either side of center circle.
SHELL 42"
(5) 1917-1922 $350
Die cut yellow porcelain Shell logo number 2 with red details and red "SHELL" across center.
SHELL 42"
(5) 1922-1929 $350
Die-cut yellow porcelain Shell logo number 4 with red details and red "SHELL" across center.
SHELL GASOLINE 36"
(5) 1929-1935 $350
Die-cut yellow porcelain Shell logo number 3 with red details and red "SHELL" arched above "GASOLINE" across center.
SHELL GASOLINE 36"
(5) 1929-1935 $350
Die-cut yellow porcelain Shell logo number 6 with red details and red "SHELL" arched above "GASOLINE" across center.
SHELL 42"
(2) 1929-1946 $350
Die cut yellow porcelain Shell logo number 6 with red details and red "SHELL" across center.
SHELL 42"
(4) 1946-1960 $350
Die-cut yellow porcelain Shell logo number 7 with red details and red "SHELL" across center.
TWO PIECE SIGN SECTIONS
NOTE: The following sign sections were assembled to form a complete die-cut shell shaped porcelain sign.
SHELL 24x48
(3) 1930s ASSEMBLED $250
Upper section of logo number 6 with red "SHELL" arched across center.
GASOLINE 24x48
(4) 1930s
Lower section of logo number 6 with red "GASOLINE" across center.
MOTOR OIL 24x48
(4) 1930s

Lower section of logo number 6 with red "MOTOR OIL" across center.

SHELL 30x60
(3) 1930s ASSEMBLED $250
Upper section of logo number 6 with red "SHELL" arched across center.

GASOLINE 30x60
(4) 1930s
Lower section of logo number 6 with red "GASOLINE" across center.

MOTOR OIL 30x60
(4) 1930s
Lower section of logo number 6 with red "MOTOR OIL" across center.

BUILDING SIGNS & LETTERS

SHELL 30"
(4) 1915-1925 $400
Die-cut and embossed three dimension white porcelain shell for rooftop mounting.

SHELL PANEL 24x120
(4) $200
Embossed yellow self framed porcelain strip with embossed red "SHELL" across strip.

SHELL PANEL 12x120
(4) $200
Embossed yellow self framed porcelain strip with embossed red "SHELL" across strip.

SHELLUBRICATION PANEL 8x144
(4) 1935-1957 $150
Embossed yellow self framed porcelain strip with embossed red "SHELLUBRICATION" across strip.

LUBRICATION 8x120
(4) 1935-1957 $100
Embossed yellow self framed porcelain strip with embossed red "LUBRICATION" across strip.

BATTERIES 8x96
(4) 1935-1957 $100
Embossed yellow self framed porcelain strip with embossed red "BATTERIES" across strip.

TIRES 8x96
(4) 1935-1957 $100
Embossed yellow self framed porcelain strip with embossed red "TIRES" across strip.

WASHING 8x120
(4) 1935-1957 $100
Embossed yellow self framed porcelain strip with embossed red "WASHING" across strip.

SHELL 8x96
(4) 1935-1957 $250
Embossed yellow self framed porcelain strip with embossed red "SHELL" across strip.

PUMP SIGNS

SHELL GASOLINE ON TAB 12x14
(5) N0 Listing
Die-cut red detailed yellow porcelain Shell logo number 3 with red "SHELL" arched above red "GASOLINE" in center of sign. Red tab for mounting on cylinder rods on visible pumps.

SHELL GASOLINE 12"
(3) 1937-1941 $800
Die-cut yellow detailed red porcelain Shell logo number 6 with black outlined yellow "SHELL" across center. Small yellow "GASOLINE" below.

SUPER SHELL GASOLINE 12"
(4) 1932-1937 $1800
Die cut yellow detailed red porcelain Shell logo number 6 with yellow "Super" diagonal above black outlined yellow "SHELL." Small yellow "GASOLINE" below.

SHELL PREMIUM GASOLINE 12"
(3) 1937-1941 $1200
Die cut yellow detailed red porcelain Shell logo number 6 with black outlined yellow "SHELL" above white "Premium" across center. Small yellow "GASOLINE" below.

SHELL ETHYL GASOLINE 12x16
(5) 1932-1937 No Listing
Die-cut black outlined yellow detailed red porcelain Shell logo number 6 with black outlined yellow "SHELL" arched across center. Small yellow "GASOLINE" below. Offset to lower left is a black outlined die-cut white circle with Ethyl (EGC) logo with yellow rays filling most of circle. Very complex die cut sign.

SILVER SHELL GASOLINE 12"
(5) 1932-1935 $1200
Die cut silver detailed red porcelain Shell logo number 6 with silver "SILVER/SHELL" across center above smaller silver "GASOLINE" below. Eastern markets only.

GASCO MOTOR FUEL 12"RD
(4) 1930-1933 $1500
Light blue outlined dark blue sign with light blue outlined white center circle. Light blue "GASCO" arched around top, "MOTOR FUEL" around bottom with light blue/dark blue soaring eagle in white center circle. Western markets only

GREEN STREAK GASOLINE A SHELL PRODUCT 12"RD
(4) 1933-1941 $800
Dark green outlined light green sign with yellow band across lower sign face. Dark green script "Green/Streak" arched across sign above green "GASOLINE" Red "A SHELL PRODUCT" on yellow band. Western markets only.

SHELL RED PLASTIC 12x12
(1) 1957-1965 $45
Red three dimensional plastic square with embossed shell logo in center with red "SHELL" across logo. White "GASOLINE" across bottom.

SHELL GOLD PLASTIC 12x12
(1) 1957-1965 $45
Gold three dimensional plastic square with embossed shell logo in center with red "SHELL" across logo. White "GASOLINE" across bottom.

SUPER SHELL 4x12
(3) 1957-1965 $25
Red plastic diagonal banner with white "SUPER SHELL" across banner.

NEON

SHELL DIE CUT EMBOSSED WITH NEON 42"
(2) 1929-1946 $1000/side
Die cut embossed yellow porcelain Shell logo number 6 with red details and red "SHELL" across center, assembled back to back with neon outline.

SHELL DIE CUT FLAT WITH NEON 42"
(2) 1929-1946 $800
Die cut yellow porcelain Shell logo number 6 with red details and red "SHELL" across center, assembled back to back on can type frame with neon outline.

CURB SIGNS

SHELL AVIATION GASOLINE 30"
(5) 1912-1917 No Listing
Black/red outlined yellow porcelain sign with red center circle. White shell scallop with black details in center circle. Red outlined yellow "AVIATION" across center of shell. Black outlined red "SHELL" arched aorund top, "GASOLINE" around bottom. Small red diamond to either side of center circle.

SHELL GASOLINE 25" $600
Shell logo number 2 with red details and red "SHELL" arched across top above red "GASOLINE" in center.

SHELL GASOLINE 25" $600
Shell logo number 3 with red details and red "SHELL" arched across top above red "GASOLINE" in center.

SHELL GASOLINE 25" $500
Shell logo number 6 with red details and red "SHELL" arched across top above red "GASOLINE" in center.

GASCO MOTOR FUEL 30"RD
(5) 1920's $750
Black outlined white porcelain sign with black outlined red center circle. Black/white eagle on center circle with red "GASCO" arched aorund top, "MOTOR FUEL" around bottom. Gasco was a west coast Shell economy grade before "Green Streak" was introduced.

SHELL MOTOR OIL 25" $550
Shell logo number 3 with red details and red "SHELL" arched across top above red "MOTOR OIL" in center.

FLANGE SIGNS

SHELL KEROSENE 15"RD
(5) $500
Red outlined yellow porcelain sign with red center circle. Black outlined red "SHELL" arched around top, "KEROSENE" around bottom with yellow Shell similar to logo number 1 in center.

SHELL MOTOR OIL 16x18
(4) 1922-1929 $350
Die cut yellow tin litho shell shaped flange signs with logo number 3. Red "SHELL" arched across top above red "MOTOR/OIL."

SHELL GASOLINE 17.5x22
(4) 1920s $350
Die cut yellow tin litho shell shaped flange signs with logo number 5. Red "SHELL" arched across top above red "GASOLINE"

SHELL MOTOR OIL 17.5x22
(4) 1920s $350
Die cut yellow tin litho shell shaped flange signs with logo number 5. Red "SHELL" arched across top above red "MOTOR/OIL"

SHELL KEROSENE 17.5x22
(4) 1920s $350
Die cut yellow tin litho shell shaped flange signs with logo number 5. Red "SHELL" arched across top above red "KEROSENE"

SHELL FURNACE OIL 17.5x22
(4) 1920s No Listing
Die cut yellow tin litho shell shaped flange signs with logo number 5. Red "SHELL" arched across top above red "FURNACE OIL"

SHELL GASOLINE 17.5x22
(4) 1929-1946 $350
Die cut yellow tin litho shell shaped flange signs with logo number 6. Red "SHELL" arched across top above red "GASOLINE."

SHELL MOTOR OIL 17.5x22
(4) 1929-1946 $350
Die cut yellow tin litho shell shaped flange signs with logo number 6. Red "SHELL" arched across top above red "MOTOR/OIL"

SHELL KEROSENE THICK LETTERS 17.5x22
(4) 1929-1946 $350
Die cut yellow tin litho shell shaped flange signs with logo number 6. Red "SHELL" arched across top above thick red "KEROSENE."

SHELL KEROSENE THIN LETTERS 17.5x22
(4) 1929-1946 $350
Die cut yellow tin litho shell shaped flange signs with logo number 6. Red "SHELL" arched across top above thin red "KEROSENE"

SHELL KEROSENE THIN LETTERS 9x12
(4) 1929-1946 $350
Die cut yellow tin litho shell shaped flange signs with logo number 6. Red "SHELL" arched across top above thin red "KEROSENE"

LUBRICATION/MOTOR OIL SIGNS

SHELL MOTOR OIL 30"
(4) 1929-1946 $500
Die cut yellow porcelain shell shaped sign with logo number 6. Thin red "SHELL" arched across top above red "MOTOR/OIL."

SHELL MOTOR OIL 30"
(4) 1929-1946 $500
Die cut yellow porcelain shell shaped sign with logo number 6. Thick red "SHELL" arched across top above red "MOTOR/OIL."

GOLDEN SHELL MOTOR OIL 30"
(4) 1929-1946 $550
Die cut yellow porcelain shell shaped sign with logo number 6. Black outlined red "Golden/SHELL" diagoanl across top above red "MOTOR/OIL."

SHELL PENN YELLOW 30"
(4) 1929-1946 $650
Die cut yellow porcelain shell shaped sign with logo number 6. Red "SHELL-PENN" arched across top above red "100%/PURE PENNSYLVANIA/MOTOR/OIL."

SHELL PENN RED 30"
(4) 1929-1946 $700
Die cut red porcelain shell shaped sign with logo number 6 and yellow details. Black outlined yellow "SHELL-PENN" arched across top above yellow "100%/PURE PENNSYLVANIA" above black outlined yellow "MOTOR/OIL."

SHELL CHASSIS & MOTOR LUBRICATION 30"
(4) 1929-1946

Die cut yellow porcelain shell shaped sign with logo number 6. Thin red "SHELL" arched across top above red "CHASSIS/AND/ MOTOR/LUBRICATION."

GOLDEN SHELL MOTOR OIL 12"
(4) 1929-1946 $450
Die cut yellow porcelain shell shaped sign with logo number 6. Black outlined red "Golden/SHELL" diagonal across top above red "MOTOR/OIL."

SHELLUBRICATION 12"
(4) 1929-1946 $750
Die cut yellow porcelain shell shaped sign with logo number 6. Red "SHELLUBRICATION" arched across top above red "THE/MODERN/UPKEEP/SERVICE."

SHELLUBRICATION 12"
(4) 1929-1946 $750
Die cut yellow porcelain shell shaped sign with logo number 6. Red "SHELLUBRICATION" arched across top above red "A/CERTIFIED/GUARANTEED/SERVICE."

SHELLUBRICATION W/ OCTAGON 30x96
(4) 1930s $1200
Red outlined yellow self framed porcelain sign with red outlined yellow and white octagon at left end. Black outlined red 'SHELLUBRICATION" above black script "The Modern/Upkeep Service" on ocatgon. Small Shell logo at bottom of octagon. Large black outlined red "SHELLUBRICATION" above black script "The Modern/Upkeep Service" to right of ocatgon.

SHELLUBRICATION OCTAGON 30x30
(4) 1930s $1200
Red outlined yellow self framed porcelain sign with red outlined yellow and white octagon at left end. Black outlined red "SHELLUBRICATION" above black script "The Modern/Upkeep Service" on ocatgon. Small Shell logo at bottom of octagon.

SECONDARY SIGNS

SHELL NO SMOKING SHELL 4x24
(4) 1920s $1500
Red outlined yellow strip with shell logo number 6 at either end. Red "NO SMOKING" across center.

MEN DIE CUT 2x8
(4) 1920s $100
Red outlined yellow porcelain strip with die-cut tab at each end. Red "MEN" across sign.

LADIES DIE CUT 2x8
(4) 1920s $100
Red outlined yellow porcelain strip with die-cut tab at each end. Red "LADIES" across sign.

MEN 2x8
(4) 1920s $75
Yellow porcelain strip with red "MEN" across sign.

LADIES 2x8
(4) 1920s $75
Yellow porcelain strip with red "MEN" across sign.

WHITE CROSS 6"RD
(4) 1930s $100
Round porcelain sign with blue band around yellow center circle. White cross in center, with white "THE SYMBOL OF" arched around top, "A HOME CLEAN REST ROOM" around bottom.

SHELL AIR 12"
(5) 1922-1929 $650
Die cut yellow porcelain shell shaped sign with logo number 4. Red "SHELL" arched across top above red "AIR."

AIR 8x10
(4) 1929-1946 $500
Red outlined yellow rectangular panel with notch at bottom for mounting. Red "AIR" across center.

SHELL AUTHORIZED DEALER 10x24
(4) 1920s $450
Red outlined yellow porcelain sign with red line across center. Red "SHELL" across top above red "AUTHORIZED DEALER."

SHELL CHARGE ACCCOUNTS ACCEPTED 10x24
(4) 1930s $450
Yellow porcealin sign with red band across center. Small Shell logo number 6 in center with red "AUTHORIZED DEALER" across top with red "Charge Accounts Accepted Here" across bottom.

SHELL CHARGE CARDS HONORED 12x24
(5) 1940s $400
Yellow outlined red porcelain vertical rectangular sign. Yellow "SHELL/CHARGE/CARDS/HONORED" across sign.

BULK PLANT SIGNAGE

SHELL KEROSENE RANGE OIL 12x24
(5) 1940s $350
Red outlined yellow porcelain sign with small Shell logo number 6 in center. Red "KEROSENE" across top, "RANGE OIL" across bottom.

TRUCK SIGNS

SHELL 5"RD
(4) 1920s $600
Dark blue porcelain sign with shell logo number 2 in center.

SHELL 5"RD
(4) 1920s $600
Dark blue porcelain sign with Shell logo number 4 in center.

CANADIAN OIL COMPANY, LTD: Created in the mid 1930's from the Canadian holdings of Cleveland, OH based National Refining Company, Canadian Oil continued to market under the White Rose brand until mergining with Shell Cnanad in 1964.

IDENTIFICATION SIGNS

WHITE ROSE 42"RD
(3) 1936-1956 $600
Black outlined white porcelain sign with black outlined yellow center circle. Detailed white rose with green leaves in ecnter circle with black outlined red "WHITE" arched around top, "ROSE" around bottom.

WHITE ROSE SHIELD ON CIRCLE 42"RD
(4) 1956-1964 $450
Yellow porcelain sign with outlined white shield logo in center. Detailed white rose with green leaves at top with red "WHITE/ROSE" below.

SECONDARY SIGNS

WHITE ROSE 18x18
(4) 1940S $150
Yellow porcelain sign with detailed white rose with green leaves in center.

NORTH STAR OIL CO, LTD - WINNIPEG, MANITOBA:

Little is known of North Star history. The company was founded in Winnepeg in 1920 and operated gasoline stations throughout western Canada until after a 1961 merger with Shell Canada.

IDENTIFICATION SIGNS

NORTH STAR 48"RD
(3) 1946-1961 $350
White outlined blue porcelain sign with white triangular area at lower right. White stars on blue area, with red star at apex of triangle. Red "NORTH/STAR" below.

SIGNAL AND SIGNAL/CHARTER BRANDS

SIGNAL OIL AND GAS - LOS ANGELES, CA In 1932 Signal Oil and Gas of Los Angeles, formerly a producer/refiner entered the retail marketplace under the Signal name. Involved with Standard of California from it's earliest days, Signal sold it's retail marketing to Standard Chevron in 1947. In 1950 they re-entered the market with the purchase of aoportion of the Bankline Oil/Norwalk & Regal Gas Stations operations. In 1958 Signal merged with Hancock Oil of Long Beach,Ca and bought the Houston based Watson Oil and rebranded the stations "Hancock." The Signal brand was operated by Chevron until about 1965 when the entire marketing network was sold to Humble Oil and stations were rebranded Enco. Hancock, Norwalk and other Signal brands were sold off beginning in 1970. Stations in the west were sold off individually, while the southeastern marketing was sold to Charter Oil/Dixie Vim.

SIGNAL OIL AND GAS - LOS ANGELES, CA

See Signal history in introduction above.

IDENTIFICATION SIGNS

SIGNAL 72"RD
(3) 1931-1962 $1000
Yellow/red outlined black porcelain sign with red outline black circle in center. Yellow "SIGNAL" arched around top, "GAS" around bottom with "GO" traffic signal design in center.

SIGNAL 42x84
(5) 1962-1966 $250
Red elongated shield shaped sign with white over yellow oval in center. Black "Signal" across color split in oval.

BUILDING SIGNS AND LETTERS

SIGNAL PANEL
(3) 19 $300
Red outlined black porcelain strip with notched corners. Yellow "SIGNAL" across center.

SIGNAL LUBRICATION PANEL
$250
Red outlined black porcelain strip with notched corners. Yellow "SIGNAL LUBRICATION" across center.

TIRES & BATTERIES PANEL
$200
Red outlined black porcelain strip with notched corners. Yellow "TIRES & BATTERIES" across center.

PUMP SIGNS

SIGNAL GASOLINE 12"RD
(2) 1946-1962 $400
White/red outlined black porcelain sign with red outlined black circle in center. Yellow "SIGNAL" arched around top, "GASOLINE" around bottom with black and white "GO" traffic signal in center.

SIGNAL ETHYL 12"RD
(2) 1946-1962 $350
White/red outlined black porcelain sign with red outlined black circle in center. Yellow "SIGNAL" across top, "GASOLINE" across bottom with red outlined white script "Ethyl" across center. Small Ethyl (EC) logo in white circle at bottom.

SIGNAL DIESEL 12"RD
(2) 1946-1962 $400
White/red outlined black porcelain sign with red outlined black circle at top. Black and white "GO" traffic signal in circle. Yellow "SIGNAL" across top, "FUEL" across bottom with white "DIESEL" in center.

SECONDARY SIGNS

SIGNAL ETHYL 12x30
(3) 1930's $350
Wood framed red tin litho sign. White "SIGNAL" over "GO" traffic signal in black diamond at left with yellow "ETHYL" with black dropshading at right.

TRUCK SIGNS

SIGNALPRODUCTS $350
Red outlined black porcelain strip with notched corners. Yellow "SIGNAL PRODUCTS" across center.

SIGNAL SMALL PANEL 4x20
(4) 1930's $400
Red outlined black porcelain strip with notched corners. Yellow "SIGNAL" across center.

SIGNAL OIL AND GAS CO 6"RD
(4) 1950-1970 $150
White outlined circle with red upper half and black lower half, split diagonally. Black oil derricks on upper red area, with black outlined white "SIGNAL" across center. White "OIL AND GAS/COMPANY" on lower black area.

HANCOCK REFINING CO. - LONG BEACH, CA

Founded in 1925, Hancock was a California based refiner/marketer with stations throughout the state. In 1958 Signal Oil and Gas merged with Hancock Refining and stations remained under the Hancock brand in California and the brand was expanded to Texas when Signal purchased the Watson operation and rebranded the stations Hancock. Hancock brand discontinued when Signal marketing was sold off in 1970.

IDENTIFICATION SIGNS

HANCOCK GASOLINE 72"RD
(4) 1946-1959 $800
White porcelain sign with orange band and stripe across top, black band and stripe across bottom. Black "HANCOCK" across center, with black and orange rooster above and orange "GASOLINE" below.

PUMP SIGNS

HANCOCK GASOLINE 12"RD
(4) 1946-1959 $500
White porcelain sign with orange band and stripe across top, black band and stripe across bottom. Black "HANCOCK" across center, with black and orange rooster above and orange "GASOLINE" below.

SECONDARY SIGNS
NO SMOKING/STOP YOUR MOTOR 8x30
(4) 1946-1959 $450
White porcelain strip with orange and black rooster in center. Black "NO SMOKING" to left, "STOP YOUR MOTOR" to right,

BANKLINE OIL COMPANY: California marketer that operated stations under the "Norwalk" brand. Purchased by Signal in 1950, Norwalk was the company's flagship brand until Hancock was purchased in 1958.
No collectible size Norwalk signs are known at this time.

REGAL SERVICE STATIONS: California retial marketer purchased by Signal in the 1950s. When Signal sold out their retail operations in 1970, the Regal chain was sold to a west coast refining operation and continued in operation through the early 1990s. Regal's gasoline grades, as shown below, were unusual in that they didn't carry the company name.
REGENT REGULAR 10x14
(5) 1946-1950 $350
White outlined red porcelain sign with white "REGENT" with blue dropshading across top. White figure of Mercury" at left, through "REGENT." White outlined blue "REGULAR" across center with white "GASOLINE" across bottom.
REGENT PREMIUM ETHYL 10x14
(5) 1946-1950 $350
White outlined blue porcelain sign with white "REGENT" with red dropshading across top. White figure of Mercury" at left, through "REGENT." Yellow "PREMIUM/ETHYL" across center with white "GASOLINE" across bottom. Yellow circle with Ethyl (EC) logo in lower left corner.
FORTUNE 12x12
(3) 195-1960 $100
White/green outlined white porcelain sign with green "FORTUNE" across top. Black outlined white horseshoe with green four leaf clover in center. Black "GASOLINE" across bottom.

WATSON OIL CO. - HOUSTON, TX
Houston based discounter purchased by Signal about 1960 and rebranded Hancock.

REGAL
California independent purchased by Signal about 1960.
No collectible size Watson signs are known at this time.

BILLUPS BROS. - GREENWOOD, MS
Founded in Greenwood, Mississippi in 1926, Billups survived as a family owned operation until 1959. In 1959 the company was split into two operating units, Billups Eastern and Billups Western. Billups Western was sold out to Signal Oil and operated by them until 1970 when Signal got out of retail gasoline marketing and sold the operation to Charter Oil. A majority of Billups Eastern was sold to Independent refiner Delhi-Taylor, who was bought out by Hess Oil and Chemical in 1963. The remaining portion of Billups Eastern was bought by Hess in 1965 and Billups stations were rebranded Hess. See Hess section for more information.
IDENTIFICATION SIGNS
FILL UP WITH BILLUPS YOUR FRIEND
(5) 1946-1965 $250
White tin litho sign with Billups hand logo in center. "Fill Up With Billups" across top, "YOUR FRIEND" across bottom.
BUILDING SIGNS AND LETTERS
BILLUPS RED PORCELAIN LETTERS 24"
(5) 1946-1980 $100
Red porcelain "BILLUPS" channel type letters.
PUMP SIGNS
BILLUPS YOUR FRIEND 12"RD
(5) 1946-1965 $300
White outlined red porcelain sign with white band across center. Series of red and blue lines forming half-circles in center. Blue "BILLUPS" in center band with white "YOUR" arched around top, "FRIEND" around bottom.
BILLUPS ETHYL 12"RD
(5) 1946-1965 $450
Black outlined red porcelain sign with white circle in center. Ethyl (EC) logo in center circle. White script "Billups" with grey inline arched around top with small airplane to either side of center circle. White "PREMIUM" arched around bottom.

SPAR - ORLANDO, FL
Originally Orlando Ice and Fuel, they were bought out by Billups Eastern soon after the Billups split in 1959. After the Billups merger with Hess, they were operated by another independent for several years.
No collectible size Spar signs are known to exist.

ROSE OIL CO. - VICKSBURG, MS
Rose Oil Co was operated by a member of the Billups family. The "MMP" stands for Mississippi Motor Products.
No collectible size Rose signs are known at this time.

SUPERTEST OIL CO. - TAMPA, FL
Postwar Florida based discounter bought out by Signal about 1960 and sold off to Charter in 1970.
PUMP SIGNS
SUPERTEST GASOLINE 10x14
(5) 1960's $75
White aluminum litho sign with red outlined elongated hexagon at top above blue band. Blue "SuperTest" in hexagon, with white "GASOLINE" in blue band.
THERE'S SOMETHING SUPER ABOUT SUPERTEST 12x18
(5) 1960's $75
Embossed red tin litho sign white gold "SUPER" across center. White "There's/Something" above "SUPER," white "about" below with white elongated hexagon at bottom. Red "SuperTest" in hexagon.
NEON SIGNS
SAVE WITH SUPERTEST 48x108
(5) 1950's No Listing
Red irreular shaped can type tin litho sign with neon sockets and chase lights around border. White "SAVE/WITH" at top above white "SUPERTEST."

SOUTHLAND OIL CO. - SAVANNAH, GA
In 1970 when Signal Oil sold off its retail marketing, "SOC"-Southland Oil Co of Savannah,Ga was sold off to it's original owners, who had remained in management since Signal's pur-

chase of SOC about 1960. They operated as an independent until about 1986 when they became a branded Texaco operation.

IDENTIFICATION SIGNS

SUPER SOC 72"RD
(5) 1960's $200
Blue/white outlined red porcelain sign with red burst in center. Blue "SOC" in center of burst, with blue outlined box just above. Red "SUPER" in box.

PUMP SIGNS

SUPER SOC 14"RD
(5) 1960's $200
Blue/white outlined red tin litho sign with red burst in center. Blue "SOC" in center of burst, with blue outlined box just above. Red "SUPER" in box.

CURB SIGNS

MIGHTY SOC Credit Cards 36"
(1) 1960's $250
Wooden cutout litho sign with character "Mighty Soc" boy at top with "CREDIT CARDS/AVAILABLE HERE" across bottom.

SOUTHERN OIL STORES/DIXIE VIM - JACKSONVILLE, FL

In 1970 Signal Oil again left the retail gasoline marketplace. Charter Oil Co of Jacksonville,FL had the year before purchased the Southern Oil Stores-Dixie Vim chain and when Billups,Supertest and the former Watson/Hancock stations were placed on the market, they added them to the Dixie Vim operation. Southern Oil Stores had been an independent marketer,one of the original trackside discount chains with approximately 50 stations. Founded about 1929 it is listed here only to aid in understanding the change in ownership of Billups and Supertest.

IDENTIFICATION SIGNS

DIXIE VIM 42x42
(5) 1930-1960 $1800
Green outlined white porcelain sign with green band across center. White "DIXIE VIM" with speed lines in green band. Green tank car in upper white area with red "From Tank Car," green 1930's automobile with red "To Car Tank" in lower white area. Red line connecting tank car and automobile.

SINCLAIR AND AFFILIATES
SINCLAIR OIL CORPORATION - NEW YORK, NY

Founded in 1916 in New York from the various holdings of founder Harry Sinclair, including the refining and marketing formerly under the Cudahy name, Sinclair was a huge operation right from the start. With oil production, pipelines, railcars, refineries, terminals, bulk plants and service stations, Sinclair was one of few oil companies to start as a completely intergrated operation. Sinclair grew larger during the prosperous 1920s in spite of Harry Sinclair's involvement in the "Teapot Dome" oil scandal of President Warren Harding's administration. With several important purchases during the early 1930s, such as Pierce, Parco, Rio Grande and Richfield of New York, Sinclair was marketing from coast-to-coast. With the 1937 reorganization of Rio Grande into Richfield of California, Sinclair direct operations on the west coast were discontinued, but Sinclair continued to market from the Rockies east. From 1932 until 1943 the corporate name was Consolidated Oil Company, changing back to Sinclair Oil Corporation in 1943. After WW2 Sinclair continued to expand direct operated and jobber marketing, building thousands of service stations along the new interstate highway system. Marketing surpassed production and the company could no longer supply all the crude oil needed for it's own branded marketing. In 1969 Sinclair merged with oil-rich and newly created Atlantic-Richfield, selling off it's east coast marketing to British Petroleum. In 1973 Atlantic-Richfield was court-ordered to sell most of Sinclair, and the operation was spun off as an independent oil refiner and marketer that today operates more than 2500 stations in 20 states.

IDENTIFICATION SIGNS

SINCLAIR OILS 30"RD
(4) 1920-1924 $450
Green outlined white porcelain sign with green outlined white center circle. Series of green vertical stripes in center with green "SINCLAIR" arched around top, "OILS" around bottom.

SINCLAIR OILS 30"RD
(3) 1924-1926 $425
Green outlined white porcelain sign with green outlined white center circle. Series of green vertical stripes in center with red "SINCLAIR" arched around top, "OILS" around bottom.

SINCLAIR GASOLINE 36"RD
(4) 1920-1926 $450
Green outlined white porcelain sign with green outlined white center circle. Series of green vertical stripes in center with green "SINCLAIR" arched around top, "GASOLINE" around bottom.

SINCLAIR GASOLINE 42"RD
(3) 1920-1926 $400

SINCLAIR GASOLINE 48"RD
(2) 1920-1926 $350
Green outlined white porcelain sign with green outlined white center circle. Series of green vertical stripes in center with red "SINCLAIR" arched around top, "GASOLINE" around bottom.

SINCLAIR HC STRIPED 48"RD
(5) 1926-1927 NO LISTING
Green outlined white porcelain sign with green outlined white center circle. Series of green vertical stripes in center, interrupted by white horizontal band. Red "H-C" in band, with red "SINCLAIR" arched around top, "GASOLINE" around bottom.

SINCLAIR AIRCRAFT 48"RD
(5) 1926-1928 $3500
Red/white outlined green porcelain sign with red/white outlined red center circle. White airplane with red details in center circle with white "SINCLAIR" arched around top, "AIRCRAFT" around bottom.

SINCLAIR AIRCRAFT RED/WHITE 48"RD
(4) 1928-1931 $2500
Black/white outlined red porcelain sign with black/white outlined red center circle. White airplane with black details in center circle with black outlined white "SINCLAIR" arched around top, "AIRCRAFT" around bottom.

SINCLAIR HC 48"
(1) 1927-1958 $350

SINCLAIR HC 72"
(1) 1935-1958 $250
Green/white outlined green porcelain sign with white outlined red center circle. White "H-C" with black dropshading in center circle, White "SINCLAIR" with black dropshading arched around top, "GASOLINE" around bottom. NOTE
Signs of this design made prior to World War 2 have baked-on painted black dropshading. THose made after WW2 have black porcelain dropshading
SINCLAIR TRIPLE CHECK ROUND 48"RD
(5) 1956-1958 $500
White porcelain sign with black trapezoidal area at top. Red, white, green check marks on trapezoidal area. Red "SINCLAIR" below.
SINCLAIR TRIPLE CHECK SHIELD 60x72
(5) 1956-1958 $500
Steel framed white porcelain trapezoidal sign with black trapezoidal area at top. Red, white, green check marks on trapezoidal area. Red "SINCLAIR" below.
SINCLAIR 48x60
(2) 1958-1972 $250
SINCLAIR 60x72
(1) 1958-PRESENT $100
Green outlined white porcelain trapezoidal shield shaped sign with green dinosaur in lower point of shield. Orange-red "Sinclair" logotype across upper sign area.

BUILDING SIGNS AND LETTERS

SINCLAIR BUILDING PANEL 10x72
(5) 1931-1959 $250
Green/red outlined flat green porcelain panel with white "SINCLAIR" across center.
SINCLAIR BUILDING PANEL 10x72
(4) 1931-1959 $250
SINCLAIR BUILDING PANEL 12x96
(2) 1931-1959 $200
SINCLAIR BUILDING PANEL 14x120
(2) 1931-1959 $100
SINCLAIR BUILDING PANEL 18x144
(4) 1931-1959 $100
Black/red outlined self framed green porcelain panel with white "SINCLAIR" with black dropshade across center.
WASHING PANEL 6x66
(3) 1931-1959 $150
Black/red outlined self framed green porcelain panel with white "WASHING" with black dropshade across center.
LUBRICATION PANEL 6x66
(3) 1931-1959 $150
Black/red outlined self framed green porcelain panel with white "LUBRICATION" with black dropshade across center.
SINCLAIR OPALINE MOTOR OIL 12x60
(2) 1929-1933 $275
Green outlined white porcelain vertical panel with red "OPALINE" vertical down center. Green "SINCLAIR" across top, "MOTOR/OIL" across bottom.
SINCLAIR MOBILINE MOTOR OIL 12x60
(4) 1929-1931 $300
Red outlined white porcelain vertical panel with red veritical rectangle down center. White "MOBILENE" with black dropshade vertical on rectangle with green "SINCLAIR/PENNSYLVANIA" across top, "MOTOR/OIL" across bottom.
SINCLAIR PENNSYLVANIA MOTOR OIL 12x60
(2) 1931-1933 $300
Green outlined white porcelain vertical panel with red veritical rectangle down center. White "PENNSYLVANIA" with black dropshade vertical on rectangle with green "SINCLAIR" across top, "MOTOR/OIL" across bottom.
SINCLAIR OPALINE MOTOR OIL 12x60
(2) 1933-1940 $300
Green outlined white self framed porcelain vertical panel with red "OPALINE" vertical down center. Green "SINCLAIR" across top, "MOTOR/OIL" across bottom.
SINCLAIR PENNSYLVANIA MOTOR OIL 12x60
(2) 1933-1940 $300
Green outlined white self framed porcelain vertical panel with red veritical rectangle down center. White "PENNSYLVANIA" with black dropshade vertical on rectangle with green "SINCLAIR" across top, "MOTOR/OIL" across bottom.

PUMP SIGNS

SINCLAIR GASOLINE
(1) 1961-1972 $35
White porcelain sign with red/white/green Sinclair shield logo at top ab
ove red "GASOLINE" across lower sign face.
SINCLAIR DIESEL
(2) 1961-1965 $45
White porcelain sign with red/white/green Sinclair shield logo at top above red "DIESEL" across lower sign face.
SINCLAIR DIESEL
(1) 1965-1972 $35
White porcelain sign with red/white/green Sinclair shield logo at top above green "DIESEL" in condensed lettering across lower sign face.

DIAL FACE INSERTS

NOTE: The following signs were used to form the lower dial face flange on bennett 2066 series pumps. All are white porcelain strips with angled ends and a single mounting hole at top center.
DINO GASOLINE 4x16
(1) 1962-1972 $35
White porcelain strip with green "DINO" beside smaller green "GASOLINE" across strip.
DINO SUPREME 4x16
(1) 1962-1972 $35
White porcelain strip with red "DINO" beside smaller red "SUPREME" across strip.
DINO GASOLINE/DINO SUPREME 4x16
(1) 1962-1972 $35
White porcelain strip with green "DINO/GASOLINE" on right half of strip and red "DINO/SUPREME" on left half.
DINO SUPREME/DINO GASOLINE 4x16
(1) 1962-1972 $35
White porcelain strip with red "DINO/SUPREME" on right half of strip and green "DINO/GASOLINE" on left half.
SINCLAIR HC 4x16
(4) 1956-1962 $50
White porcelain strip with green "SINCLAIR/HC/GASOLINE" across center.
SINCLAIR POWER X 4x16

(4) 1956-1962 $50
White porcelain strip with red "SINCLAIR/POWER-X/GASOLINE" across center.

NEON SIGNS

Sinclair used some building panel signs with neon overlays as rooftop signs. None are of collectible size.

CURB SIGNS

SINCLAIR OILS 24"RD
(5) 1921-1930 $450
Green outlined white porcelain sign with green outlined white center circle. Series of green vertical stripes in center with red "SINCLAIR" arched around top, "OILS" around bottom.

SINCLAIR GASOLINE 24"RD
(5) 1921-1926 $425
Green outlined white porcelain sign with green outlined white center circle. Series of green vertical stripes in center with red "SINCLAIR" arched around top, "GASOLINE" around bottom.

SINCLAIR HC 24"RD
(4) 1927-1940 $375
Green/white outlined green porcelain sign with white outlined red center circle. White "H-C" with black dropshading in center circle, White "SINCLAIR" with black dropshading arched around top, "GASOLINE" around bottom.

SINCLAIR OPALINE MOTOR OIL w/CAN 24"RD
(5) 1921-1925 No Listing
Red outlined white porcelain sign with detailed green and white striped vertical gallon can in center.

SINCLAIR OPALINE MOTOR OIL 24"RD
(5) 1930-1936 $350
Green/white outlined green porcelain sign with white band across center. Red "OPALINE" with green dropshading across center with white "SINCLAIR" with black dropshading arched around top, "MOTOR OIL" around bottom.

SINCLAIR OPALINE MOTOR OIL 24"RD
(4) 1936-1946 $1250
White outlined green porcelain sign with red circle positioned above center. White outlined green "SINCLAIR" logotype above red "OPALINE/MOTOR OIL" on upper red circle with Green outlined white dinosaur standing on red mound below. White "MELLOWED/80 MILLION YEARS" on mound.

SINCLAIR PENNSYLVANIA MOTOR OIL 24"RD
(4) 1936-1946 $1250
White outlined red porcelain sign with white circle positioned above center. Black outlined white "SINCLAIR" logotype above red "PENNSYLVANIA/MOTOR OIL" on upper white circle with white outlined black dinosaur standing on black mound below. White "MELLOWED/100 MILLION YEARS" on mound.

LUBRICATION/MOTOR OIL SIGNS

DRAIN AND REFILL WITH NEW OPALINE 24x36
(2) 1920s $450
Red/white/green outlined white porcelains ign with red "NOW" across top. Green "while you wait/Drain(red) and Refill(red) with/New" across center and red "OPALINE" across bottom.

AUTHORIZED DEALER SINCLAIR OPALINE 20x48
(3) 1922-1929 $450
Red/white outlined green porcelain sign with white/green outlined white panel offset to left. Green/white striped gallon vertical can at right, with green "SINCLAIR" over red "OPALINE" over green "MOTOR OIL" on white panel. White "Authorized Dealer" across top, "Seals Power at every Degree of Wear" across bottom.

SINCLAIR OPALINE 20x48
(4) 1922-1929 $850
Red/white outlined green porcelain sign with white/green outlined white panel offset to left. Green/white striped gallon vertical can at right, with green "SINCLAIR" over red "OPALINE" over green "MOTOR OIL" on white panel. White "Seals Power at every Degree of Wear" across bottom.

SINCLAIR OPALINE F MOTOR OIL
AUTHORIZED DEALER 20x48
(4)1920-1926 $925
Green outlined white porcelain sign with red line across bottom. Green/white Sinclair striped logo at right end of line. Green "Made Expressly for FORD
(red) cars" arched across top above "to keep them silent/and smooth." Green "SINCLAIR" above red "OPALINE F" above green "MOTOR OIL" across right end of sign with green "Authorized Dealer" across bottom below red line.

SINCLAIR OPALINE 12x12
(4) 1922-1925 $450
Red porcelain sign with green outlined white circular logo in center with green outlined white center circle. Series of green vertical stripes in center with green "SINCLAIR" arched around top, "OPALINE" around bottom.

SINCLAIR OPALINE MOTOR OILS 12x17
(4) 1920-1925 $450
White/red outlined white porcelain sign with red square at top. Green outlined white circular logo in center with green outlined white center circle. Series of green vertical stripes in center with green "SINCLAIR" arched around top, "OPALINE" around bottom. Green "MOTOR/OILS" across bottom.

SINCLAIR OILS 6"RD
(3) 1925-1959 $525
Green outlined white porcelain sign with green outlined white center circle. Series of green vertical stripes in center with green "SINCLAIR" arched around top, "OILS" around bottom.

SINCLAIR OILS 10"RD
(3) 1925-1959 $600
Green outlined white porcelain sign with green outlined white center circle. Series of green vertical stripes in center with green "SINCLAIR" arched around top, "OILS" around bottom.

SINCLAIR OILS 12"RD
(3) 1925-1959 $600
Green outlined white porcelain sign with green outlined white center circle. Series of green vertical stripes in center with green "SINCLAIR" arched around top, "OILS" around bottom.

SINCLAIR OPALINE MOTOR OIL 10"RD
(5) 1930-1936 $400
Green/white outlined green porcelain sign with white outlined red band across center. White "OPALINE/MOTOR OIL" with green dropshading across center with white "SINCLAIR" with black dropshading arched around top.

SINCLAIR OPALINE MOTOR OIL 16"RD
(5) 1930-1936 $450
Green/white outlined green porcelain sign with white outlined red band across center. White "OPALINE/MOTOR OIL" with

Miscellaneous pump plate signs. *Bruce Miller.*

Miscellaneous pump plate signs. *Bruce Miller.*

Miscellaneous pump plate signs. *Bruce Miller.*

Miscellaneous pump plate signs. *Bruce Miller.*

Miscellaneous small signs. *Bruce Miller.*

D-X hanging neon sign.

Sohio hanging neon sign.

Skelly hanging neon sign.

Four different Shell flange tin signs.

Pure porcelain credit card sign.

Shell Motor Oil 42" porcelain sign.

Phillips 66 hanging neon sign.

green dropshading across center with white "SINCLAIR" with black dropshading arched around top.

SINCLAIR OPALINE MOTOR OIL 10"RD
(4) 1936-1946 $1100
White outlined green porcelain sign with red circle positioned above center. White outlined green "SINCLAIR" logotype above red "OPALINE/MOTOR OIL" on upper red circle with Green outlined white dinosaur standing on red mound below. White "MELLOWED/80 MILLION YEARS" on mound.

SINCLAIR OPALINE MOTOR OIL 16"RD
(4) 1936-1946
White outlined green porcelain sign with red circle positioned above center. White outlined green "SINCLAIR" logotype above red "OPALINE/MOTOR OIL" on upper red circle with Green outlined white dinosaur standing on red mound below. White "MELLOWED/80 MILLION YEARS" on mound.

SINCLAIR OPALINE MOTOR OIL 16" SQ
(4) 1936-1946 $1200
White outlined green porcelain sign with red circle positioned above center. White outlined green "SINCLAIR" logotype above red "OPALINE/MOTOR OIL" on upper red circle with Green outlined white dinosaur standing on red mound below. White "MELLOWED/80 MILLION YEARS" on mound.

SINCLAIR PENNSYLVANIA MOTOR OIL 10"RD
(5) 1932-1936 $1100
White outlined red porcelain sign with white "SINCLAIR" logotype with black dropshade across top. Black standing dinosaur, with white band superimposed across dinosaur with black "PENNSYLVANIA/MOTOR OIL" on band. White "MELLOWED A HUNDRED/MILLION YEARS" on black mound below dinosaur.

SINCLAIR PENNSYLVANIA MOTOR OIL 16"RD
(5) 1932-1936 $1200
White outlined red porcelain sign with white "SINCLAIR" logotype with black dropshade across top. Black standing dinosaur, with white band superimposed across dinosaur with black "PENNSYLVANIA/MOTOR OIL" on band. White "MELLOWED A HUNDRED/MILLION YEARS" on black mound below dinosaur.

SINCLAIR PENNSYLVANIA MOTOR OIL 10"RD
(3) 1936-1940 $1200
White outlined red porcelain sign with white circle positioned above center. Black outlined white "SINCLAIR" logotype above red "PENNSYLVANIA/MOTOR OIL" on upper white circle with white outlined black dinosaur standing on black mound below. White "MELLOWED 100/MILLION YEARS" on mound.

SINCLAIR PENNSYLVANIA MOTOR OIL 16"RD
(4) 1936-1940 $1200
White outlined red porcelain sign with white circle positioned above center. Black outlined white "SINCLAIR" logotype above red "PENNSYLVANIA/MOTOR OIL" on upper white circle with white outlined black dinosaur standing on black mound below. White "MELLOWED/100 MILLION YEARS" on mound.

SINCLAIR PENNSYLVANIA MOTOR OIL 16"SQ
(4) 1936-1940 No Listing
White outlined red porcelain sign with white circle positioned above center. Black outlined white "SINCLAIR" logotype above red "PENNSYLVANIA/MOTOR OIL" on upper white circle with white outlined black dinosaur standing on black mound below. White "MELLOWED/100 MILLION YEARS" on mound.

SINCLAIR OILS LUBRICATNS OF LONG LIFE 20x48
(3) 1930s $850
White porcelain sign with red band across upper right. Green dinosaur at left, with white "SINCLAIR/OILS" across red band, green "LUBRICANTS OF LONG LIFE" across bottom.

SINCLAIR EXTRA DUTY MOTOR OIL 18x24
(3) 1953-1956 $75
Rectangular tin litho sign with red upper background and green lower background with fine white lines. White trapezoid superimposed over background with green "SINCLAIR" across top and "MOTOR/OIL" across bottom. Red "EXTRA/DUTY" in center.

SECONDARY SIGNS

SINCLAIR GASOLINE 12x19.5
(5) 1916-1922 No Listing
Green outlined cream embosssed tin litho sign with red "SINCLAIR/GASOLINE" across center.

SINCLAIR KEROSENE 12x19.5
(5) 1916-1922 No Listing
Green outlined cream embosssed tin litho sign with green "SINCLAIR/KEROSENE" across center.

SINCLAIR OPALINE MOTOR OIL 12x19.5
(5) 1916-1922 $600
Red outlined cream embosssed tin litho sign with green "SINCLAIR/OPALINE/MOTOR OIL" across center.

SINCLAIR OPALINE MOTOR OIL 12x19.5
(5) 1922-1925 $850
Red outlined cream embosssed tin litho sign with green "SINCLAIR/OPALINE/MOTOR OIL" across offset to right. Green and white striped can at left.

NOTICE - FILLING SERVICE... 18x24
(5) 1916-1922 $450
White porcelain sign with red "NOTICE" across top. Green "FILLING SERVICE/WILL NOT BE RENDERED/WHILE MOTORS ARE RUNNING/OCCUPANTS SMOKING/OR LAMPS BURNING" below. Reference to lamps indicates the relative age of this sign - automobiles originally had acetylene headlights.

SINCLAIR GASOLINE STRIPED FLANGE 16"RD
(4) 1920-1926 $350
Green outlined white tin litho flange sign with green outlined white center circle. Series of green vertical stripes in center with red "SINCLAIR" arched around top, "GASOLINE" around bottom. White flange area.

SINCLAIR OILS STRIPED FLANGE 16"RD
(4) 1920-1926 $350
Green outlined white tin litho flange sign with green outlined white center circle. Series of green vertical stripes in center with red "SINCLAIR" arched around top, "OILS" around bottom. White flange area.

DIE CUT DINO 12x16
(5) 1934-1940 No LIsting
Die cut green porcelain dinosaur. Used on top of mounting brackets for hanging credit card signs.

APPROVED DEALER STATION 14x18
(5) 1928-1934 $400

Red outlined white embossed tin litho sign with red "APPROVED DEALER STATION" across top above black "SINCLAIR" logotype. Red "CREDIT CARDS" above green "HONORED HERE" below.

APPROVED DEALER SINCLAIR CREDIT CARDS HONORED
(2) 1934-1946 $350
Green tin litho sign with white band across bottom. White "APPROVED DEALER" across top above white "SINCLAIR" logotype with black dropshading. Red "CREDIT CARDS/HONORED" on white band below.

SINCLAIR CREDIT CARDS HONORED
(1) 1946-1962 $150
Green outlined white porcelain sign with green "Sinclair" across top, "HONORED" across bottom and red "CREDIT CARDS" across center.

CLEAN REST ROOMS/WINDSHIELD SERVICE 30x36
(3) 1956-1959 $225
Green outlined white porcelain sign wtih red "CLEAN/REST/ROOMS" across top. Smaller green "WINDSHIELD SERVICE" over red "A SINCLAIR DEALER SERVICE" across bottom.

CLEAN REST ROOMS/WINDSHIELD SERVICE 30x36
(3) 1959-1972 $225
Green outlined white porcelain sign wtih red "CLEAN/REST/ROOMS" across top. Smaller green "WINDSHIELD SERVICE" over red "A Sinclair DEALER SERVICE" (new logotype) across bottom.

MEN 4x14
(4) 1934-1942 $100
White porcelain strip with cut corners. Green "MEN" across center.

LADIES 4x14
(4) 1934-1942 $100
White porcelain strip with cut corners. Green "LADIES" across center.

ULTRAVIOLET RAY SANATIZED REST ROOMS $125
Irregular shaped green porcelain sign with white circular area at top. Green center circle with whtie cross in upper circle with green "ULTRAVIOLET" arched around top, "RAY" around bottom. White script "Sanatized" diagonal above white "REST ROOMS" on lower sign area. Refers to a process invented in the 1930's to sterilize rest room fixtures using ultraviolet light.

SINCLAIR TRUCK STOP 30x60
1960s $300
Red outlined white porcelain sign with red "Sinclair" logotype above red "TRUCK/STOP" across sign.

FENCE PANEL SIGNS

SINCLAIR OPALINE MOTOR OIL 30x60
(5) 1930-1936 No Listing
Black/white outlined green tin litho sign with white band across center. White "SINCLAIR" with black dropshade across top, "MOTOR OIL" across bottom. Red "OPALINE" with black dropshade across center white band.

SINCLAIR PENNSYLVANIA MOBILENE MOTOR OIL 30x60
(5) 1930-1932 No Listing
Black/white outlined white tin litho sign with red band across center. Green "SINCLAIR/PENNSYLVANIA" across top, "MOTOR OIL" across bottom. Green outlined white "MOBILENE" across center red band.

SINCLAIR PENNSYLVANIA MOTOR OIL 30x60
(5) 1932-1936 No Listing
Red tin litho sign with white "SINCLAIR" with black dropshade across top. Black standing dinosaur to right, with white bnad across dinosaur. Black "PENNSYLVANIA/MOTOR OIL" on band.

SINCLAIR OPALINE TAMPER PROOF CANS 30x60
(5) 1936-1940 No Listing
Red tin litho sign with green dinosaur on mound across bottom. White "SINCLAIR/PENNSYLVANIA MOTOR/OIL" with black dropshade across top with white "can" at right. Green script "in/Tamper/Proof/Cans" on can. White "MELLOWED 100 MILLION YEARS" on mound.

SINCLAIR PENNSYLVANIA TAMPER PROOF CANS 30x60
(5) 1936-1940 No Listing
Red tin litho sign with green dinosaur on mound across bottom. White "SINCLAIR/OPALINE MOTOR/OIL" with black dropshade across top with white "can" at right. Green script "in/Tamper/Proof/Cans" on can. White "MELLOWED 80 MILLION YEARS" on mound.

BULK PLANT SIGNAGE

SINCLAIR OILS GREEN STRIP 12x96
(5) 1910s $150
Green/white outlined grren porcelain horizontal strip with white "SINCLAIR OILS" across center.

SINCLAIR OILS WHITE STRIP 12x96
(5) 1910s $150
White/green outlined white porcelain horizontal strip with green "SINCLAIR OILS" across center.

SUPER FLAME OIL FOR HEAT
(5) 1940s $550
White outlined green porcelain sign with large white outlined red flame in center. White "SuperFlame" with black dropshade across top, "OIL FOR HEAT" across bottom.

TRUCK SIGNS

SINCLAIR ROOFTOP SIGN 4x48
(5) 1930s No Listing
Black/red outlined self framed green porcelain panel with white "SINCLAIR" with black dropshade across center.

SINCLAIR HC ON SQUARE 14x14
(2) 1926-1935 $450
Black/red outlined green porcelain sign with green/white outlined green circlular logo with white outlined red center circle. White "H-C" with black dropshading in center circle, White "SINCLAIR" with black dropshading arched around top, "GASOLINE" around bottom.

FUEL OIL 12x12
(5) 1930s $100
Green/white outlined red porcelain sign with white "FUEL/OIL" with black dropshade across center.

SINCLAIR DINO 5x7
(4) 1959-1970 $225
Green outlined white porcelain trapezoidal shield shaped sign with green dinosaur in lower point of shield. Orange-red "Sinclair" logotype across upper sign area. NOTE: This sign is easily confused with those which have been cut from porcelain pump dial faces. Look carefully at the edges of the sign to determine if it is a factory cut or a homemade fake.

ADVANCE SIGNS

SINCLAIR AHEAD 72x96
(4) 1959-1972 $100
White tin litho sign with green trapezoidal Sinclair logo at top above red "AHEAD."

PIERCE PETROLEUM CORPORATION - ST. LOUIS, MO

Primarily a production and refining affiliate of the Standard Oil trust. Operated as Waters-Pierce Oil Company prior to 1920. Had extensive gasoline marketing in Mexico. Entered gasoline marketing in the south central states in the early 1920s. Purchased by and rebranded Sinclair in 1930. Marketing under the "Red Rooster" Productos Sinclair brand continued in Mexico until Mexico nationalized their petroleum industry in 1938.

IDENTIFICATION SIGNS

PENNANT GASOLINE 12x24
(5) 1920s No Listing
Wood framed yellow outlined green tin litho sign with yellow flag on upper sign face. Green "PENNANT" on flag, with black outlined yellow "GASOLINE" below. Black outlined yellow "PIERCE PETROLEUM CORPORATION" across bottom.

PIERCE OIL CORPORATION PENNANT 48"RD
(5) 1925-1930 $600
White/blue outlined red porcelain sign with blue outlined white center circle. Blue and white flag with white "PENNANT" in center above blue "GASOLINE/AND/OILS." White "PIERCE OIL" arched around top, "CORPORATION" around bottom.

PRODUCTOS SINCLAIR GASOLINA 48x54
(5) 1931-1938 No Listing
Used in Mexico only. Keyhole shaped die cut porcelain sign. Upper round section has red/white outline ring around green outer band. White/red outlined white center circle with red rooster in center. White "PRODUCTOS" with black dropshade arched around top, "SINCLAIR" around bottom. Lower area is red/white outlined green with white "GASOLINA" with black dropshade across center.

LUBRICATION/MOTOR OIL SIGNS

PIERCE OIL CORPORATION PENNANT OILS 12"RD
(5) 1925-1930 $450
White/blue outlined red porcelain sign with blue outlined pink center circle. Blue and white flag with white "PENNANT" in center above blue "OILS." White "PIERCE OIL" arched around top, "CORPORATION" around bottom.

PIERCE PENNANT MOTOR OIL 12x24
(4) 1920s $125
Green outlined cream tin litho sign wtih green flag on green pole in center. Orange "PENNANT" across flag, with green "PIERCE" above and "MOTOR OIL" below.

PIERCE PENNANT MOTOR OIL GREASES GASOLINE 12x24
(5) 1920s $125
Green outlined cream tin litho sign wtih green flag on green pole in center. Orange "PENNANT" across flag, with green "PIERCE" above and "MOTOR OIL/GREASES - GASOLINE" below.

PRODUCERS AND REFINERS OIL COMPANY - INDEPENDENCE, KS

Rocky Mountain refiner/marketer partially owned by Prairie Oil and Gas. When Sinclair purchased Prairie in the early 1930s, they also acquired part ownership of PARCO. They purchased the remaining portions in 1934 and the operation was rebranded Sinclair.

CURB SIGNS

PARCO ETHYL GASOLINE 30"RD
(5) 1926-1930 $475
Black outlined red porcelain sign with black outlined white center circle. White "PARCO" arched aorund top, "GASOLINE" around bottom with large Ethyl (EGC) logo in center.

LITTLE AMERICA REFINING COMPANY - LITTLE AMERICA, WY

A Wyoming refiner/marketer that had ties to Fletcher Oil Co. The Larco brand was purchased by Sinclair in 1955. Larco marketed under the Covey brandname in the 1950s.
No collectible size Larco or Covey signs are known at this time.

SIMPSON OIL CO. - ST. LOUIS, MO

Missouri marketer purchased by Sinclair in 1961. The Simpson "SUPER A" brand survived as a Sinclair secondary brand until 1965.

IDENTIFICATION SIGNS

SIMPSON OILS QUALITY SERVICE 42"RD
(5) 1940s $150
Red porcelain sign with white outlined red band around red and white quadrants of center circle. White "SIMPSON" arched around top, "OILS" around bottom with red "QUALITY/SERVICE" in white quadrants in center.

STOLL REFINING CO. - LOUISVILLE, KY

Louisville refiner/marketer founded in 1896 and involved in Kerosene and lubricants in the early years of this century. Entered gasoline marketing in the 1920s and continued suppling stations until the Stoll "Golden Tip" brand until after the 1948 purchase of Stoll refining by Sinclair. The "Golden Tip" stations were rebranded Sinclair about 1952.

IDENTIFICATION SIGNS

GOLDEN TIP GASOLINE MOTOR OIL 30x60
(5) 1930S $700
Blue porcelain sign with yellow band across top. Blue and yellow Stoll "S" logo in top band, with yellow arroe below. Blue "GOLDEN TIP" in arrow, with yellow GASOLINE-MOTOR OIL" below.

SILVER TIP GASOLINE 30x60
(5) 1930s $700
Yellow outlined blue porcelain sign with white script "Silver Tip" diagonal across center. White "GASOLINE" below.

GOLDEN TIP GASOLINE 46x84
(5) 1935-1952 $500
Maroon outlined yellow die cut arrow shaped porcelain sign with maroon "GOLDEN TIP" across center above maroon "GASOLINE."

PUMP SIGNS

GOLDEN TIP GOES GASOLINE 7x9
(4) 1946-1952 No Listing

Wolf's Head tin sign with hanger.

Small Shell porcelain directional sign. *Bruce Miller.*

Shell Dealer porcelain sign. *Bruce Miller.*

Shell Penn porcelain sign. *Bruce Miller.*

Shell porcelain rest room sign. *Bruce Miller.*

Miscellaneous Shell signage. *Bruce Miller.*

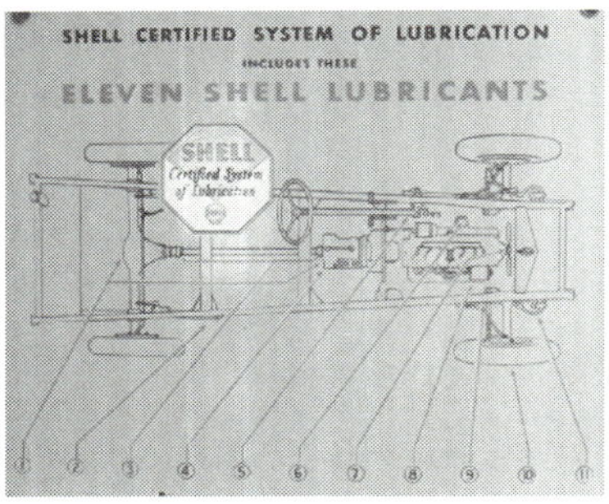

Rare porcelain Shell Lube sign. *Bruce Miller.*

Pennzoil 36"x12" porcelain sign.

Pennzoil metal flange sign.

Miscellaneous Shell signage. *Bruce Miller.*

Miscellaneous rare Shell signs. *Bruce Miller.*

Rare Shell Air 12" porcelain sign.

Standard of New Jersey 30" porcelain sign.

White porcelain sign with yellow band across upper sign face. Dark blue "GOLDEN TIP/GOES" above farm scene landscape on upper yellow band. Red "GASOLINE" below.

GOLDEN TIP GOES DIESEL 7x9
(4) 1946-1952 No Listing
White porcelain sign with yellow band across upper sign face. Dark blue "GOLDEN TIP/GOES" above farm scene landscape on upper yellow band. Red "DIESEL" below.

GOLDEN TIP GOES LAMP OIL 7x9
(4) 1946-1952 No Listing
White porcelain sign with yellow band across upper sign face. Dark blue "GOLDEN TIP/GOES" above farm scene landscape on upper yellow band. Red "LAMP OIL" below.

NEON SIGNS

GOLDEN TIP GASOLINE 48x96
(5) 1935-1952 $1500
Maroon outlined yellow die cut arrow shaped porcelain can type sign with maroon "GOLDEN TIP" across center above maroon "GASOLINE."

CURB SIGNS

STOLL 30"RD
(5) 1925-1930 $1200
Yellow outlined blue porcelain sign with large yellow winged "S" logo in center. Blue "STOLL" diagonal on center stroke of "S."

STOLL ETHYL GASOLINE 30"RD
(5) 1926-1935 $900
Black outlined yellow porcelain sign with balck outlined white center circle. Black "STOLL" arched around top, "GASOLINE" around bottom with large Ethyl
(EGC) logo in center. Small Stoll "S" logo on black circle on either side of center on yellow outer area.

LUBRICATION/MOTOR OIL SIGNS

VISCOYL MOTOR OIL 16x16
(5) 1935-1950 $350
Blue outlined yellow porcelain sign with blue band across bottom. Yellow arrow on blue band with blue "GOLDEN TIP/GASOLINE" across arrow. Blue "VISCOYL/MOTOR OIL" across top of yellow area above small blue "PARTNER IN POWER TO."

RICHFIELD OIL CO. OF NEW YORK

Although they are Sinclair affiliates, Richfield of New York, Sherwood/Betholine and Rio Grande are listed under the RICHFIELD heading for clarity.

SKELLY OIL CO., TULSA, OK

GETTY OIL COMPANY, TULSA, OK

SKELLY OIL CO. - TULSA, OK Skelly Oil Company was founded in Tulsa, Oklahoma in 1919 from the holdings of oil trader William Grove Skelly. The company was involved in gasoline marketing from the beginning and actively purchased chains of service stations throughout the mid-continent area during the early years. During the depression Skelly defaulted on loans and control of the company fell into the hands of Standard Oil of NJ (Exxon) who formed the Mission Corporation to operate it's interest in Skelly and Tidewater Oil. J. Paul Getty, in attempting to gain control of Tidewater, purchased control of Mission in 1937, only to find out he had also gained control of Skelly. Getty retained majority interest in Skelly until the 1960s when the remaining interests were purchased and Skelly was merged into Getty, along with Tidewater. The company continued to operate somewhat independently though throughout the 1970s and in 1983 Skelly began a gradual changeover to the Getty brand. Getty's purchase by Texaco halted the rebranding and most Skelly locations finally ended rebranded Texaco by 1985. See the Flying A section for more information.

IDENTIFICATION SIGNS

SKELLY 42"
(3)1930-1960 $300
SKELLY 60"
(3)1930-1960 $250
SKELLY 72"
(1)1946-1984 $100
White/red outlined blue porcelain square diamond shaped sign with white "S" in center with white band superimposed across "S." Red "SKELLY" in white band.

SIGN POLES

SKELLY "DIAMOND S" POLE
(4) 1930'S
Pipe pole structure with apex shaped mast-arm for installationof diamond shaped sign. "S" shaped bracket between pole and arm.

BUILDING SIGNS & LETTERS

SKELLY 24x24
(5)1920's No Listing
White/red outlined blue painted glass square diamond shaped sign with white "S" in center with white band superimposed across "S." Red "SKELLY" in white band Designed for installation in building facade.

SKELLY OIL COMPANY 12x120
(4)1924-1946 $125
Self framed red outlined white porcelain strip sign with black outlined red "SKELLY OIL COMPANY" across sign.

SKELLY PRODUCTS STRIP 12x120
(4)1924-1946 $125
Self framed red outlined white porcelain strip sign with black outlined red "SKELLY PRODUCTS" across sign.

TIRES & BATTERIES STRIP 12x120
(4)1924-1946 $75
Self framed red outlined white porcelain strip sign with black outlined red "TIRES & BATTERIES" across sign.

LUBRICATION SERVICE STRIP 12x120
(4)1924-1946 $75
Self framed red outlined white porcelain strip sign with black outlined red "LUBRICATION SERVICE" across sign.

AUTO LAUNDRY STRIP 12x120
(4)1924-1946 $75
Self framed red outlined white porcelain strip sign with black outlined red "AUTO LAUNDRY" across sign.

"SKELLY" 18"
(4)1946-1984 $100
Red porcelain individual letters "SKELLY"

NEON:

SKELLY 72"
(4)1946-1958 $750

White/red outlined blue porcelain can type square diamond shaped sign with white "S" in center with white band superimposed across "S." Red "SKELLY" in white band. Neon outline and lettering.

CURB SIGNS

SKELLY GASOLINE BUY IT HERE 28x39.5
(2) 1924-1940 $300
Red outlined tombstone shaped white porcelain sign with large red/white/blue Skelly diamond logo at top. Red "GASOLINE" over blue "Buy it Here" below diamond.

SKELLY TAGOLENE BUY IT HERE 28x39.5
(4) 1924-1940 $300
Red outlined tombstone shaped white porcelain sign with large red/white/blue Skelly diamond logo at top. Blue "TAGOLENE" over smaller blue "Buy it Here" below diamond. Red outlined white tag hanging from "G" in Tagolene.

SKELLY TAGOLENE OILS & GREASES 28x39.5
(4) 1924-1940 $275
Red outlined tombstone shaped white porcelain sign with large red/white/blue Skelly diamond logo at top. Red "TAGOLENE/ OILS & GREASES" over blue "Buy Them Here" below diamond.

SKELLY AROMAX ETHYL 30"RD
(3) 1926-1940 $450
White outlined red porcelain sign with white outlined blue band across center. Skelly diamond logo at top, with white "AROMAX" in band and Ethyl
(EGC) with yellow rays logo at bottom.

SKELLY AROMAX ETHYL 30"RD
(3) 1926-1940 $450
White porcelain sign with large red diamond area covering most of sign. White band across center with Skelly diamond logo above and white circle with Ethyl
(EGC) logo below. Blue "AXOMAX" in center band.

FORTIFIED TAGOLENE MOTOR OIL 30"RD
(3) 1940-1955 $150
White tin litho sign with red/white/blue Skelly diamond at top. Gold script "Fortified" over red "TAGOLENE" with blue drop-shading. Blue "MOTOR OIL" across bottom.

SKELCO MOTOR OIL 30"RD
(3) 1940-1955 $150
White outlined red tin litho sign with small Skelly diamond logo at top. White "SKELCO" across center with "MOTOR OIL" below.

SKELLY TAGOLENE MOTOR OIL 30"RD
(4) 1940s $150
White outlined blue tin litho sign with white outlined red band across center. Red/white/blue Skelly diamond logo at top, with white "TAGOLENE" in band and white "MOTOR OIL" arched around bottom.

GREASE MASTER FOR A SAFE GREASE JOB 30"RD
(4) 1940s $150
White outlined blue tin litho sign with white outlined red band across center. Red/white/blue Skelly diamond logo at top, with white "GREASE/MASTER" in band and white "FOR A SAFE GREASE JOB" with stars arched around bottom.

SECONDARY SIGNS

SKELLY COUPONS AND CREDIT CARDS 10x24
(5)1930's $125
White tin litho sign with Skelly diamond at left. Blue "WE HONOR" across top above black outlined red "SKELLY COUPONS." Blue "AND" in center above black outlined red "CREDIT CARDS."

SKELLY COUPONS/CREDIT CARDS 12x18
(5)1940's $125
White tin litho sign with red band across top and bottom. Small Skelly diamond logo in center, with two parallel blue lines to either side. Blue "ACCEPTED" to left, "HERE" to right with black outlined "COUPONS" arched around top, "CREDIT CARDS AROUND BOTTOM."

SKELLY CREDIT CARDS HONORED 12x18
(5)1950's $125
White tin litho sign with red band across top and bottom. Small Skelly diamond logo in center, with black outlined red "CREDIT CARDS" arched across top and "ACCEPTED HERE" flat across bottom.

BULK PLANT SIGNAGE

NOTE: Skelgas was Skelly Oil Company's LPG division. Dealers supplied LPG/propane and sold gas equipment. The division continues to operate independantly today.

SKELGAS DIE CUT 12x20
(5) 1950's No Listing
Blue outlined white angled oval die-cut porcelain sign with blue outlined white band across center. Red logotype "Skelgas" in band. Blue gas bottle vertical through center of sign.

SKELGAS NEON 24x60
(5) 1950's No Listing
Blue can type porcelain sign with red horizontal lines at top and bottom of faces and white "SKELGAS" across center.

SURFCO

Surfco was a 1960s Skelly secondary brand.
No collectible size Surfco signs are known at this time.

SOHIO AND AFFILIATES

STANDARD OIL CO. OF OHIO - CLEVELAND, OH

Sohio is, of course, Standard Oil of Ohio, which traces its roots to the original formation of Standard Oil by John D. Rockefeller in Cleveland, Ohio in 1870. Although Standard of New Jersey was to become the "main office" as such after the formation of the Standard Oil Trust in 1882, Standard of Ohio was always considered to be "the original." After the break up of Standard Oil in 1911, Standard of Ohio was assigned only marketing in Ohio and a small refinery. Being primarily a marketing company, Sohio was a pioneer in service stations, operating a chain of company owned stations statewide by 1920. The brandname "SOHIO" was introduced in 1928 to identify all the company's marketing. In their first marketing venture outside Ohio, Sohio purchased Spears and Riddle of Wheeling, West Virginia, a large regional marketer selling gasoline under the Fleet-Wing name. Fleet-Wing Corporation was formed as a subsidiary and the brand was offered to jobbers in the states surrounding Ohio. Sohio opened the first company owned locations outside Ohio in Newport, Kentucky in 1956 under the Boron name. Boron had first been used as a brandname for Sohio's super-premium grade of gasoline in 1954. Little was done with the Boron brand until stations were opened in western Pennsylvania in 1962. In

1969 British Petroleum purchased a controlling interest in Sohio and BP's U.S. marketing subsidiary was transferred to Sohio. About that time Sohio also purchased Scot Service Stations, Inc. of McLean, Virginia. Boron and BP became Sohio's primary non-Ohio brandnames. Fleet-Wing was sold to Pennzoil about 1970. In 1990 BP, having purchased the remaining portions of Sohio, began rebranding stations under the Boron, Wm. Penn, Gas-n-Go, and Gulf (southeast, purchased in 1985) to BP. Sohio stations were rebranded BP beginning in 1991 and now the entire company markets under the BP name.

IDENTIFICATION SIGNS

RED CROWN GASOLINE 24"RD
(3) 1915-1928 $650
Green outlined white porcelain flange sign with green outlined white center circle and white flange area. Green outlined red "RED CROWN" arched around top, "GASOLINE" around bottom with retailed red and white crown in center.

RED CROWN GASOLINE RED & GREEN 30"RD
(4) 1920-1928 $600
Green outlined white porcelain flange sign with green outlined white center circle. Green outlined red "RED CROWN" arched around top, "GASOLINE" around bottom with retailed red and white crown in center.

RED CROWN GASOLINE RED & BLUE 30"RD
(5) 1928-1930 $550
Blue outlined white porcelain flange sign with blue outlined white center circle. Blue outlined red "RED CROWN" arched around top, "GASOLINE" around bottom with retailed red and white crown in center.

SOHIO OVAL W/DIAMOND 24x48
(4) 1930-1962 $550

SOHIO OVAL W/DIAMOND 30x60
(4) 1930-1962 $550
Blue/white outlined porcelain sign with series of blue and white horizontal lines across center. Red outlined red flat diamond superimposed over lines with white band across center. Red "SOHIO" across band.

SOHIO OVAL "STANDARD OIL DEALER" 32x48
(4) 1930-1962 $575
Blue/white outlined porcelain sign with series of blue and white horizontal lines across center. Red outlined red flat diamond superimposed over lines with white band across center. Red "SOHIO" across band. Blue outlined white band across bottom with red "STANDARD OIL DEALER" across band.

SOHIO NEW OVAL 42x72
(4) 1962-1989 $100
Blue ovalized shield sign with white oval in center. Red "SOHIO" on center oval. While many single sided versions of this sign exist, having been used as a station facade sign in the 1960's, a few double sided examples were used for identification signs in rural areas.

BUILDING SIGNS AND LETTERS

THE STANDARD OIL COMPANY STRIP 10x60
(5) 1930-1935 $200
Red outlined white strip with red "The STANDARD OIL Co." across strip.

SOHIO BUILDING PANEL 42x72
(2) 1962-1989 $100
Blue ovalized shield sign with white oval in center. Red "SOHIO" on center oval.

NEON SIGNS

SOHIO OVAL W/DIAMOND 30x60
(4) 1930-1962 $1000
Blue/white outlined porcelain sign faces assembled back-to-back on can-type frame. Series of blue and white horizontal lines across center. Blue outlined red flat diamond superimposed over lines with white band across center. Blue outlined red "SOHIO" across band.

CURB SIGNS

OILS/POLARINE/GREASES 30"RD
(4) 1920-1928 $350
Blue outlined white porcelain sign with blue outlined white center circle. Blue outlined red "OILS" arched around top, "GREASES" around bottom with "Polarine" across center of sign.

RED CROWN AVIATION GASOLINE 30"RD
(5) 1922-1926 No Listing
Blue outlined white porcelain sign with blue outlined white center circle. Blue outlined red "RED CROWN" arched around top, "GASOLINE" around bottom with "Aviation" across center of sign.

HIGH TEST SOHIO ETHYL BANNER 28x30
(4) 1928-1930 $375
White outlined red die-cut vertical "banner" shaped porcelain sign. White script "HighTest" across top with blue outlined white "SOHIO" just below. Blue outlined white circle in center with Ethyl
(EGC) logo with blue rays in circle. Black script "Anti" to left, "knock" to right. Blue outlined white "GASOLINE" arched around bottom.

SOHIO ETHYL GASOLINE BANNER 28x30
(4) 1930-1940 $375
White outlined red die-cut vertical "banner" shaped porcelain sign. Blue outlined white "SOHIO" across top. Blue outlined white circle in center with Ethyl
(EGC) logo with blue rays in circle. Black script "Anti" to left, "knock" to right. Blue outlined white "GASOLINE" arched around bottom.

SOHIO MOTOR OIL BANNER 28x30
(3) 1930-1940 $300
Die cut vertical "banner" shaped porcelain sign with red upper band, white center and blue bottom. White "SOHIO" across top, "OIL" across bottom with blue "MOTOR" in center.

LUBRICATION/MOTOR OIL SIGNS

SOHIO MOTOR OIL PANEL 36x60
(3) 1930-1940 $250
White porcelain sign with red and blue wavy border lines top and bottom. Large blue "SOHIO/MOTOR OIL" across sign face with smaller red "DEPENDABLE and ECONOMICAL" below.

BORON: Sohio secondary brand introduced in 1956 for use in non-Ohio locations where "Standard Oil" & it's related tradeemarkes were owned by others .

IDENTIFICATION SIGNS

BORON OVAL 42x72

(4) 1961-1990 $100
Blue ovalized shield sign with white oval in center. Red "BORON" on center oval.

BUILDING SIGNS & LETTERS

BORON BUILDING PANEL 42x72
(3) 1961-1990 $100
Blue ovalized shield sign with white oval in center. Red "BORON" on center oval.

CALDWELL & TAYLOR - CINCINNATI, OH

Cincinnati based marketer purchased by Sohio in 1928.

SECONDARY SIGNS

CALDWELL & TAYLOR FLANGE 12x18
(5) 1915-1920 No Listing
Red porcelain flange sign with black outlined white "CALDWELL/AND/TAYLOR'S/ORIGINAL/BENZOL GAS" on upper sign face. Detailed black and white "BENZOL GAS" logo in center at bottom with black outlined white "MORE MILEAGE" across bottom.

SPEARS & RIDDLE - WHEELING, WEST VIRGINIA

A large regional wholesale & lubricants operation founded in the 1870s, they began selling gasoline under the Fleet-Wing brandname about 1925. Sohio purchased Spears & Riddle in 1928 and created the Fleet-Wing Corporation as a marketing subsidiary to brand jobbers in Ohio and surrounding states. Pennzoil purchased Fleet-Wing from Sohio about 1970 and continues to operate about a dozen stations under the Fleet-Wing name.

IDENTIFICATION SIGNS

FLEET WING GASOLINE 30x60
(4) 1950S $100
White tin-litho rectangular sign with red bands across upper and lower sign face. Red soaring eagle at top above red "FLEET-WING." Small red "GASOLINE" below.

PUMP SIGNS

FLEET WING FLITE-X 7x8
(4) 1940-1960 $125
Blue outlined beige tin litho sign with red soaring eagle at top. Red "FLEET-WING" across center with blue "Flite-X" below.

FLEET WING SUPER ETHYL EC 8x14
(4) 1950-1960 $150
Blue outlined beige tin litho sign with black outlined white circle extending from bottom edge. Red soaring eagle at top, above red "FLEET-WING." Red "SUPER" with blue dropshading below. Large Ethyl (EC) logo in white circle at bottom.

CURB SIGNS

FLEET WING MOTOR OIL OVAL 18x30
(5) 1930-1935 $350
Beige oval tin litho sign with blue "FLEET-WING" superimposed over red soaring eagle with speed lines below. Blue "MOTOR/OIL" below.

FLEET WING MOTOR OIL OVAL 18x30
(5) 1935-1940 $375
Beige oval tin litho sign with blue "FLEET-WING" superimposed over red soaring eagle with speed lines below. Blue "MOTOR/OIL" below. Small PGCOA seal at lower left.

FLEET WING ETHYL 30"RD
(4) 1926-1930 $375
Beige/blue outlined beige tin litho sign with blue "FLEET-/WING" arched across top above large Ethyl (EGC) logo with blue rays.

SECONDARY SIGNS

FLEET WING ETHYL 12x42.5
(4) 1926-1930 $200
Blue outlined beige tin litho sign with blue "FLEET WING" across strip with large Ethyl (EGC) logo with blue rays at right end.

FLEET WING COURTESY CARDS ACCEPTED 10x14
(5) 1935-1940 $200
Blue tin litho sign with white oval/red soaring eagle logo at lower left corner. White "FLEET-WING" above "Courtesy Cards/Accepted/Here."

FLEET WING CREDIT CARDS HONORED 18x24
(5) 1950-1960 $150
Beige tin litho flange sign with red "FLEET-WING" across top above blue "NATIONAL/CREDIT CARDS/HONORED."

THE REFINERS OIL COMPANY - DAYTON, OH

Refiners was a Dayton, Ohio based marketer that operated around 300 retail stations in Ohio and Indiana. When the General Motors Chemical Company developed the antiknock additive Tetraethyl Lead in early 1923, Dayton based Refiners was chosen to test market the new product. The test was so successful that in September of that year Standard Oil of Indiana signed on to be an exclusive distributor of the new additive, "Ethyl," through its marketing territory. Standard Oil of Ohio purchesed Refiners Oil company in 1930 and rebranded the stations Sohio.

IDENTIFICATION SIGNS

GASOLINE REFINERS MOTOR OIL 30"RD
(4) 1926-1930 $450
Black border area around orange sunburst centered porcelain sign with white band across center. Orange "GASOLINE" arched around top, "MOTOR OILS" around bottom with black script "Refiners" logotype across center.

CURB SIGNS

REFINERS PIONEER DISTRIBUTORS ETHYL 30"RD
(4) 1926-1930 $375
Yellow outlined white porcelain sign with black script "Refiners" logotype across upper sign face. Black band in center with white "PIONEER DISTRIBUTORS" on band. Large Ethyl (EGC) logo with yellow rays at bottom.

LUBRICATION/MOTOR OIL SIGNS

HIGH SPEED MOTOR OIL 24x36
(5) 1920-1925 $200
Black/orange outlined orange tin litho sign with black "HIGH SPEED/MOTOR OILS" across upper sign face. Smaller black "The Refiners Oil Company" across bottom.

CANFIELD OIL COMPANY - CLEVELAND, OHIO

Founded in Cleveland in 1886, Canfield was primarily a lubricants manufacturer, developing, among others, the "Wm. Penn" motor oils. Canfield became involved in gasoline marketing in the late 1910s, operating stations in and around

Cleveland. When Standard Oil purchased Canfield in 1945, there were over 200 branded stations at that time. The Sohio brandname gradually replaced the Canfield name, although Canfield and Wm. Penn motor oils continued in production. In the 1970s, after British Petroleum purchased a controlling interest in Standard Oil, many dealer operated BP stations in the east were rebranded "Wm. Penn," continuing as such until the BP name was selected to replace all Sohio affiliate brandnames in 1990.

IDENTIFICATION SIGNS

CANFIELD OVAL 28x72
(5) 1946-1960 $275
White porcelain oval with green outlined red "Canfield" script logotype across sign face.

BUILDING SIGNS AND LETTERS

CANZOL DEALER STRIP 10x54
(5) 1946-1960 $175
White porcelian strip with green outlined red "CANZOL DEALER" across sign face.

NEON SIGNS

CANZOL 48x84
(5) 1935-1946 $600
Elaborate die-cut porcelain faces on can-type sign. Green outline with green outlined red "Canzol" across center. Neon outline and Canzol.

CURB SIGNS

WM PENN MOTOR OILS 24x26
(5) 1930s $850
Red tin litho keyhole shaped sign with white charachature of William Penn at top. White "WM PENN" across center, above white "MOTOR OIL" and other lettering.

LUBRICATION/MOTOR OIL SIGNS

WM PENN MOTOR OILS PURE PENNSYLVANIA 18x19
(5) 1920-1926 $750
Green porcelain flange sign with red circular center. Green band diagonal across sign with large white script "MotorOils" across band. White script "Wm.Penn" in upper red area with green "Pure/Pennsylvania" in lower. Small white "THE CANFIELD OIL CO" arched around bottom edge.

CANFIELD OIL CO WM PENN MOTOR OILS 18"RD
(5) 1926-1935 $650
Green porcelain sign with yellow/black/white outlined red shield in center. GreenBlack outlined yellow oval at top of shield with black outlined red "Canfield/Oil Co." across oval. Yellow "QUALITY & EXPERIENCE SINCE 1886" on red Canfield underline. Black outlined yellow "Wm.Penn/MOTOR OILS" across center of shield with black PGCOA seal in yellow area at bottom. Yellow "PERMIT NO.29" just above seal.

CANFIELD OIL CO WM PENN MOTOR OILS 18"RD
(5) 1926-1935 $650
Green porcelain sign with yellow/black/white outlined red shield in center. Green outlined yellow oval at top of shield with green William Penn sillouette in oval. Green outlined yellow "Wm.Penn/MOTOR OILS" across center of shield with green PGCOA seal in yellow area at bottom. Yellow "PERMIT NO.29" just above seal.

BP-BRITISH PETROLEUM-CLEVELAND, OH:

British Petroleum, founded in 1912 as Anglo-Persian Oil, entered North American markets in 1958 with several stations in Canada. In 1964 they purchased Cities Service's Canadian marketing operations, greatly expanding their station count. Next, in 1967, came Supertest, whcih BP operated as a secondary brand until the early 1970's. Reaching for the lucrative U.S. gasoline market, they purchased marketing operations along the eastern seaboard that Atlantic-Richfield was forced to sell with their 1969 purchase of Sinclair Oil. BP signs replaced Sinclair at stations from Maine to Florida. In 1970 BP traded a share of their ever-increasing Alaskan oil production for minority interest in Standard Oil of Ohio. In 1972 BP withdrew from branded marketing in the southeast, selling the former Sinclair outlets to Fina. By the early 1980's BP had gained control of Sohio and in 1985 purchased the Gulf marketing in the southeast when Chevron was forced to sell in the Chevron-Gulf merger. In 1989 Gulf stations in the southeast began rebranded to BP and in 1990 Sohio and Boron outlets in the north central states began converting to BP. They have bought their way into several othother markets and now operate in select markets nationwide.

IDENTIFICATION SIGNS

BP SHIELD ON WHITE WITH RED STRIPE
(4) 1969-1980 $100
Steel framed white porcelain sign with green shield logo in center. Yellow "BP" on shield. Red stripe across below white area.

PUMP SIGNS

BP SHIELD ON WHITE 7x7
(1) 1969-1975 $25
White porcelain sign with green shield logo in center. Yellow "BP" on shield.

BULK PLANT SIGNAGE

BP SHIELD ON WHITE
(3) 1969-1975 $75
White porcelain sign with green shield logo in center. Yellow "BP" on shield. SUPERTEST

SUPERTEST: Canada's "All Canadian" company, BP purchased Supertest in 1967.

IDENTIFICATION SIGNS

SUPERTEST 30x72
(5) 1920s $300
Orange porcelain self framed sign with white outlined black "SUPERTEST" across upper sign face. Other lettering below, details not available.

SUPERTEST 48"RD
(3) 1936-1957 $450
Black/red outlined white porcelain sign with large green outlined orange maple leaf in center. White outlined black "SUPERTEST" across leaf, with two white banners across leaf below. Upper banner has black "CANADA'S," lower has "ALL CANADIAN COMPANY."

SPUR AND AFFILIATES

MURPHY OIL CO. - EL DORADO, ARK
Murphy traces it's orgins to the banking, timber and oil interests of founder C.H.Murphy, brought together in 1907. Oil

production began in earnest in the 1920s the the company shifted almost entirely to oil exploration after WW2. In 1957 Murphy entered the retail market with the purchase of River States Oil Company of Evanston,IL. Murphy then introduced the Royal brandname and was expanding successfully when the company purchased Spur Distributing in 1960. By 1962 all the former Royal stations had rebranded Spur and the company added several other markers to the collection, replacing that operation's brand with Spur. The company brands over 500 Spur stations in 14 states today.

RIVER STATES OIL CO. - EVANSTON, IL

River States, marketing in Wisconsin, Minnesota, Illinois, and Michigan, was a combination of several north central marketing firms brought together about 1955. In 1957 River States was purchased as the first marketing division of Murphy Oil and the "ROYAL" brand was introduced, replacing the earlier "RIVER STATES" brand. In 1962 the Spur brandname replaced Royal at several hundred stations in the north central states.

IDENTIFICATION SIGNS

RIVER STATES OVAL 60x84
(5) 1955-1957 $100
Pole framed oval porcelain sign with white band across center. "RIVER STATES" in center band.

RS ROYAL OVAL 60x84
(5) 1957-1962 $250
Pole framed red outlined white porcelain oval sign with red "ROYAL" across center superimposed over blue/white striped "RS."

WEBB CUT PRICE GAS AND OIL - DULUTH, MN

Founded in 1923, Webb was truly the pioneer discount gasoline operation. They operated stations in Minnesota, Wisconsin, and Michigan until the mid 1960s when they either merged with or rebranded Spur.

IDENTIFICATION SIGNS

WEBB 72x72
(5) 1960's $250
Porcelain can type die cut triangular sign with white outline area around yellow center. Large "W" in upper red area with red "WEBB" below.

SECONDARY SIGNS

WEBB CUT PRICE GAS ? MILES TIN 24x24
(4) 1950's $85
White tin litho sign with large black outlined red "W" on black outlined yellow bar on sign. White "WEBB/CUT PRICE" on "W" with "GAS" on yellow bar. Black "MILES" below with space for mileage to be painted in.

SPUR DISTRIBUTING, NASHVILLE, TN

Founded by John Mason Houghland in Nashville in 1928 as the original trackside gasoline chain, Spur had 348 stations in 21 states when they were purchased by Murphy Oil in September,1960. Murphy, who had only recently entered the retail marketing field, marketed under the various regional names (Royal, Spur, Ingram) until 1962 when the Spur brandname was chosen as the company's only brand and the others were replaced with the new oval and spur logo. The company supplies about 500 branded stations in 14 states today.

IDENTIFICATION SIGNS

SPUR GAS FOR LESS 60x84
(5) 1940's $150
Steel framed white sign with red "SPUR" at top above "GAS/FOR/LESS." "FOR" is smaller than other lettering.

SPUR POWERFUL/KNOCKLESS 60x84
(5) 1950's $100
Steel framed white sign with red "SPUR" at top above space for pricing numbers to be attached. Red band across bottom with white "POWERFUL-KNOCKLESS" with blue dropshading on band.

SAVE WITH SPUR/FREE PREMIUMS 60x84
(5) 1950's $100
Steel framed white sign with red "SAVE" at top and "SPUR" across center. Blue "WITH" in band across between "SAVE" & "SPUR." Blue "And Receive/FREE/PREMIUMS" below.

SPUR OVAL 42x98
(3) 1962-1978 $150
White/blue outlined white porcelain oval can type sign with red field to right. Blue "SPUR" on white area with white rowel in red area.

BUILDING SIGNS & LETTERS

SPUR SMALL MODUCAL
(4) 1976-1980 $50

SPUR MEDIUM MODUCAL
(2) 1976-1980 $35

SPUR LARGE MODUCAL
(3) 1976-1980 $35
White aluminum litho strip with red panel at right end. Red "SPUR" on white area. Also available in fiberglass.

SPUR W/STAR SMALL MODUCAL
(4) 1980-1990 $50

SPUR W/STAR MEDIUM MODUCAL
(2) 1980-1990 $35

SPUR W/STAR LARGE MODUCAL
(3) 1980-1990 $35
Same as above with white rowel on red area.

PUMP SIGNS

SPUR FORMED OVAL 7x15
(1) 1962-1969 $45
Embossed white porcelain oval sign with keyhole slot mounting. Blue "SPUR" to left with red field to right. White rowel on red area.

REGULAR PANEL 15.5x29
(3) 1962-1966 $25
Blue porcelain tapered panel with white "REGULAR" across center below area for mounting formed oval sign.

PREMIUM PANEL 15.5x29
(3) 1962-1966 $25
Red porcelain tapered panel with white "PREMIUM" across center below area for mounting formed oval sign.

POWER BLEND PANEL 22x27
(4) 1965-1969 $35
Tapered porcelain pump panel with red blue area to left and white area to right with white progressive vertical stripes down center separating colors.

SMALL BLUE PANEL 14x26
(4) 1969-1974 $50

Nu-Pro Oils embossed tin sign.

Atlantic pump plate porcelain sign.

Rare Richfield metal oil rack signs.

United Motors wood-framed tin sign.

Miscellaneous porcelain pump plates. *Bruce Bednar.*

Conoco pump plates. *Bruce Bednar.*

Mobil porcelain plates - rare pair. *Bruce Bednar.*

Sinclair H-C 24" porcelain sign and matching globe. *Bruce Bednar.*

Oilzum porcelain sign. *Bruce Bednar.*

Hood tires - self-framed rare porcelain sign.

LARGE BLUE PANEL 22x27
(4) 1969-1974 $50
Tapered blue porcelain pump panel with white area at bottom. Small red/white/blue Spur oval logo on lower white area. Used on regular pumps.
SMALL RED PANEL 14x26
(4) 1969-1974 $50
LARGE RED PANEL 22x27
(4) 1969-1974 $50
Tapered red porcelain pump panel with white area at bottom. Small red/white/blue Spur oval logo on lower white area. Used on premium pumps.
APG PANEL 14x26
(5) 1971-1974 No Listing
Blue tin litho panel with small red/white/blue Spur logo at bottom. Yellow "APG" on upper panel with white "UNLEADED/GASOLINE" below.
RED/BLUE PANEL 22x27
(3) 1969-1974 $50
Tapered porcelain pump panel with blue/red area split diagonal over white area at bottom. Small red/white/blue Spur oval logo on lower white area. Used on blending pumps.
SAVE PANEL 22x27
(4) 1974-1980 $35
Tapered blue porcelain pump panel with white area at bottom. Small red/white/blue Spur oval logo on lower white area. Yellow "SAVE" on upper blue area.

CURB SIGNS

SPUR OIL W/BOOT 24"RD
(5) 1930-1935 No Listing
White porcelain sign with large boot and spur logo in center. White "SPUR" lettering on boot. Large "OIL" below. "ONE OF THE WORLD'S FINEST MOTOR OILS" arched around top, "CUTS COSTS IN HALF" around bottom.
SPUR PRICE SIGN W/BOOT 24"RD
(5) 1930-1935 No Listing
White porcelain sign with small boot and spur logo at top. White "SPUR" on boot. "GAS" at bottom with "HIGH" to left and "TEST" to right.

LUBRICATION/MOTOR OIL SIGNS

SPUR SUPREME OIL RACK SIGN 6x20
(3) 1960's $25
White tin litho sign with Spur oval logo to left. Blue "SUPREME" over "MOTOR OIL" to right.

SECONDARY SIGNS

MENS REST ROOM SHIELD 7x11
(4) 1946-1962 $35
White porcelain sign with blue "colonial" style border with blue "MENS/REST ROOM" across center.
LADIES REST ROOM SHIELD 7x11
(4) 1946-1962 $35
White porcelain sign with blue "colonial" style border with blue "LADIES/REST ROOM" across center.
REST ROOMS WHITE/BLACK 8x15
(2) 1976-1992 $15
Black porcelain flange sign with white "REST/ROOMS" across center.
CLEAN REST ROOMS/PRICE PANEL 36x41
(4) 1962-1969 $25
White tapered porcelain sign with blue script "Clean" over blue "REST ROOMS." Reverse side had place to attach blue porcelain price numbers with blue "REGULAR" across bottom. Designed for reversible use.
PREMIUM ATTACHMENT TO PRICE PANEL 12x30
(5) 1962-1969 $10
White porcelain attachment to above sign with blue "PREMIUM" on front and blue porcelain reverse with white script "No Stamps."
MURPHY OIL COMPANY PLACARD 9x20
(3) 1960s $25
Aluminum framed white plastic sign with blue "MURPHY" over smaller blue "OIL COMPANY" across sign. Red rowel above "U" in "MURPHY."
MURPHY OIL CORPORATION PLACARD 9x20
(3) 1970s $25
Aluminum framed white plastic sign with blue "MURPHY" over smaller blue "OIL CORPORATION" across sign. Red rowel above "U" in "MURPHY."
MURPHY CORPORATION PLACARD 9x20
(3) 1980s $25
Aluminum framed white plastic sign with blue "MURPHY" over smaller blue "CORPORATION" across sign. Red rowel above "U" in "MURPHY."

WOOD RIVER OIL COMPANY - WOOD RIVER, IL

Formed in 1939 as a refiner/marketer. The Ingram family was involved with Wood River from its founding and eventually Wood River ooperations in the south were spun off andrebranded Inngram.

IDENTIFICATION SIGNS

WOOD RIVER EARLY 72"RD
(5) 1939-1956 $100
Blue/red outlined white porcelain sign with blue band across center. White "WOOD RIVER" on blue center band.
WOOD RIVER LATER 72"RD
(4) 1956-1980 $100
Red/white outlined red porcelain sign with white band across center. Blue "WOOD RIVER" on white center band.

INGRAM OIL - NASHVILLE, TN

The Ingram brand was introduced at Wood River stations in the south in 1956. Murphy Oil purchased the Ingram marketing, 201 stations, in 1961 and rebranded the stations Spur.

IDENTIFICATION SIGNS

INGRAM 72"RD
(4) 1956-1961 $75
Red/white outlined red porcelain sign with white band across center. Blue "INGRAM" on white center band.

PUMP SIGNS

INGRAM 11"RD
(4) 1956-1961 $100
Red/white outlined red porcelain sign with white band across center. Blue "INGRAM" on white center band.

HEMPSTEAD OIL STORAGE CO. - MOBILE, AL

Purchased by Murphy Oil in 1962 and rebranded Spur.
No collectible size Hosco signs known at this time.

STANDARD AND UTOCO

STANDARD OIL CO. OF INDIANA - CHICAGO, IL

Formed in 1889 as a refining and marketing division of the Standard Oil Trust, Standard of Indiana had established an extensive marketing organization by the time of the 1911 breakup of Standard. Primarily a marketer, Standard established an extensive network of gasoline stations throughout the midwest. Involvement with Pan American and Amoco, detailed in the Amoco section in this book, explains Standard's purchase of Pan-Am in order to gain crude oil production. Midwest Oil Refining, listed below, was another producer purchased by Standard in order to keep its extensive refining and marketing operation going. In 1961 Standard shifted all domestic marketing to the subsidary American Oil Company. Stations in former Amoco, Pan-Am and Utoco territory were rebranded American, while stations in the Standard territory remained Standard. In 1974 the Amoco brand replaced American and began to replace Standard, but many stations retain the Standard brand today. Standard Oil of Indiana became Amoco Corporation in 1984.

IDENTIFICATION SIGNS

RED CROWN GASOLINE 42"RD
(3) 1915-1934 $550

RED CROWN GASOLINE 48"RD
(2) 1925-1934 $525

Blue outlined white porcelain sign with blue outlined white center circle. Red, white, and blue detailed crown in center circle. Red "RED CROWN" arched around top, "GASOLINE" around bottom.

POLARINE MOTOR OIL THE PERFECT MOTOR OIL 42"RD
(3) 1920-1934 $450

POLARINE MOTOR OIL THE PERFECT MOTOR OIL 48"RD
(4) 1920-1934 $400

Red outlined white porcelain sign with blue line triangle in center. Red logotype "Polarine" with white "THE PERFECT MOTOR OIL" in underline across top of triangle with blue "CONSULT CHART" in lower triangle.

STANDARD SERVICE 36x60
(3) 1934-1946 $225

STANDARD SERVICE 48x84
(2) 1934-1946 $175

Red/white outlined blue porcelain sign with white "STANDARD/SERVICE" across sign face.

STANDARD TORCH & OVAL 36x48
(4) 1946-1961 $300

Die-cut porcelain STANDARD torch and oval with red flame with white details.

STANDARD TORCH & OVAL 54x72
(3) 1946-1961 $150

Steel framed oval porcelain STANDARD torch and oval with detachable one piece porcelain red flame bolted to frame.

STANDARD TORCH & OVAL W/ GLASS FLAME
(1) 1946-1961 $225

Pole framed oval porcelain STANDARD torch an oval with bolt on torch bowl and internally lit plastic or glass flame.

STANDARD torch & oval 60x100
(3) 1961-1974 $100

Steel framed porcelain sign with formed porcelain two-piece flame. Red, white & blue. 1961 logo.

STANDARD torch & oval 30x54
(3) 1961-1974 $100

Tin litho diecut sign. Red, white & blue. 1961 logo.

STANDARD torch & oval 60x100
(5) 1971-1974 $100

Steel framed porcelain sign with formed porcelain two-piece flame. Red, white & blue. 1971 logo.

BUILDING SIGNS AND LETTERS

STANDARD "SERVICE" CERAMIC CREST 24"RD
1926-1940 $350

Blue/white outlined blue cast ceramic disc with white "SERVICE" across center. Gold torch with white flame superimposed over "E" in service.

STANDARD "RED CROWN" CERAMIC CREST 24"RD
(4) 1940-1961\ $300

Blue outlined white cast ceramic disc with red detailed "Red Crown" logo in center.

STANDARD OIL PRODUCTS STRIP 12x96
(1) 1934-1961 $125

Blue/white outlined embossed blue porcelain sign with white "STANDARD OIL PRODUCTS" across sign face.

STANDARD OIL DEALER STRIP 12x96
(1) 1934-1961 $125

Blue/white outlined embossed blue porcelain sign with white "STANDARD OIL DEALER" across sign face.

STANDARD OIL PRODUCTS LETTERS 10"&16"
(3) 1940-1955 $150

Art deco style blue porcelain letters large (16") "STANDARD" and smaller (10")"OIL PRODUCTS."

SERVICE LETTERS 16"
(3) 1940-1955 $100

Art deco style blue porcelain letters large "SERVICE."

PUMP SIGNS

STANDARD RED CROWN 12x13
(1) 1948-1955 $125

White/blue outlined white porcelain sign with blue outlined red "STANDARD/RED CROWN" across upper sign face above red/white/ blue Standard torch and oval logo. Blue circle at either bottom corner with interlocked "SO" in center. Blue "STANDARD OIL COMPANY" between circles.

STANDARD WHITE CROWN 12x13
(1) 1948-1955 $125

White/blue outlined white porcelain sign with blue outlined red "STANDARD/WHITE CROWN" across upper sign face above red/white/ blue Standard torch and oval logo. Blue circle at either bottom corner with interlocked "SO" in center. Blue "STANDARD OIL COMPANY" between circles.

CURB SIGNS

RED CROWN GASOLINE 30"RD
(2) 1915-1934 $550

Blue outlined white porcelain sign with blue outlined white center circle. Red, white, and blue detailed crown in center circle. Red "RED CROWN" arched around top, "GASOLINE" around bottom.

RED CROWN ETHYL A GENERAL MOTORS PRODUCT 30"RD
(5) 1923-1925 $750

Black outlined white porcelain sign with black outlined white center circle. Black triangular early Ethyl logo in center with blue rays. Yellow "ETHYL" across top above detailed red/white/blue crown. Yellow banner below, with black "GEN-ERAL/MOTORS/PRODUCT" on banner. Red "RED CROWN" arched around top, "GASOLINE" below. This is the first Ethyl curb sign.

RED CROWN ETHYL GASOLINE 30"RD
(4) 1926-1932 $525
Blue outlined white porcelain sign with blue outlined white center circle. Large Ethyl
(EGC) logo with blue rays in center circle. Red "RED CROWN" arched around top, "GASOLINE" around bottom.

SOLITE GASOLINE W/ETHYL 30"RD
(5) 1932-1936 $500
White outlined red porcelain sign with blue 1/3 left side. Blue outlined white "SOLITE/GASOLINE" with red inlines over script white "with" across sign above small blue & white Eth-yl
(EGC) logo at lower right.

POLARINE THE PERFECT MOTOR OIL TRIANGLE 28"
(4) 1920s $375
Red painted steel frame around white porcelain die-cut trian-gle. Red "Polarine" logotype across top above red "THE PER-FECT/MOTOR/OIL."

POLARINE THE PERFECT MOTOR OIL CONSULT CHART 30"RD
(1) 1920s $350
Red outlined white porcelain sign with blue line triangle in center. Red logotype "Polarine" with white "THE PERFECT MOTOR OIL" in underline across top of triangle with blue "CONSULT CHART" in lower triangle.

POLARINE MOTOR OIL STURDY & RICH 30"RD
(2) 1920s $350
Red outlined white porcelain sign with blue line triangle in center. Red logotype "Polarine" with white "MOTOR OIL" in underline across top of triangle with blue "STURDY/ AND/RICH" in lower triangle.

ISO-VIS STANDARD OIL COMPANY 30"RD
(3) 1930-1935 $450
Red outlined white porcelain sign with red outlined white cen-ter circle. Red circle in center with white "SERVICE" and torch (identical to Standard corporate emblem) with blue "STAN-DARD OIL COMPANY" arched around top and "INDIANA" around bottom. Blue "ISO=VIS" arched around top of sign, "MOTOR OIL" around bottom.

LUBRICATION/MOTOR OIL SIGNS

POLARINE CRANKCASE SERVICE PANEL 30x60
(4) 1924-1928 $350
Red outlined white porcelain sign with red outlined circle in center. Red "Polarine" logotype across center with blue "CRANKCASE" arched around top, "SERVICE" around bot-tom.

TOP QUALITY ISO VIS "D"MOTOR OIL 16x60
(3) 1928-1935 $150
Blue/white/red outlined self framed blue porcelain sign with white "ISO/VIS" diagonal in center. White outlined red "TOP/QUALITY" across top, "D" above white "MOTOR OIL" across bottom.

ANTI-SLUDGE ISO VIS "D"MOTOR OIL 16x60
(3) 1928-1935 $175
Blue/white/red outlined self framed blue porcelain sign with white "ISO/VIS" diagonal in center. White outlined red "ANTI-/SLUDGE" across top, "D" above white "MOTOR OIL" across bottom.

TIRES ATLAS BATTERIES GUARANTEED 16x60
(3) 1930-1940 $125
Blue outlined self framed red porcelain sign wtih white out-lined blue diagonal stripe at upper right and lower left corners. White "ATLAS" vertical on sign, with white "TIRES" across top and "BATTERIES" across bottom. White "GUARANTEED BY" arched above red/white/blue Standard corporate "SERVICE" logo at bottom.

ATLAS TIRES REPAIRS GUARANTEED 16x60
(3) 1930-1940 $125
Blue outlined self framed red porcelain sign wtih white out-lined blue diagonal stripe at upper right and lower left corners. White "ATLAS" vertical on sign, with white "TIRES/REPAIRS" across bottom. White "GUARANTEED BY" arched above red/white/blue Standard corporate "SERVICE" logo at bot-tom.

LUBESTER SIGNS

POLARINE THE PERFECT MOTOR OIL 7"RD
(4) 1920-1935 $250
Red outlined white porcelain sign with blue line triangle in center. Red logotype "Polarine" with white "THE PERFECT MOTOR OIL" in underline across top of triangle with blue "CONSULT CHART" in lower triangle.

ISO-VIS MOTOR OIL STANDARD SERVICE 7"RD
(4) 1926-1935 $250
Red outlined white porcelain sign with red outlined white cen-ter circle. Red circle in center with white "SERVICE" and torch (identical to Standard corporate emblem) with blue "STAN-DARD OIL COMPANY" arched around top and "INDIANA" around bottom. Blue "ISO=VIS" arched around top of sign, "MOTOR OIL" around bottom.

SECONDARY SIGNS

POLARINE FROST & CARBON PROOF OIL 9x19.5
(4) 1911-1915 $450
White outlined blue tin litho sign with white/blue/orange gal-lon square can at left side. Red outlined white "POLARINE" across top above white "THE FROST AND CARBON PROOF." Red outlined white "Motor Car Oil" across bottom.

EUREKA HARNESS OIL 14x19.5
(5) 1911-1915 $350
Blue outlined orange tin litho sign with blue outlined green circle with multicolor hose picture bottom center. Blue "EURE-KA/HARNESS OIL" across top, with blue "Lengthens/the life" to left and "of your/Harness" at right.

RED CROWN/GASOLINE/POWER-ECONOMY-SERVICE 24x60
(5) 1915-1920 $350
Black outlined white tin litho sign wtih red "RED CROWN/ GASOLINE" across top above black "POWER-SERVICE-ECON-OMY" across bottom.

POLARINE THE PERFECT MOTOR OIL TIN 24x60
(5) 1915-1920 $350
Black outlined white tin litho sign with red script "Polarine" lo-gotype above blac "THE PERFECT MOTOR OIL" across sign.

SOLITE A PERFECT GASOLINE FLANGE 14x18
(5) 1922-1926 $325

Red outlined white tin litho flange sign. Blue outlined white circle with blue outlined white center circle. Red outlined blue triangle in center circle with white "A/PERFECT" across. Red "SOLITE" arched around top "GASOLINE" around bottom

RED CROWN GASOLINE POWER MILEAGE 24x72
(4) 1925-1930 $350

White porcelain sign with blue outlined white circle logo with blue outlined white center circle. Red, white, and blue detailed crown in center circle. Red "RED CROWN" arched around top, "GASOLINE" around bottom. Blue "POWER" to left, "MILEAGE" to right.

APPROVED CREDIT CARDS HONORED 12x18
(4) 1946-1961 $100

Blue porcelain sign with large white "CREDIT CARDS" across center. Smaller "APPROVED" across top, "HONORED" across bottom.

STANDARD CREDIT CARDS HONORED 24x42
(4) 1961-1965 $100

White outlined red vertical rectangular porcelain sign with American logo at top over white "CREDIT/CARDS/HONORED"

THERMOMETER PANELS

All are large wooden backed self framed porcelain thermometer panels for outdoor use.

RED CROWN GASOLINE FOR POWER MILEAGE POLARINE THE PERFECT MOTOR CONSULT CHART $900

Blue outlined white porcelain thermometer panel. Blue outlined white circle at top with blue outlined white center circle. Red and blue crown in center circle with red "RED CROWN" arched around top, "GASOLINE" around bottom. Red "for/POWER/MILEAGE" across top of white area above red and blue text and thermometer bulb. Blue/white/red outlined white circle at bottom with blue triangle in center. Red script "Polarine" across triangle above blue "CONSULT CHART" .

RED CROWN GASOLINE FOR POWER MILEAGE POLARINE THE PERFECT MOTOR OIL MADE IN FIVE GRADES CONSULT CHART $900

Blue outlined white porcelain thermometer panel. Blue outlined white circle at top with blue outlined white center circle. Red and blue crown in center circle with red "RED CROWN" arched around top, "GASOLINE" around bottom. Red "for/POWER/MILEAGE" across top of white area above red and blue text and thermometer bulb. Red/white/blue outlined white triangle at bottom with red script "Polarine" across above blue "MADE IN/FIVE/GRADES." Red "CONSULT" & "CHART" across bottom.

RED CROWN GASOLINE FOR POWER SERVICE ECONOMY POLARINE THE PERFECT MOTOR OIL MADE IN FIVE GRADES CONSULT CHART $900

Blue outlined white porcelain thermometer panel. Blue outlined white circle at top with blue outlined white center circle. Red and blue crown in center circle with red "RED CROWN" arched around top, "GASOLINE" around bottom. Red "for/POWER/SERVICE/ECONOMY" across top of white area above red and blue text and thermometer bulb. Red/white/blue outlined white triangle at bottom with red script "Polarine" across above blue "MADE IN/FIVE/ GRADES." Red "CONSULT" & "CHART" across bottom.

SAVES THE MOTOR/POLARINE MOTOR OIL/STURDY & RICH/ISO VIS STANDARD SERVICE DOES NOT THIN OUT $900

Blue outlined white porcelain thermometer panel. Blue/white/red outlined white circle at top with blue triangle in center. Red script "Polarine" across triangle above blue "CONSULT CHART". Blue outlined white circle ISO-VIS logo at bottom. White "SAVES THE/MOTOR" across above Polarine logo, "DOES NOT/THIN OUT" below ISO-VIS logo. Red and blue text around thermometer bulb.

BULK PLANT SIGNAGE

STANDARD OIL COMPANY (INDIANA) 12x96
1934-1961 $125

Blue/white outlined embossed blue porcelain sign with white "STANDARD OIL COMPANY" across sign face. Small white "(INDIANA)" center across bottom.

STANDARD OIL COMPANY (INDIANA) 12x96
1950s $125

Black outlined embossed white porcelain sign with red "STANDARD OIL COMPANY" across sign face. Small red "(INDIANA)" center across bottom.

STANDARD OIL/DIVISION AMERICAN 24x96
(2) 1961-1971 $100

Red outlined white porcelain sign with red band across bottom. Red/white/blue 1961 torch and oval logo at left with blue "STANDARD OIL" across white area. White "DIVISION OF AMERICAN OIL COMPANY" across bottom red band.

ADVANCE SIGNS

STANDARD SERVICE X MILES 30x42
(5) 1934-1946 $150

Three piece porcelain sign assembled on post. Top panel is red/white outlined blue porcelain rectangle with white "STANDARD/SERVICE" across fac. Lower section is blue outlined white panel with blue "MILES" across face. Center section is die-cut white outlined red porcelain number, changable for appropriate distance.

STANDARD OIL OF NEBRASKA

A marketing operation founded in 1906 and long supplied by Standard of Indiana, they merged with Standard of Indiana in August, 1939. Stations retained the Nebraska identification until after WW2.

Std NEB

IDENTIFICATION SIGNS

RED CROWN/GASOLINE 18x22
(5) 1920-1925 $1300

Blue outlined white porcelain sign with geartooth edge around white/blue border. Red and black crown with "RED CROWN GASOLINE" in upper center with blue "RED CROWN" arched above and "GASOLINE" across below. Smaller blue "STANDARD OIL COMPANY/OF NEBRASKA" across bottom.

CURB SIGNS

RED CROWN GASOLINE 30"RD
(4) 1915-1934 $675

Black outlined white porcelain sign with black outlined white center circle. Red and black detailed crown in center circle. Red outlined black "RED CROWN" arched aorund top, "GASOLINE" around bottom.

LUBRICATION/MOTOR OIL SIGNS

POLARINE MOTOR OIL FLANGE 18x19

(5) 1920-1925 $1100

White outlined blue porcelain sign with geartooth edge around blue/white border. Red outlined white script "Polarine" above "MOTOR OIL" on upper sign face with white "STANDARD OIL COMPANY/OF NEBRASKA" across bottom of sign.

THERMOMETER PANELS

RED CROWN GASOLINE POLARINE THE PERFECT MOTOR OIL FOR SALE HERE $900

Blue outlined white porcelain thermometer panel with white circle at top. Red and black crown in upper circle with "RED CROWN GASOLINE" across crown. Red "RED/CROWN/ GASOLINE" in upper white area above thermometer bulb and red and blue text. Blue script "Polarine" above red "THE PERFECT/MOTOR OIL" above blue "FOR SALE HERE" across bottom.

MIDWEST REFINING CO. - CHEYENE, WY

Midwest was a Rocky Mountain producer/refiner when Standard of Indiana purchased controlling interest in the company in 1917. In 1928, Midwest entered the gasoline market, purchasing Arro Oil and Refining of Lewiston, Montana, Aero Oil Co. of Cheyene, Wyoming and 75 gas stations in Colorado and New Mexico from Vickers. The Midwest brandname replaced the individual brands. In April 1930, Standard of Indiana purchased the entire Midwest marketing and rebranded the stations Standard. The Midwest subsidiary, Utah Oil Refining Co., having ventured into gasoline marketing, continued to market under the Utah Oil brands.

No collectible size Midwest signs are known at this time.

ARRO OIL CO. - LEWISTON, MT: Purchased by Standard's Midwest subsidiary and rebranded Midwest in 1929. See more details above.

No collectible size Arro signs known at this time.

UTAH OIL REFINING CO. - SALT LAKE CITY, UT: Founded as a subsidiary of Midwest Refining in 1928, the company marketed motor oils under the Vico name and gasoline under the Pep 88 brand in Utah and Idaho until after WW2. The "UTOCO" brand was introduced at the end of WW2 and gradually replaced Vico/Pep 88. In 1961, when all of Standard's U.S. marketing was assigned to the American Oil Company subsidiary, the American brandname replaced Utoco. In 1974 Amoco replaced American.

Utoco

IDENTIFICATION SIGNS

VICO MOTOR OIL/PEP 30"RD

(5) 1928-1930 $400

Round porcelain sign with black upper half and orange lower half. Color split is a swirl across sign. White "VICO/MOTOR OIL" in upper black area with black "PEP" over white band on lower orange area. Black "The GAS with a PUNCH" on white band.

VICO MOTOR OIL/PEP 88 42"RD

(5) 1930-1934 $350

Round porcelain sign with black upper half and orange lower half. Color split is a swirl across sign. White "VICO/MOTOR OIL" in upper black area with white outlined black "PEP 88/GASOLINE" on lower orange area.

VICO MOTOR OIL/PEP 88 48"RD

(5) 1934-1947 $200

Round porcelain sign with blue upper half and red lower half. Color split is a swirl across sign. White "VICO/MOTOR OIL" in upper blue area with white outlined blue "PEP 88/GASOLINE" in lower red area.

UTOCO 54x72

(5) 1947-1961 $300

Steel framed oval porcelain UTOCO torch and oval with detachable one piece porcelain red flame bolted to frame.

UTOCO LARGE W/ GLASS FLAME

(5) 1947-1961 $250

Pole framed oval porcelain UTOCO torch an oval with bolt on torch bowl and internally lit plastic or glass flame.

PUMP SIGNS

VICO/PEP 88 10"RD

(4) 1935-1947 $250

Round porcelain sign with blue upper half and red lower half. Color split is a swirl across sign. White "VICO/MOTOR OIL" in upper blue area with blue outlined white "PEP 88/GASOLINE" in lower red area.

ULTRA POWER WHITE 8x13

(4) 1955-1961 $175

White porcelain embossed sign with red "ULTRA/Power" diagonal across sign.

ULTRA POWER RED 8x13

(4) 1955-1961 $175

Red porcelain embossed sign with white "ULTRA/Power" diagonal across sign.

UTOCO 12x13

(4) 1947-1961 $250

Embossed die-cut porcelain sign in form of torch and oval logo. Red upper third, blue lower with white band across center. Blue "UTOCO" on white band. Red flame torch.

CURB SIGNS

ISO VIS MOTOR OIL UTAH OIL 30"RD

(5) 1930-1932 $325

Red outlined white porcelain sign with blue outlined white center circle. Blue circle in center with white "SERVICE" and torch

(identical to Standard corporate emblem) with red "UTAH OIL REFINING" arched around top and "COMPANY" around bottom. Blue "ISO=VIS" arched around top of sign, "MOTOR OIL" around bottom.

LUBRICATION/MOTOR OIL SIGNS

WE SELL VICO OILS & GREASES

$200

White/black outlined white porcelain sign with black "WE SELL" at top. Black outlined red "VICO" logotype in center with black "Paraffin Base" below. Black outlined red "OILS and GREASES" across bottom, with black outlined red "U" at top right.

STANAVO: Formed in 1929 by Standard of Indiana, New Jersey and California as a joint venture to research aviation petroleum needs and market those products through their individual marketing outlets. Stanavo first marketed aviation gasoline in 1931. With the withdrawal of Standard of New Jersey in 1937, the other two participants decided to discontinue the effort and Stanavo was dissolved in June 1938.

Stanavo

IDENTIFICATION SIGNS

STANAVO PANEL 6x18
(5) 1931-1938 $250
Blue outlined red porcelain sign with blue outlined white "STANAVO" across center.

TRUES OIL CO. - SPOKANE, WA

Independent marketer with stations in Washington, purchased by Standard in 1959. Stations remained under Rainbow brand until after 1961 when most were rebranded American. Standard continues to market under "Rainbow" name as a secondary brand today.

Rainbow

No collectible size Rainbow signs are known at this time.

STANDARD OIL/KYSO

STANDARD OIL COMPANY "INCORPORATED IN KENTUCKY" - LOUISVILLE, KY

Standard Oil of Kentucky was, for 50 years, the world's largest gasoline "jobber." One of the Standard divisions spun off with the 1911 break up of Standard Oil, Standard of Kentucky (commonly called Kyso) was assigned marketing in Kentucky, Florida, Georgia, Alabama annd Mississippi, but had no oil refineries, oil production or exploration, or oil pipeline connections. It was simply a markekting group with retail service stations in five southeast states. As a result of this, Kyso contracted with Standard of New Jersey (Esso/Exxon) to supply their needs and simply sold Esso and Esso Extra gasoline under the Crown and Crown Extra names until 1961. When Esso introduced the super-premium "Golden Esso Extra" in 1956, Kyso followed with "Super Crown Extra." This neat arrangement ended in 1961 when Standard of California (Chevron) purchased Kyso outright and cancelled the long-standing Esso supply agreements. Esso scrambled to establish markekts in the Kyso territory and Chevron promptly sued them to limit the use of the Standard trademark "Esso" in the five Kyso states. Although Esso had unsuccessfully tried to market under the Esso brandname in Indiana Standard territory (St. Louis 1935-1938), they had long since began using alternate brandnames in other territories. s. The southeast, however, was different with their already established brand (Kyso sold Esso motor oils) and a court battle ensued. In 1967 Esso lost the rights to the trademark in Kyso territory and service stations signs were covered with tiger decals until they could be replaced by Enco signs. After a Supreme Court ruling in 1969 confirming this limitation, Esso began the name search process that eventually led to the Exxon brandname. (See Esso/Exxon section in this book for more on this). In the meantime, with the 1961 purchase by Chevron, stations were completely re-imaged and signage was updated. The Chevron name replaced "STANDARD" at most Kyso stations in 1977, although a few stations remain Standard to protect the long held brand.

IDENTIFICATION SIGNS

STANDARD OIL COMPANY INCORPORATED 42x96
(5) 1928-1935 $250
Framed white outlined blue porcelain rectangle with rounded corners. White "STANDARD/OIL/COMPANY" on sign face. Small white "INCORPORATED" below, with white bar to either side of "OIL."

STANDARD OIL PRODUCTS 24x30
(4) 1935-1945 $300
STANDARD OIL PRODUCTS 34x42
(4) 1935-1945 $300
STANDARD OIL PRODUCTS 58x72
(5) 1935-1945 $250
Blue porcelain rectangle with rounded corners. White "STANDARD/OIL/PRODUCTS" on sign face.

STANDARD OIL DEALER 30x72
(3) 1945-1949 $350
Blue embossed porcelain die-cut sign faces assembled back-to-back. Two red stripes across top with single red stripe across bottom of sign face. White "STANDARD" across center, with "OIL" below in notch area. White "DEALER" in area below bottom stripe.

STANDARD OIL 28x72
(3) 1949-1962 $350
Blue embossed porcelain die-cut sign faces assembled back-to-back. Two red stripes across top with single red stripe across bottom of sign face. White "STANDARD" across center, with "OIL" below in notch area.

BUILDING SIGNS AND LETTERS

STANDARD RED PORCELAIN 12"
(4) 1945-1962 $150
Red porcelain "STANDARD" lettering for facade or rooftop mounting.

NEON SIGNS

STANDARD OIL PRODUCTS 58x72
(5) 1935-1945 $650
Blue porcelain rectangle with rounded corners. White "STANDARD/OIL/PRODUCTS" on sign face.

STANDARD OIL DEALER 42x108
(4) 1945-1962 $750
Blue embossed porcelain die-cut sign faces assembled back-to-back on steel frame with curved endcaps. Two red stripes across top with single red stripe across bottom of sign face. White "STANDARD" across center, with "OIL" below in notch area. White "DEALER" on removable panel below bottom stripe.

FLANGE SIGNS

CROWN GASOLINE "KYSO" BLACK 22x22
(4) 1915-1920 $1600
White porcelain sign with red outlined black elaborate crown design covering most of sign face. White "CROWN/GASOLINE" covering on crown, with red "KYSO" in oval at bottom of crown. Black "STANDARD OIL COMPANY" arched above "KYSO," "INCORPORATED IN KENTUCKY" arched below.

CROWN GASOLINE BLACK 22x30
(4) 1920-1923 $1600
White porcelain sign with red outlined black elaborate crown design covering most of sign face. Red outlined white "CROWN/GASOLINE" covering on crown, with red "STAN-

Three very different pump plates. *Spencer Nelson.*

Gulf Refining Co. porcelain sign. *Pete Clark.*

Vico Oils porcelain sign. *Spencer Nelson.*

Richfield neon die-cut sign, rare. *Jim Steeves.*

Purol Motor Gasoline rare metal sign with space for dealer name.

Miscellaneous wall of porcelain signs. *Dennis McClellan.*

Texaco keyhole porcelain sign.

Texaco Diesel Fuel #2 pump plate sign.

Polly Gas sign. *Jeff Decker.*

Oilzum 24" porcelain sign. *Leo Mathieu.*

Firestone tin and Hyvis porcelain signs.

Harris Oils porcelain hanging sign. *Leo Mathieu.*

Mohawk porcelain neon surrounded by miscellaneous signs. *Dennis Leith.*

Penzene hanging tin sign. *Robert Bridges.*

DARD OIL COMPANY" arched above black "INCORPORATED" in oval at base of crown.

CROWN GASOLINE BLUE 22x30
(4) 1923-1928 $1600
White porcelain sign with red outlined blue elaborate crown design covering most of sign face. Red outlined white "CROWN/GASOLINE" covering on crown, with red "STANDARD OIL COMPANY" arched above black "INCORPORATED" in oval at base of crown.

SUN OIL COMPANY - PHILADELPHIA,PA

Sun Oil, a refiner-marketer founded by the Pew family in Philadelphia inn 1890, began retail gasoline marketing in 1920. Primarily a northeastern marketer, Sunoco had stations from Maine to the Chicago area south to Virginia by the early 1930s, as well as marketing in Florida. For many years Sunoco marketed only a premium grade of gasoline "Blue Sunoco," but for a brief time after WW2 added a super premium "Sunoco Dynafuel." In 1956 Sunoco, in cooperation with the Wayne Pump Company began experimenting with a multiple-grade blending system "CUSTOM BLENDING" that sold nine grades of gasoline through a single pump, drawing from two tanks and mixing in proper proportion. As the new equipment was installed, stations were usually re-imaged and by 1959 the Custom-Blenders were in place at all Sunoco locations. In 1968 Sun merged with Sunray DX and for the next twenty years marketed under the Sunoco name in the northeast and east while retaining the DX name in the mid-continent area. About 1988 Sun began re-imaging in the midwest, replacing the DX brand with Sunoco at high volume locations and selling off or closing the others. Some pockets of DX branding remain, but this is primarily from jobbers who just haven't removed old signage. DX and related brand signs can be found under the listing DX in this book.

IDENTIFICATION SIGNS

SUNOCO GAS OIL 36x92
(3) 1926-1938 $750
Blue outlined yellow porcelain die cut flat diamond shaped sign faces assembled can style. Blue "SUNOCO" in bowtie lettering across center, with "GAS" abovve and "OIL" below. Blue outlined red porcelain arrow pieces attached to each side of frame at diamond point.

SUNOCO GAS OIL 36x92
(5) 1930-1938 $750
Steel framed blue outlined yellow porcelain die cut flat diamond shaped sign. Blue "SUNOCO" in bowtie lettering across center, with "GAS" above and "OIL" below. Blue outlined red porcelain arrow pieces attached to each side of frame at diamond point.

SUNOCO DIAMOND CUT 42x96
(3) 1935-1959 $650
Steel framed blue outlined yellow porcelain die cut flat diamond shaped sign. Blue "SUNOCO" in bowtie lettering across center. Blue outlined red porcelain arrow pieces attached to each side of frame at diamond point.

SUNOCO SQUARE CUT 42x96
(3) 1946-1959 $650
Steel framed blue outlined yellow porcelain die cut flat diamond shaped sign with squared ends allowing for "S" and "O." Blue "SUNOCO" in bowtie lettering across center. Blue outlined red porcelain arrow pieces attached to each side of frame at diamond point.

SUNOCO DIAGONAL ARROW 60x96
(5) 1959-1968 $500
Pole framed blue outlined yellow porcelain die cut flat diamond shaped sign with squared ends allowing for "S" and "O." Blue "SUNOCO" in bowtie lettering across center. White outlined red porcelain arrow pieces attached to top and bottom of frame to appear diagonal through sign.

BUILDING SIGNS AND LETTERS

SUNOCO BLUE PORCELAIN LETTERS
(5) 1932-1950 $225
Blue formed porcelain lettering "SUNOCO." Size varies as lettering is bowtie style.

CADUCES 30x42
(5) 1932-1950 $400
Yellow porcelain formed torch with red flame and yellow snakes wrapped around torch shaft. Blue formed porcealin wings on either side of torch. The caduces is associated with Mercury in Greek mythology and Sun Oil used the element mercury in their oil refining process, hence the association.

PUMP SIGNS

STRIPES 2"WIDE
(2) 1946-1959 $35
Yellow porcelain stripe with blue lines dividing the yellow into multiple stripes. Custom sizes for different makes of pumps as these were desinged to install vertically above and below diamond shaped pump signs listed below. Some upper short stripes are curved at the top to match contour of pumps.

BLUE SUNOCO 8x12
(1) 1946-1956 $75
Dark yellow porcelain diamond shaped die-cut sign with small blue "BLUE" over larger "SUNOCO" across center. Blue lettering is shaded with lighter blue.

BLUE SUNOCO WITH BORDER 8x12
(5) 1956-1959 $200
Bright yellow porcelain diamond shaped die-cut sign with blue outline and small blue "BLUE" over larger "SUNOCO" across center. Blue lettering is shaded with lighter blue. First signs used on custom blencer pumps.

DYNAFUEL 8x12
(3) 1946-1950 $125
Bright yellow porcelain diamond shaped die-cut sign with yellow/red outline. Blue "DYNALUBE" across center. Blue lettering is shaded with lighter blue.

BLUE SUNOCO 200 15x20.5
(2) 1956-1958 $250
Elaborate die-cut porcelain sign with white outlined blue diamond area across center with red arrow parts pointing right to either side of diamond. Yellow "BLUE/SUNOCO" across diamond. Large yellow vertical triangle extensions above and below diamond logo. Blue "200" on lower yellow triangle. Studs welded on back for mounting.

BLUE SUNOCO 200X 15x20.5
(3) 1958-1959 $250
Same as above, with black "x" decal below "200."

BLUE SUNOCO 19x22
(1) 1959-1974 $225

Elaborate die-cut porcelain sign with white outlined yellow diamond area across center with red arrow parts pointing right to either side of diamond. Blue "BLUE/SUNOCO" across diamond. Large blue vertical triangle extensions above and below diamond logo. Studs welded on back for mounting.
BLUE SUNOCO 190 15x21
(5) 1962-1965 $150
Elaborate die-cut porcelain sign with white outlined yellow diamond area across center with red arrow parts pointing right to either side of diamond. Blue "BLUE/SUNOCO" across diamond. Large blue vertical triangle extensions above and below diamond logo. Yellow "190" on lower blue triangle. Studs welded on back for mounting. Very thin porcelain over what appears to be galvanized metal.

NEON SIGNS

SUNOCO GAS OIL 32x96
(5) 1930s $1200
Building mounted blue outlined yellow porcelain die cut flat diamond shaped sign. Blue "SUNOCO" in bowtie lettering across center, with "GAS" above and "OIL" below. Blue outlined red porcelain arrow pieces attached to each side of frame at diamond point. Neon "SUNOCO" lettering and diamond outline.

CURB SIGNS

AUTHORIZED SUNOCO DEALER 22x31.5
(4) 1925-1935 $425
Blue outlined yellow porcelain die-cut flat diamond sign with blue "SUNOCO" in bowtie lettering across center. Blue "AUTHORIZED" above "SUNOCO," "DEALER" below.
OILS SUNOCO GREASES 22x31.5
(4) 1925-1935 $400
Blue outlined yellow porcelain die-cut flat diamond sign with blue "SUNOCO" in bowtie lettering across center. Blue "OILS" above "SUNOCO," "GREASES" below.
MERCURY MADE SUNOCO OIL 22x31.5
(5) 1925-1935 $400
Blue outlined yellow porcelain die-cut flat diamond sign with blue "SUNOCO" in bowtie lettering across center. Blue caduces design with red script "Mercury Made" above "SUNOCO," "OIL" below.

LUBRICATION/MOTOR OIL SIGNS

SUNOCO THE DISTILLED OIL SUNSET 19.5x26
(5) 1915-1918 $475
Blue porcelain flange sign with white outlined yellow "SUNOCO" over "MOTOR OIL" on lower sign face. Red outlined yellow diamond with red "SUN OILS" on upper sign face with yellow sunset/water scene
SUNOCO THE DISTILLED OIL DIAMOND 19.5x26
(2) 1918-1925 $350
Similar to above with larger "SUNOCO" logotype, smaller diamond and solid blue background without sunset.
DISTILLED SUNOCO MOTOR OIL 15x15
(3) 1925-1935 $350
Yellow/blue outlined yellow porcelain sign with red caduces design at top. Red outlined diamond at top with red "THE/DISTILLED/OIL" in diamond.
DISTILLED SUNOCO MOTOR OIL 10x12
(3) 1925-1935 $300
Yellow/blue outlined yellow porcelain sign with red caduces design at top. Blue outlined diamond at top. "DISTILLED" in diamond.
MERCURY MADE SUNOCO MOTOR OIL 10x12
(3) 1925-1935 $300
Yellow/blue outlined yellow porcelain sign with red caduces design at top. Blue "MERCURY/MADE" superimposed over caduces with blue "SUNOCO/MOTOR OIL" below.
SUNOCO MOTOR OILS FLANGE
(5) 1935-1940 $300
Blue outlined yellow porcelain die-cut flat diamond sign with blue "SUNOCO" in bowtie lettering across center. Blue "MOTOR" above "SUNOCO," "OIL" below. Black flange area.
SUNOCO A to Z LUBRICATION die cut $450
Elaborate die cut tin litho sign. Blue/white outlined blue diamond sign with red outlined "A to Z" across center. Blue outlined yellow banner above and below diamond with blue "SUNOCO" on upper banner and "LUBRICATION" on lower banner.

SECONDARY SIGNS

REST ROOMS 6x14
(3) 1946-1964 $225
Yellow outlined blue porcelain die-cut rounded rectangle sign with yellow/blue outlined white circle at either end. Blue silouette of lady with parasol on one circle, man with cane on other. White script "Rest/Rooms" in blue area between circles.
MEN 2.5x5
(3) 1946-1964 $75
Yellow outlined blue porcelain die cut rectangular sign rounded on left end. Yellow/blue outlined white circle on rounded end with sillouette of man with cane. White script "Men" on blue area.
LADIES 2.5x5
(3) 1946-1964 $75
Yellow outlined blue porcelain die cut rectangular sign rounded on left end. Yellow/blue outlined white circle on rounded end with sillouette of lady with parasol. White script "Ladies" on blue area.
CHARGE ACCOUNTS HONORED HERE 18x24
(4) 1930's $150
Blue outlined white porcelain die-cut banner shaped sign with blue "CHARGE ACCOUNTS" contoured around top above small "OF." Larger "SUN OIL COMPANY" above "HONORED HERE" below.
SUNOCO CHARGE ACCOUNTS HONORED $150
Blue outlined white porcelain die-cut banner shaped sign with blue band across center. White "CHARGE ACCOUNTS" on blue banner with blue "SUNOCO" in bowtie lettering above and blue "HONORED" below.
SUNOCO JUST AHEAD STRIP 22x70
(2) 1950's $175
Blue tin litho sign with yellow Sunoco diamond logo at right. White outlined red arrow to lower left with yellow "JUST/AHEAD..." above.
SUNOCO A TO Z LUBRICATION STRIP 22x70
(4) 1930's $150
Blue tin litho strip sign with red outline arrow across center. Yellow "SUNOCO" logotype to upper leftt with white "A/TO Z" in arrowhead. Blue "LUBRICATION" below.
SUNOCO OIL CRANKCASE SERVICE STRIP 22x70

(4) 1930's $150
Yellow tin litho strip sign with blue "SUNOCO/OILS" to upper left and red box to right. White "CRANKCASE/SERVICE" on red box, with blue band across bottom of sign. Yellow "KEEPS MOTORS KNOCKLESS" on band.
POSITIVELY NO SMOKING SUN OIL COMPANY 8x20
(4) 1930's $100
White porcelain sign with red outline and red "POSITIVELY/ NO SMOKING across center. Small red "SUN OIL COMPANY" across bottom.

TENNECO - HOUSTON, TX

Tenneco was formed in 1961 when Tennessee Gas Transmission (a pipeline company) merged with Bay Petroleum of Denver, Colorado. Bay had been in the process of buying out regional private brands throughout the southeast.

BAY PETROLEUM - DENVER, CO.

Bay was founded in McPherson, Kansas and dates back to the late 1930s, although little is known about the early history. By the mid-1950s, the operation had shifted to Denver, Colorado and in 1961 Bay Petroleum merged with Tennessee Gas Transmission to form Tenneco. During the 1960s, Tenneco purchased several private brand operations that it continued to operate company direct and a short-lived network of branded Tenneco dealers was created but most of the "Tenneco" stations were branded "Bay." During the 1950s, Bay had purchased several southern independents that had been converted to Bay and when Tenneco decided to get out of retail gasoline marketing, the remaining portion of the Bay brand, mostly in Georgia and Florida, was sold to Dixie Oil of Tifton, Georgia, a Tenneco supplied independent that continues to operate Bay branded stations. A note of trivia here: The Bay shield logo, last revised about 1946, is the oldest logo ever to be associated with a major brand that is still in use today, without change.

IDENTIFICATION SIGNS

BAY SHIELD 48x60
(5) 1938-1950 $200
Steel framed black outlined red three point shield with black outlined white band diagonal from upper left to lower right. Black "BAY" in diagonal band.
BAY SHIELD 72x84
(1) 1946-PRESENT $75
Pole framed black outlined red three point shield with black outlined white band diagonal from upper left to lower right. Black "BAY" in diagonal band.

PUMP SIGNS

BAY REGULAR 10.5x12.5
(4) 1946-1950 $300
Irregular die cut porcelain sign. Upper area is black outlined three point red shield with black outlined white band diagonal from upper left to lower right. Black "BAY" in diagonal band. Lower area is white outlined black rectangle attached to shield at bottom point. White "REGULAR" across black band.
92+ OCTANE 12x17
(2) 1960-1974 $25
Blue outlined white porcelain sign with large blue "92+" over smaller blue "OCTANE."
100 OCTANE 12x17
(2) 1960-1974 $25
Red outlined white porcelain sign with large red "100" over smaller red "OCTANE."

CITIZENS OIL COMPANY - TALLAHASSEE, FL

Citizens had originally operated as a franchise of the Atlanta based Greyhound stations and had introduced the Citizens 77 brand logo in 1937. About 1950 Citizens merged with Atlanta based Speed Oil Co. Citizens 77 operated dealer stations while the Speed brand was used in their direct operated stations. With the 1957 purchase by Bay Petroleum, the Speed name was retained for several years in a transitional Bay shaped "SPEED" shield.

IDENTIFICATION SIGNS

GREYHOUND MOTOR FUEL 24x40
(5) 1936-1946 $500
Red porcelain sign with white running greyhound across center of sign face. Grey "SUPER" above white"GREYHOUND" logotype above dog, white "MOTOR FUEL" below. Small grey "SPEED" to left of dog, "POWER" to right.
GREYHOUND MOTOR FUEL 24x40
(3) 1946-1950 $350
White porcelain sign with grey running greyhound across center of sign face. Grey "SUPER" above red "GREYHOUND" logotype above dog, red "MOTOR FUEL" below. Small grey "DISTRIBUTOR OR DEALER IS OPERATING UNDER CONTRACT WITH GREYHOUND OIL COMPANY" across bottom.
CITIZENS 77 72"RD
(4) 1939-1957 $300
Blue outlined white porcelain sign with red "CITIZENS" logotype arched across upper half of sign. Blue running greyhound below, with red "77" in quotes at bottom.

SPEED OIL CO. - ATLANTA, GA

Atlanta based discounter that merged with Citizens Oil Company about 1950.

IDENTIFICATION SIGNS

SPEED SHIELD 72x84
(5) 1957-1961 $100
Pole framed black outlined red three point shield with black outlined white band diagonal from lower left to upper right. Black "SPEED" in diagonal band.

GULF COAST OIL CO. - NEW ORLEANS, LA

Gulf Coast was bought out by Bay Petroleum in the late 1950s and rebranded Bay.
No collectible size Gulf Coast signs are known to exist.

TENNECO

Tenneco brandname introduced in 1961 and used on a group of dealer operated "Class A" stations in the southeast. In 1970 Tenneco began experimenting with combination gasoline/ convenience stores and built a network of stores in southern metropolitan markets, eventually expanding the store concept to their Direct and Red Diamond locations. In the late 1980s Tenneco sold off the gasoline marketing to TOC Retail and Tenneco branded locations were rebranded "Majik Market."

IDENTIFICATION SIGNS

TENNECO SHIELD 48x108
(5) 1962-1970 $100
Die-cut shield shaped porcelain can type sign with outline area red on the left and blue on the right. White teardrop oval in center with blue "TENNECO" across center.

PUMP SIGNS

TENNECO SHIELD 9.5x15
(4) 1962-1965 $65
Die-cut shield shaped embossed porcelain sign with outline area red on the left and blue on the right. White teardrop oval in center with blue "TENNECO" across center.

DIRECT OIL CO. - NASHVILLE, TN

Direct Oil Co. was founded in Nashville in 1946 by Calvin Houghland, the son of John Mason Houghland, founder of Spur Distributing. In the early years Direct acted much like a Spur jobber, distributing products through stations in eight states. Tenneco purchased the operation, totalling 132 stations, in 1962 and as of this writing in 1993, TOC retail, successors to the Tenneco marketing, continue to operate Direct stations throughout the southeast.

IDENTIFICATION SIGNS

DIRECT SHIELD 48x108
(3) 1962-PRESENT $100
Yellow porcelain can type die-cut trapezoidal sign with red trapezoid across center of sign face. White script "Direct" on band with blue "arrow' design above and below red area.

BUILDING SIGNS AND LETTERS

DIRECT OIL COMPANY 10x36
(3) 1946-1989 $75
Red porcelain strip with white script "Direct Oil Company" across center.

PUMP SIGNS

92+ OCTANE 12x17
(2) 1962-1974 $25
Blue outlined white porcelain sign with large blue "92+" over smaller blue "OCTANE."
100 OCTANE 12x17
(2) 1962-1974 $25
Red outlined white porcelain sign with large red "100" over smaller red "OCTANE."

RED DIAMOND OIL COMPANY - PICKENS, SC

Trackside discounter that was operating about 25 stations in North and South Carolina when purchased by Tenneco in the 1970s. Most of the original Red Diamond locations remain branded Red Diamond today.
No collectible size Red Diamond signs are known to exist.

TRUST/Q-S PETROLEUM-NASHVILLE, TN: Another regional southern marketer that Tenneco purchased part of in the 1970's. Semi-independent, the company survived the TOC Retail breakup and continues to operate today.

PUMP SIGNS

TRUST REGULAR GASOLINE 14x20
(3) 1970-PRESENT $50
White porcelain trapezoidal panel with blue/white outlined blue oval at top. White "TRUST" in oval. Red "REGULAR/ GASOLINE/ SELF SERVE" below oval.
TRUST PREMIUM GASOLINE 14x20
(3) 1970-PRESENT $50
White porcelain trapezoidal panel with blue/white outlined blue oval at top. White "TRUST" in oval. Red "PREMIUM/GASOLINE/ SELF SERVE" below oval.
TRUST UNLEADED GASOLINE 14x20
(3) 1970-PRESENT $50
White porcelain trapezoidal panel with blue/white outlined blue oval at top. White "TRUST" in oval. Red "UNLEADED/ GASOLINE/ SELF SERVE" below oval.

DIXIE OIL CO. - TIFTON, GA

Dixie Oil Co. of Tifton, Georgia is an independent company supplied by Tenneco beginning in the late 1960s. When Tenneco got out of retail gasoline marketing in the late 1980s, Dixie Oil purchased most of the former Bay branded operation in Georgia and Florida and continues to operate the Bay stations today.

IDENTIFICATION SIGNS

DIXIE FLAG 32x42
(5) 1947-1955 $350
Yellow porcelain sign with red triangles in upper and lower areas forming yellow diagonal band through center of sign. Green "DIXIE" in diagonal band.
DIXIE SHIELD 72x84
(3) 1955-PRESENT $300
Pole framed blue outlined three point shield with red and white vertical stripes in center. Blue "DIXIE" in band across stripes, with series of white stars in outline area

PUMP SIGNS

SUPER DIXIE GAS PANEL 15x28
(3) 1955-1970 $100
Porcelain trapezoidal panel with blue outline area and series of red and blue/white vertical stripes in center. Red/white red bands at top of panel with white "SUPER" and "GAS" on red bands and blue "DIXIE" on white band between.
DIXIE SHIELD PANEL 15x28
(2) 1970-PRESENT $100
Porcelain trapezoidal panel with blue outline area and series of red and blue/white vertical stripes in center. Dixie shield logo near top.

SUNSET

Sunset is a Dixie secondary operation.
No collectible size Sunset signs are known.

TEXACO

THE TEXAS COMPANY - NEW YORK, NY

Texaco. Throughout this country and around the world the first gasoline pump most people ever saw was a Texaco pump. Founded in 1902 in Port Arthur, Texas, Texaco was early on a worldwide producer, refiner and marketer. Their early "sales offices," set up to market oils and greases, specialty products, asphalts and even roofing shingles were to become the first organized network of gasoline filling staions. Promoting gasoline very heavily before 1910, they became the first marketer to establish a brand image-in 1915-and went on to become the first oil company to operate gas stations in all 48 (and later all

50) states. They continued to build and operate stations everywhere and with the coming of the Interstate Highway System in 1956 and the many new locations offered by it, Texaco eventually branded more than 40,000 gas stations in the United States. The oil shortages of the 1970s forced Texaco to pull back, moving out of some entire areas of the country by the mid-1980s. Texaco's purchase of Getty in 1984 added some former Skelly locations in areas Texaco had withdrawn from, but forced the company to the edge of bankruptcy when Pennzoil sued over the Getty purchase and won. Forced to scale back even more and to team up with a group of overseas investors, Texaco survived and currently markets through more than 12,000 stations in 46 states.

IDENTIFICATION SIGNS

THE TEXAS COMPANY FILLING STATION 42"RD
(3) 1915-1923 $2000
Red outlined porcelain sign with black outlined white center circle. Large red star in center without a border around a green "T" in center of star with black "TEXACO" across upper area of star, "MOTOR OIL" across lower area of star. White "GASOLINE" arched around top, "FILLING STATION" around bottom.

THE TEXAS COMPANY FILLING STATION 42"RD
(3) 1915-1923 $2200
Red outlined porcelain sign with black outlined white center circle. Large red star in center with black outlined green "T" in center of star with black "TEXACO" across upper area of star, "MOTOR OIL" across lower area of star. White "GASOLINE" arched around top, "FILLING STATION" around bottom.

TEXACO GASOLINE/MOTOR OIL 42"RD
(3) 1923-1936 $450
Red outlined two sided porcelain sign with black outlined white center circle. Large red star in center with black outlined green "T" in center of star with black "TEXACO" across upper area of star. White "GASOLINE" arched around top, "MOTOR OIL" around bottom. Black "MADE BY THE TEXAS COMPANY USA" arched under star.

TEXACO GASOLINE 42"RD
(3) 1934-1936 $500
Red outlined porcelain sign with black outlined white center circle. Large red star in center with black outlined green "T" in center of star with black "TEXACO" across upper area of star. White "TEXACO" arched around top, "GASOLINE" around bottom.

TEXACO 72"RD
(1) 1936-1967 $150
Black outlined white porcelain sign with large red star in center. White outlined green "T" in center of star with black "TEXACO" across upper area of star. Most likely the most repeated commercial identification ever to appear along our nation's highways.

TEXACO HEXAGON 42x96
(2) 1962-1982 $100

SIGN POLES

CURVED POLE
(4) 1925-1936 $200
Pipe pole rolled into "shepherd's crook" shape for hanging 42" diameter signs.

DENVER POLE
(5) 1928-1936 $500
Fluted cast column with clamp on mast arm for installation of 42" signs.

BANJO POLE
(1) 1936-1968 $100
The classic Texaco pole. Welded sheet metal and rolled angle iron assembly.

SIDEMOUNT POLE
(2) 1946-1968 $100
Sidemount version of the Classic Texaco pole. Welded sheet metal construction with "V" fork curving out to offset sign frame from center line of pole.

BUILDING SIGNS AND LETTERS

DEALER NAMEPLATE 10x42
(4) 1925-1937 $75
Embossed brown porcelain strip sign with dealer or station name in gold across center. No mention of Texaco and very uncharacteristic colors.

TEXACO RED PORCELAIN LETTERS 18"
(4) 1937-1955 $350
Red porcelain channel letters "TEXACO" for attachment to building facade.

RED PORCELAIN STARS 16"
(3) 1937-1955 $150
Red porcelain formed three-dimensional star, for attachment to building facade.

TEXACO LEADED GLASS INSERT 22"RD
(3) 1922-1937 $1000
Lead framed and channeled stained glass Texaco sign for installation in building portico.

FLAT PUMP SIGNS

DRAIN AND REFILL WITH CLEAN, CLEAR, GOLDEN TEXACO MOTOR OIL 15" $275
White outlined black porcelain sign with white band across upper sign face. Gold stream of motor oil on left side with white "Clean, Clear" across top of black area above yellow "GOLDEN." Grey outlined white "TEXACO/MOTOR/OIL" below. Black "DRAIN/AND REFILL WITH" across upper white area.

TEXACO 7"RD
(1) 1930-1944 $250
TEXACO 8"RD
(1) 1930-1944 $250
TEXACO 10"RD
(1) 1930-1944 $250
TEXACO 15"RD
(1) 1930-1944 $250
Black outlined white porcelain sign with large red star in center. Blacke outlined green "T" in center of star with black "TEXACO" across upper area of star.

TEXACO 8"RD
(2) 1944-1950 $250
Black outlined white porcelain sign with large red star in center. White outlined green "T" in center of star with black "TEXACO" across upper area of star.

FIRE CHIEF 12x18
(1) 1940-1967 $75

FIRE CHIEF 10x15
(3) 1955-1967 $150
FIRE CHIEF 8x12
(1) 1940-1967 $175
White/black outlined white porcelain sign with detailed red/black/gold fire hat in center. Small black outlined white Texaco star emblem to left of hat. Black outlined red "FIRE CHIEF" across top of sign above black "GASOLINE." All known originals are dated in either lower left or lower right corner.
SKY CHIEF 12x18
(1) 1940-1954 $75
SKY CHIEF 8x12
(1) 1940-1954 $150
Green porcelain sign with black band across bottom and series of white lines and white band across top. Red "Sky Chief" on upper white band. Black outlined red wings with Texaco logo at center in center of sign. White "GASOLINE" across bottom in black area.
SKY CHIEF SUPERCHARGED WITH PETROX 12x22.25
(1) 1955-1958 $125
SKY CHIEF SUPERCHARGED WITH PETROX 12x18
(3) 1955-1958 $150
Green porcelain sign with black band across lower area and series of white lines and white band across top. Red "Sky Chief" on upper white band. Black outlined red wings with Texaco logo at center in center of sign. White "GASOLINE" across bottom in black area. White band just below black area and red band across bottom. Red "Super-Charged/WITH" on white band, white "PETROX" on red band. Lower two bands give appearance of an add-on to basic sign design.
PETROX ADD-ON 4.25x12
(4) 1955-1958 $35
Painted tin add on panel to bolt below basic Sky Chief sign. White band above red band with red "Super-Charged/WITH" on white band, white "PETROX" on red band.
SKY CHIEF SU-PREME 12x18
(1) 1959-1967 $65
SKY CHIEF SU-PREME 10x15
(3) 1959-1967 $100
SKY CHIEF SU-PREME 8x12
(3) 1959-1967 $150
Green porcelain sign with white band across upper area and white band across bottom. Red "Sky-Chief" above black "Supreme"on upper white band, black "SUPER-CHARGED WITH" over red "PETROX" on lower white band with black outlined red wings with Texaco logo at center in center of sign with white "GASOLINE" below.
INDIAN 12x18
(2) 1940-1942 $250
INDIAN 8x12
(3) 1940-1942 $325
Porcelain sign with green area over blue. Red/white/ yellow/black Indian beadwork design on upper green area with white "INDIAN" over green "GASOLINE" across lower blue area.
DIESEL CHIEF 12x18
(5) 1940-1946 $500
Porcelain sign with white area over red lower band. Center of red band has projecting injector nozzle with thick red lines radiating up from nozzle. Texaco logo (white outlined "T") superimposed over lines. Black "DIESEL/CHIEF" over white "(DIESEL FUEL)" on red band.
DIESEL CHIEF 12x18
(1) 1946-1967 $200
DIESEL CHIEF 8x12
(3) 1946-1967 $400
Porcelain sign with white area over red lower band. Center of red band has projecting injector nozzle with red lines radiating up from nozzle. Texaco logo
(white outlined "T") superimposed over lines. White "DIESEL/CHIEF" over black "(DIESEL FUEL)" on red band.
DIESEL CHIEF SHIELD 10x15
(2) 1967-1980 $100
White outlined green tin lith hexagon shaped sign with white hexagon across center. Small black outlined red star Texaco logo at bottom center. Black "Diesel Chief" across center.
DIESEL CHIEF L 12x18
(4) 1960's $450
Porcelain sign with white area over red lower band. Center of red band has projecting injector nozzle with red lines radiating up from nozzle. Texaco logo
(white outlined "T") superimposed over lines. White "DIESEL/CHIEF L" over black "(DIESEL FUEL)" on red band.
DIESEL FUEL 2 12x18
(2) 1946-1967 $300
DIESEL FUEL 2 8x12
(4) 1946-1967 $450
Porcelain sign with white area over light green lower band. Center of green band has projecting injector nozzle with green lines radiating up from nozzle. Texaco logo
(white outlined "T") superimposed over lines. White "DIESEL/FUEL 2" on green band.
MARINE WHITE 12x18
(2) 1940-1967 $750
MARINE WHITE 8x12
(5) 1940-1967 $950
Porcelain sign with green area over black. Red and white ship's wheel around black outlined white Texaco logo with red star/white outlined green "T"on upper green area. Black lined white "rope" "Marine/White" script over red "GASOLINE" on black area across bottom.
MARINE WHITE SHIELD 10x15
(4) 1967-1980 $125
Red outlined green tin lith hexagon shaped sign with white hexagon across center. Small black outlined red star Texaco logo at bottom center. Black "Marine White" across center.
SKY CHIEF MARINE 12x18
(4) 1946-1955 $225
SKY CHIEF MARINE 8x12
(5) 1946-1955 $250
Green porcelain sign with black band across bottom and series of white lines and white band across top. Red "Sky Chief /Marine" on upper white band. Black outlined red wings with Texaco logo at center in center of sign. White "GASOLINE" across bottom in black area.
SKY CHIEF MARINE SUPER CHARGED WITH PETROX 12x22.25
(3) 1955-64 $225

Pencoil porcelain bottle rack sign. *Bruce Bednar.*

Pencoil 7" porcelain lubester sign.

Havoline F sign. Indian Oil Refining. *Leelan Clanzer.*

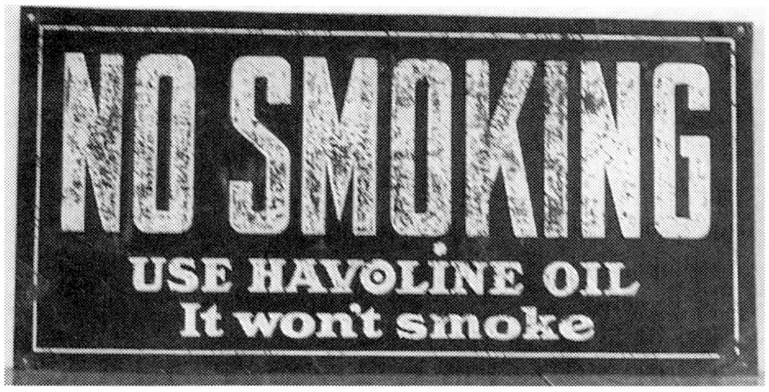

Havoline No Smoking Indian Oil Refining rare tin.

Havoline porcelain flange sign. *Leelan Glanzer.*

Harbor porcelain hanging sign. *Bruce Bednar.*

Pair of Gulf and Texaco Marine signs. *Bruce Bednar.*

Penn Drake tin sign. *John Phippen.*

Miscellaneous porcelain pump plates. *Bruce Bednar.*

Polly Gas die-cut porcelain sign. *Dennis Leith.*

Three rare porcelain signs. *Bruce Bednar.*

Union 76 outboard pump sign. *Dunbar's Gallery.*

SKY CHIEF MARINE SUPER CHARGED WITH PETROX 12x18
(4) 1955-1964 $250
SKY CHIEF MARINE SUPER CHARGED WITH PETROX 8x12
(4) 1955-1964 $250
Green porcelain sign with black band across lower area and series of white lines and white band across top. Red "Sky Chief" over black "MARINE" on upper white band. Black outlined red wings with Texaco logo at center in center of sign. White "GASOLINE" across bottom in black area. White band just below black area and red band across bottom. Red "Super-Charged/WITH" on white band, white "PETROX" on red band. Lower two bands give appearance of an add-on to basic sign design.
SEA CHIEF SHIELD 10x15
(4) 1967-1980 $150
White outlined gold tin lith hexagon shaped sign with white hexagon across center. Small black outlined red star Texaco logo at bottom center. Black "Sea Chief" across center.
FUEL CHIEF 1 DIESEL FUEL 12x18
(5) 1963-1967 $800
Porcelain sign with white area over yellow lower band. Center of yellow band has projecting injector nozzle with thick yellow lines radiating up from nozzle. Texaco logo (white outlined "T") superimposed over lines. Black "FUEL/CHIEF 1" on yellow bqand over red "(DIESEL FUEL)."
AUTO DIESEL 12x18
(5) 1963-1967 $250
Tin litho sign with white area over red lower band. Center of red band has projecting injector nozzle with thick red lines radiating up from nozzle. Texaco logo
(white outlined "T") superimposed over lines. White "AUTO/DIESEL" on red band.
AUTO DIESEL 12x18
(5) 1967-1970 $225
Tin litho sign with white area over red lower band. Center of red band has projecting injector nozzle with thick red lines radiating up from nozzle. Red Hexagon Texaco logo superimposed over lines. White "AUTO/DIESEL" on red band.

CURVED PUMP SIGNS

DRAIN AND REFILL WITH CLEAN,CLEAR,GOLDEN TEXACO MOTOR OIL 15" $300
White outlined black porcelain sign with white band across upper sign face. Gold stream of motor oil on left side with white "Clean, Clear" across top of black area above yellow "GOLDEN." Grey outlined white "TEXACO/MOTOR/OIL" below. Black "DRAIN/AND REFILL WITH" across upper white area.
TEXACO 7"RD
(1) 1930-1944 $225
TEXACO 8"RD
(1) 1930-1944 $225
TEXACO 10"RD
(1) 1930-1944 $225
TEXACO 15"RD
(1) 1930-1944 $225
Curved black outlined white porcelain sign with large red star in center. Blacke outlined green "T" in center of star with black "TEXACO" across upper area of star.

FIRE CHIEF 12x18
(1) 1940-1967 $65
FIRE CHIEF 8x12
(1) 1940-1947 $100
White/black outlined curved white porcelain sign with detailed red/black/gold fire hat in center. Small black outlined white Texaco star emblem to left of hat. Black outlined red "FIRE CHIEF" across top of sign above black "GASOLINE." All known originals are dated in either lower left or lower right corner.
SKY CHIEF 12x18
(1) 1940-1955 $65
SKY CHIEF 8x12
(2) 1940-1955 $100
Green porcelain sign with black band across bottom and series of white lines and white band across top. Red "Sky Chief" on upper white band. Black outlined red wings with Texaco logo at center in center of sign. White "GASOLINE" across bottom in black area.
INDIAN 12x18
(2) 1940-1946 $200
INDIAN 8x12
(3) 1940-1946 $275
Porcelain sign with green area over blue. Red/white/ yellow/black Indian beadwork design on upper green area with white "INDIAN" over green "GASOLINE" across lower blue area.
MARINE WHITE 12x18
(4) 1946-1966 $400
Porcelain sign with green area over black. Red and white ship's wheel around black outlined white Texaco logo with red star/white outlined green "T"on upper green area. Black lined white "rope" "Marine/White" script over red "GASOLINE" on black area across bottom.

CURB SIGNS

TEXACO ETHYL 30"RD
(4) 1930-1939 $375
Red porcelain sign with black outlined white Texaco red star and green "T" logo
(black outlined "T") at top over black outlined white circle at bottom with Ehtyl
(EGC) logo with yellow rays.
FREE CRANKCASE SERVICE 30"SQ
(4) 1928-1934 $350
Red outlined white porcelain sign with black/white outlined vertical rectangle with cut corners in center. Silver can pouring gold oil through center of black area. White "Clean,/Clear" above yellow "GOLDEN" and small red outlined Texaco star logo beside oil. Black "FREE CRANKCASE SERVICE" above red "REFILL WITH" above black "TEXACO/MOTOR OIL" across sign below black area.
FREE CRANKCASE SERVICE 30"SQ
(4) 1928-1934 $350
Red outlined white porcelain sign with black/white outlined vertical rectangle with cut corners in below center. Silver can pouring gold oil through center of black area. White "Clean,/Clear" above yellow "GOLDEN" and large red outlined Texaco star logo beside oil. Black outlined red "FREE CRANKCASE SERVICE" above black "REFILL WITH" above

black outlined red "TEXACO/MOTOR OIL" across sign above black area.

TEXACO MOTOR OIL CLEAN,CLEAR,GOLDEN 30"SQ
(3) 1928-1934 $350
White porcelain sign with cut corners. Black outlined red "TEXACO/MOTOR OIL" across top of sign above black/white outlined vertical rectangle with cut corners in center. Silver can pouring gold oil through center of black area. White "Clean,/Clear" above yellow "GOLDEN" and large red outlined Texaco star logo beside oil.

INSULATED NEW TEXACO MOTOR OIL 30"SQ
(5) 1934-1937 $350
Black porcelain sign with cut corners. Large red/white/black "NEW TEXACO MOTOR OIL" can in center with white outlined red "INSULATED" arched across top.

TEXACO CERTIFIED SERVICE 30"SQ
(4) 1928-1937 $250
White porcelain sign with cut corners. Red band across center with black outlined white "CERTIFIED" across band. Black outlined red "TEXACO" across top, "SERVICE" across bottom with smaller writing across bottom.

HAVOLINE MOTOR OIL 30"SQ
(5) 1934-1937 $350
White over blue porcelain sign with blue/white outlined red dot logo extending up from blue area. Red "KEEPS YOUR ENGINE CLEAN" across top with white "HAVOLINE/MOTOR OIL" on blue area below dot. Red "DISTILLED INSULATED" across bottom.

TEXACO SKY CHIEF 30"RD
(5) 1939-1942 $500
Green porcelain sign with black band across bottom and series of white lines and white band across top. Red "Sky Chief" on upper white band. Black outlined red wings with Texaco logo at center in center of sign. White "GASOLINE" across bottom in black area.

TEXACO AVIATION 24"RD
(5) 1925-1935 NO LISTING
Red outlined two sided porcelain sign with black outlined white center circle. Large red star in center with black outlined green "T" in center of star with black "TEXACO" across upper area of star. Black outlined white "AVIATION" arched around top, "PRODUCTS" around bottom.

LUBRICATION/MOTOR OIL SIGNS

EASY POUR CAN 2 QTS TEXACO MOTOR OIL 13x13
(5) 1922-1925 $850
White/black outlined white porcelain sign green/white/red two quart can on square with black "EASY POUR/CAN above, Red "2" black "QTS" above black "TEXACO/MOTOR OIL" below.

EASY POUR CAN 2 QTS TEXACO MOTOR OIL 13x13
(5) 1922-1925 $850
White/black outlined white porcelain sign black/white/red two quart can on square with black "EASY POUR/CAN above, Red "2" black "QTS" above black "TEXACO/MOTOR OIL" below.

TEXACO CERTIFIED LUBRICATION 9x39
(4) 1920s $300
White porcelain sign with Texaco logo
(black outline "T") at either end. Black outlined red "TEXACO/CERTIFIED LUBRICATION" above black "IN ACCORDANCE WITH APPROVED PRACTICE" between logos.

TEXACO CRANKCASE SERVICE 22x28
(4) 1928-1930 $250
Black square dotted border around white porcelain sign. Black outlined Texaco logo
(black border "T") at top over black "CRANKCASE/SERVICE" red "Let us Drain your Crankcase and/Refill with Clean, Clear, Golden (in yellow)" across center above black "TEXACO MOTOR OIL."

CLEAN,CLEAR,GOLDEN TEXACO MOTOR OIL 10"SQ
(5) 1928-1934 $450
White outlined black porcelain sign with gold stream of motor oil on left side. White "Clean,/ Clear" across top of black area above yellow "GOLDEN." Grey outlined white "TEXACO/MOTOR/OIL" below.

CLEAN,CLEAR,GOLDEN TEXACO MOTOR OIL 14"SQ
(3) 1928-1934 $350
White outlined black porcelain sign with gold stream of motor oil on left side. White "Clean,/ Clear" across top of black area above yellow "GOLDEN." Grey outlined white "TEXACO/MOTOR/OIL" below.

MARFAK LUBRICATION 24x40
(2) 1946-1960 $175
Black outlined white tin litho sign with red band diagonal across center. White "MARFAK" on band with Texaco logo above and black script "Lubrication" below.

LET US MARFAK YOUR CAR 24x40
(2) 1946-1960 $175
Black outlined white tin litho sign with red band diagonal across center. White "MARFAK" on band with black script "Let us" above and black script "your car" below.

NEW TEXACO MOTOR OIL FURFURAL'd 18x30
(5) 1937-1938 $550
Detailed sign with attendant holding dip stick with red and black "NEW TEXACO MOTOR OIL" can to right side. Black "STAYS FULL(red) LONGER" across center with white "The FURFURAL'd FILM does it!" on blue band across bottom. Originals are dated "2/37."

LUBESTER SIGNS

TEXACO MOTOR OIL 5"RD
(3) 1915-1934 $350
Red outlined porcelain sign with black outlined white center circle. Large red star in center with black outlined green "T" in center of star with black "TEXACO" across upper area of star. White "TEXACO" arched around top, "MOTOR OIL" around bottom. Black "MADE BY THE TEXAS COMPANY U.S.A." arched under star.

TEXACO MOTOR OIL FORDS 3x12
(4) 1915-1930 $225
Black outlined white porcelain sign with black "TEXACO MOTOR OIL/FORDS" across sign.

FLANGE SIGNS

EASY POUR CAN 12x18
(4) $850
Red porcelain flange sign with black outlined white square at top. Green/white/red two quart can on square with black "EASY POUR/CAN above, "TWO QUARTS" below. Black outlined white "TEXACO/MOTOR OIL" above smaller writing below square.

CLEAN, CLEAR, GOLDEN TEXACO MOTOR OIL 12x16
(4) 1928-1933 $850
Black porcelain flange tombstone shaped sign with gold stream of motor oil on left side. White "Clean, Clear" across top of black area above yellow "GOLDEN." Grey outlined white "TEXACO/MOTOR/OIL" below. Small red outlined Texaco star logo at top.

CLEAN, CLEAR, GOLDEN TEXACO/
TEXACO MOTOR OIL 12x18
(5) $500
Red outlined white porcelain sign with black/white outlined vertical rectangle with cut corners in center. Silver can pouring gold oil through center of black area. White "Clean,/Clear" above yellow "GOLDEN" and small red outlined Texaco star logo beside oil. Black "TEXACO/MOTOR OIL" across sign below black area.

NEW TEXACO MOTOR OIL CAN $250
Die-cut red/white/black porcelain flange sign replica of 1935 "NEW TEXACO MOTOR OIL" can.

HAVOLINE MOTOR OIL WAXFREE CAN $250
Die-cut red/white/blue porcelain flange sign replica of 1935 "HAVOLINE MOTOR OIL WAXFREE" can.

THE TEXAS COMPANY, USA PETROLEUM PRODUCTS TEXACO MOTOR OIL 23x18 $900
Die cut black porcelain flange sign with large red bordered white Texaco red star/black outlined green "T" logo at top. Black outlined white "THE TEXAS COMPANY, U.S.A." arched around top on red band, "PETROLEUM PRODUCTS" around bottom. White "TEXACO/MOTOR OIL" on black area below logo.

SECONDARY SIGNS

TEXACO NO SMOKING WT/BLK 6x24
(4) 1915-1930 $1600
White outlined black porcelain sign with red outlined white Texaco "THE TEXAS COMPANY, USA PETROLEUM PRODUCTS" logo at each end. White "No Smoking" across between logos.

NO SMOKING THE TEXAS COMPANY 3x15
(5) 1930-1935 $200
Black outlined white porcelain sign with black "NO SMOKING" across sign. Small black "THE TEXAS COMPANY U.S.A." across bottom.

TEXACO NO SMOKING BLK BORDER T 4x23
(2) 1935-1944 $250
White/black outlined white porcelain sign with black outlined Texaco red star and green "T"
(black border) at either end. Black "NO SMOKING" across center.

TEXACO NO SMOKING WT BORDER T 4x23
(1) 1944-1962 $225
White/black outlined white porcelain sign with black outlined Texaco red star and green "T"
(white border) at either end. Black "NO SMOKING" across center.

TEXACO NO SMOKING NEW LOGO 4x23
(3) 1962-1982 $150
White porcelain sign with Hexagon Texaco logo at left end. Black "NO SMOKING" across center.

LOCKED FOR YOUR PROTECTION 3.5x3.5
(3) 1946-1968 $100
White/black outlined white porcelain sign with Hexagon Texaco logo at left end. Black "LOCKED FOR/YOUR/PROTECTION/ATTENDANT/HAS KEY" across center.

TEXACO MAIL PORT 17x23
(4) 1950s $350
White porcelain sign with yellow and blue/white/red marine signal flags on pole at left. Red "TEXACO" over black "MAIL/PORT" across sign face.

BUY THE BEST BUY TEXACO 24x40
(3)1946-1960 $225
Green outlined tin litho sign with red band across top and white band across bottom. Green trapezoid superimposed over right end of upper red band. Black script "Buy the Best" across red and green upper area with red "Buy beside black outlined red "TEXACO" on bottom white area.

REGISTERED REST ROOM
A TEXACO DEALER SERVICE $100
White tin litho sign with green irregular green band across top and bottom. White "REGISTERED" in upper band, "A TEXACO DEALER SERVICE" in lower band with black "REST ROOM" across center.

MEN 3x8
(2) 1937-1968 $50
White porcelain flange sign with three green stripes across bottom. Green "MEN" above stripes.

LADIES 3x8
(2) 1937-1968 $50
White porcelain flange sign with three green stripes across bottom. Green "MEN" above stripes.

MEN DIE CUT 3x8
(2) 1968-1982 $35
Off white tin litho die-cut hexagon shaped flange sign with green "MEN" across center.

LADIES DIE CUT 3x8
(2) 1968-1982 $35
Off white tin litho die-cut hexagon shaped flange sign with green "LADIES" across center.

TEXACO KEROSINE CLEAN BURNING 12x20
(3) 1958-1965 $200
White self framed tin litho sign with blue band across bottom. Series of blue flames across top of blue band. Red "TEXACO" over blue "KEROSINE" on upper white area, white "Clear" red "burning" on bottom blue band.

TEXACO KEROSENE CLEAN BURNING 12x20
(3) 1958-1965 $200
White self framed tin litho sign with blue band across bottom. Series of blue flames across top of blue band. Red "TEXACO" over blue "KEROSENE" on upper white area, white "Clear" red "burning" on bottom blue band.

BULK PLANT SIGNAGE

TEXACO FARM LUBRICANTS SOLD HERE 30x42
(4) 1946-1960 $250
Green porcelain sign with green outlined white band across bottom. Texaco logo in center of upper green area with black "FARM LUBRICANTS" over red "SOLD HERE" across lower white area.

TEXACO MARINE LUBRICANTS W/ BOATS 15x30
(3) 1947-1961 $2800
White porcelain sign with green band across top and bottom. Black outlined white seagulls on upper green band with red/black/green boats on lower green band. Black outlined red "TEXACO" arched across top over black "MARINE LUBRICANTS." Texaco logo to either side below upper green band.
TEXACO MARINE LUBRICANTS 3 STRIPES 15x30
(5) 1961-1968 $1500
White porcelain sign with three color diagonal band at left. Black/white boats on each of the three color bands, with small Texaco logo at upper left. Black outlined red "TEXACO" above black "MARINE/LUBRICANTS" across sign face.
TEXACO INDUSTRIAL LUBRICANTS 3 STRIPES 15x30
(5) 1961-1968 $1500
White porcelain sign with three color diagonal band at left. Black/white trucks and equipment on each of the three color bands, with small Texaco logo at upper left. Black outlined red "TEXACO" above black "INDUSTRIAL/LUBRICANTS" across sign face.
TEXACO MARINE LUBRICANTS 3 STRIPES 15x30
(5) 1968-1971 $1200
White porcelain sign with three color diagonal band at left. Black/white boats on each of the three color bands, with small Texaco hexagon logo at upper left. Black outlined red "TEXACO" above black "MARINE/LUBRICANTS" across sign face.
TEXACO INDUSTRIAL LUBRICANTS 3 STRIPES 15x30
(5) 1968-1971 $1200
White porcelain sign with three color diagonal band at left. Black/white trucks and equipment on each of the three color bands, with small Texaco hexagon logo at upper left. Black outlined red "TEXACO" above black "INDUSTRIAL/LUBRICANTS" across sign face.
TEXACO MARINE PRODUCTS 24x96
(5) 1946-1960 $650
Red porcelain sign with black band across bottom. Black outlined white Texaco logo with white outlined green "T" on red star at either end above black band with black otulined white "TEXACO" between. White "MARINE PRODUCTS" on lower black band.
TEXGAS SOLD HERE 16x16
(5) 1968-1970 $100
White outlined green tinlitho sign with white band across bottom. Black and white bottle gas tank with red Texaco hexagon logo on upper green area. Black "TEXGAS/SOLD HERE" below.

TRUCK SIGNS

TEXACO AGENT THE TEXAS COMPANY 11x12.5
(3) 1920s-1944 $350
Black outlined white "keyhole" shaped sign with Texaco logo (black outlined "T") at top over black outlined white base area. Black Jobber's name on lower area above black "AGENT/THE TEXAS/COMPANY."
TEXACO DISTRIBUTOR THE TEXAS COMPANY 11x12.5
(3) 1920s-1944 $350
Black outlined white "keyhole" shaped sign with Texaco logo (black outlined "T") at top over black outlined white base area. Black Jobber's name on lower area above black "DISTRIBUTOR/THE TEXAS/COMPANY."
TEXACO DISTRIBUTOR 11x12.5
(2) 1920s-1944 $350
Black outlined white "keyhole" shaped sign with Texaco logo (black outlined "T") at top over black outlined white base area. Black Jobber's name on lower area above black "DISTRIBUTOR."
TEXACO CONSIGNEE 11x12.5
(2) 1944-1968 $350
Black outlined white "keyhole" shaped sign with Texaco logo (black outlined "T") at top over black outlined white base area. Black Jobber's name on lower area above black "CONSIGNEE."
TEXACO DISTRIBUTOR 11x12.5
(2) 1944-1968 $350
Black outlined white "keyhole" shaped sign with Texaco logo (white outlined "T") at top over black outlined white base area. Black Jobber's name on lower area above black "DISTRIBUTOR."
TEXACO LETTERING 14"
(4) 1925-1960 $250
Black outlined white individual die-cut letters "TEXACO."
TOUR WITH TEXACO 10/8"
(5) 1930-1960 $100
Black outlined white individual die-cut letters "TOUR WITH TEXACO."

ADVANCE SIGNS

TEXACO AHEAD 30x60
(3) 1946-1968 $100
White self framed tin litho sign with red "TEXACO/AHEAD" across sign face.

INDIAN REFINING COMPANY - LAWRENCEVILLE, IL

Indian was a refiner/marketer with branded gasoline stations in 19 northeast and north central states. They are best remembered for the motor oil brand Havoline, introduced in 1906 by Havoline Oil Co of New York which Indian Refining purchased in the late 1910s. Texaco continues to use the Havoline brand today. Texaco purchased Indian in 1931 in order to acquire the Havoline process patents and Indian stations were rebranded Texaco, while "Indian Gas" became a Texaco economy grade gasoline marketed until 1942.

IDENTIFICATION SIGNS

INDIAN GAS 36x45
(5) 1924-1931 $650
Blue porcelain die-cut sign with blue/white outlined red dot circle extending from top of sign. White "INDIAN" across top of blue area with white "GAS" offset to right. Small red "INDIAN REFINING CO./INCORPORATED" in lower left corner.
HAVOLINE SERVICE STATION 36x45
(5) 1924-1931 $650
Blue porcelain die-cut sign with blue/white outlined red dot circle extending from top of sign. White "HAVOLINE/SERVICE STATION" across center of blue area with red "AUTHORIZED DEALER" across bottom of sign.

CURB SIGNS

HAVOLINE WAXFREE OIL 12x16
(5) 1930-1935 $750
Blue porcelain die-cut sign with blue/white outlined red dot

circle extending from top of sign. White "HAVOLINE" across top of blue area above white "WAXFREE OIL."

LUBRICATION/MOTOR OIL SIGNS

BLUE GRASS AXLE GREASE 6.5x14
(5) 1900-1910 $375
Red/white outlined blue porcelain sign with white outlined red "BLUE GRASS/AXLEXLE GREASE" arched through center of sign. White "USE" above, with white "FOR" over white outlined red "TIRED WHEELS" below. Small red outlined arrowhead at lower left with black/white Indian head in center and white "INDIAN REFINING COMPANY/(INCORPORATED)" across bottom.

HAVOLINE OIL "IT MAKES A DIFFERENCE" $450
Blue/white outlined blue porcelain sign with white "HAVOLINE/OIL" across center. White "It makes a difference" across bottom of sign.

HAVOLINE OIL 24x48 $475
Wood framed blue tin litho sign with white "HAVOLINE/OIL" across center. Smaller red "INDIAN REFINING COMPANY INC." across bottom.

HAVOLINE VERTICAL SRTIP 12x72
(5) 1920s $300
Red outlined blue porcelain sign wtih white "HAVOLINE" vertical down sign. Red dot under "H" & "A" at top.

HAVOLINE HORIZONTAL STRIP 12x72
(5) 1920s $300
Red outlined blue porcelain sign wtih white "HAVOLINE" across sign. Red and white concentric circles form "O" in "HAVOLINE."

HAVOLINE F 10x12
(5) 1920s $225
Blue porcelain die-cut flange sign with blue/white outlined red dot circle extending from top of sign. White "HAVOLINE" across top of blue area with white "OIL" offset to right. Small red "INDIAN REFINING CO./INCORPORATED" in lower left corner.

HAVOLINE THE POWER OIL 20x24
(5) 1920s $300
White band over blue square porcelain sign with blue/white outlined red dot circle extending from top of blue area. White "HAVOLINE/ the power oil" across center of blue area with red "INDIAN REFINING COMPANY, INC." across bottom of sign.

FLANGE SIGNS

HAVOLINE THE POWER OIL 16x18
(5) 1925-1934 $350
Blue porcelain die-cut flange sign with blue/white outlined red dot circle extending from top of sign. White "HAVOLINE/the power oil" across center of blue area with red "INDIAN REFINING COMPANY, INC." across bottom of sign.

HAVOLINE OIL FLANGE 16x18
(5) 1925-1934 $350
Blue porcelain die-cut flange sign with blue/white outlined red dot circle extending from top of sign. White "HAVOLINE" across top of blue area with white "OIL" offset to right. Small red "INDIAN REFINING CO./INCORPORATED" in lower left corner.

SECONDARY SIGNS

NO SMOKING USE HAVOLINE OIL 10x24
(5) 1915-1920 $225
Blue/white outlined blue tin litho sign with white "NO SMOKING" across center. Small white "USE HAVOLINE OIL/It won't smoke" across bottom.

CALIFORNIA PETROLEUM COMPANY - LOS ANGELES, CA:

LA basin refiner marketer purchased by and converted to Texaco in 1928 to create the first marketing presence Texaco had in California.

CURB SIGNS

CALPET 24x24
(5) 1920-1928 $350
Red bordered white porcelain flange sign with white center diamond and black band across center. Black outlined white "CALIFORNIA PETROLEUM COMPANY" around border area with red outlined white "CALPET" on band across center.

PUMP SIGNS

CALPET 10x16
(5) 1920-1928 $300
Red bordered white porcelain flange sign with white center diamond and black band across center. Black outlined white "CALIFORNIA PETROLEUM COMPANY" around border area with red outlined white "CALPET" on band across center. White mounting tab and base area with red "GASOLINE" across below diamond.

PARAGON REFINING CO - LONG ISLAND CITY, NY:

Suburban NYC marketer, primarily involved in fuel oil distribution, that was purchased by Texaco in 1959.

CURB SIGNS

PARAGON GASOLENE 24"RD
(5) 1920-1925 $1600
White porcelain sign with large red triangle, point down, in center. Black outlined green "PARAGON" arched across upper triangle with black outlined white "GASOLENE" below. Black/white tank truck at top of sign with black/white oil derrick to left and refinery scene at right. Small "PRODUCERS REFINERS" at bottom of sign.

PARAGON MOTOR OILS 24"RD
(5) 1920-1925 $1600
White porcelain sign with large green triangle, point down, in center. Black outlined red "PARAGON" arched across upper triangle with black outlined white "MOTOR/OIL" below. Black/white tank truck at top of sign with black/white oil derrick to left and refinery scene at right. Small "PRODUCERS REFINERS" at bottom of sign.

McCOLL FRONTENAC - TORONTO, ONTARIO:

Created in 1927 in a merger between McColl Brothers, Ltd, dating back to 1873, and Frontenac Oil Refineries, founded in 1925. The company purchased Texaco Canada in 1939 in a stock swap that eventually gave Texaco control of the company. Conversion to Texaco brands began in 1947.

IDENTIFICATION SIGNS

McCOLL FRONTENAC PRODUCTS 42"RD
(4) 1930-1947 $650
Black outlined white porcelain sign with black outlined red and black center circle. Red and black Indian head on white circle on center circle. Black McCOLL-FRONTENAC" arched around top, "PRODUCTS" around bottom.

PUMP SIGNS

DIE CUT INDIAN 12"RD

(4) 1935-1947 $600

Irrengular round die cut porcelain sign. Black outlined white circle with red and black Indian in full headdress offset to right.

LUBRICATION/MOTOR OIL SIGNS

RED INDIAN MOTOR OILS 17x24

(5) 1930s $500

Black/white outlined self framed red and black porcelain sign with black outlined white circle with red and black Indian in full headdress offset to right in center. Black outlined white "RED INDIAN" across top, "MOTOR OILS" across bottom.

RED INDIAN AVIATION MOTOR OIL 11x14

(5) 1930s $850

White outlined red porcelain sign with black band across bottom. White rectangle diagonal in center with black "Aviation" above red "MOTOR OIL" on rectangle. White "RED INDIAN" across top, "NONE BETTER" across bottom.

TOTAL AND AFFILIATES

TOTAL PETROLEUM - ALMA, MI/DENVER, CO

Total, a unit of the French government, entered the US market in 1970 with the purchase of Leonard Refineries. The more than 700 Leonard stations in Michigan were rebranded Total by 1971 and the company remained a one-state marketer until 1978 when they purchased APCO. They have since acquired Vickers Petroleum, from Esmark in 1980, and the Road Runner stations of the Truman Arnold Corp. Until 1993 they marketed in Michigan and surrounding states under the Total brand, and throughout the midwest under Vickers, Apco, and Road Runner. Beginning that year Apco, Vickers and Road Runner stations began rebranding Total. Small areas of Vickers marketing was sold off and a few stations remain with the brand. With Apco being primarily a jobber operation, conversion to Total has been limited so far to high volume and high visibility locations.

TOTAL PETROLEUM - DENVER, CO

No collectible size Total signs known to exist.

MC CLANAHAN OIL COMPANY - ALMA, MI

Founded in the early 1930s in Michigan's oil region. McClanahan entered the gasoline market about 1935 and the company was reorganized as Leonard Refineries in 1936. Gasoline was rebranded Leonard in 1938.

No collectible size McClanahan signs known at this time.

LEONARD REFINERIES - ALMA, MI

McClanahah Oil Company was reorganized in 1936 as Leonard Refineries. The Leonard gasoline brand was introduced in 1938. In 1970 Total Petroleum purchased Leonard Refineries and rebranded the stations "TOTAL." They continue to operate under the Total brand in Michigan, Ohio and Wisconsin.

IDENTIFICATION SIGNS

LEONARD FLASH 42x84

(4) 1950-1957 $300

Steel framed red porcelain sign with white burst in center. Blue band across center, with white "LEONARD" in band.

NEON SIGNS

LEONARD FLAG SIGN 72x72

(5) 1938-1950 $550

Yellow porcealin can type sign with red flag in center of sign face. White "burst" on flag, with blue banner across burst. White "LEONARD" on banner. Blue "GASOLINE" below flag. Neon outline, flag outline and "LEONARD."

LEONARD FLASH 42x84

(5) 1950-1957 $400

Red porcelain can-type sign with white burst in center. Blue band across center, with white "LEONARD" in band. Neon outline, band outline and "LEONARD"

SECONDARY SIGNS

YOU CAN DEPEND ON LEONARD 30x48

(4)1950-1957 $100

Blue tin litho sign with red band across bottom. White outlined red Leonard oval logo with white flash situated on color split. White outlined red "You can/depend on" above logo. Reverse is solid red litho with logo at top and white italic "your sign/of happier/motoring" below.

YOU CAN DEPEND ON LEONARD 30x48

(4) 1957-1964 $100

Blue tin litho sign with red oval at top, white oval in center and blue outlined white Leonard oval logo with red flash situated on color split. White "You can" in red oval, red "depend on" in white logo.

ROOSEVELT OIL CO. - MT. PLEASANT, MI

Michigan marketer operating under the Roosevelt brand before WW2 and branding Power- Flight from 1946 until 1955. Purchased by and rebranded Leonard in 1955.

No collectible size Roosevelt signs are known at this time.

ROWLEY AND CHURCH - ST. LOUIS, MICHIGAN

Private brand jobber that operated under the "Michigan Maid" brand until it was rebranded ROC Power Flight in 1946. No collectible size Rowley & Church signs ae known at this time.

MID-WEST REFINING - MT. PLEASANT, MI

Michigan refiner/marketer that marketed under the Mid-West brand until after WW2 when they licensed the "White Rose" trademark from National Refining. Stations were then rebranded Mid-West White Rose. Leonard purchased Mid-West in 1955 and continued to operate the White Rose stations as a secondary brand until after Total purchased Leonard in 1970. Stations were then rebranded Total.

IDENTIFICATION SIGNS

EN-AR-CO WHITE ROSE DEALER 54x54

(5) 1946-1950 $150

Green porcelain diamond shaped with cream colored band across center. Cream "EN-AR-CO" in upper green area, "DEALER" below with green "WHITE/ROSE" in center.

(See also National White Rose in the Ashland section. Many types of signs were common to both operations.)

CAVALIER SERVICE STAS, INC. - FLINT, MI

Michigan regional marketer. cavalier sold out to Leonard Re-

fineries in 1967. In 1968 the operation was rebranded "BEST" and when Leonard sold out to Total in 1970, Best was retained as a secondary brand. Gradually changed to Total with only three Best stations in operation as of this writing.
No collectible size Cavalier or Best signs known at this time.

ANDERSON PRICHARD - OKLAHOMA CITY, OK

Founded in Oklahoma City in 1921, Anderson-Prichard was marketing gasoline under the Col-Tex brand in several midwestern states prior to 1930. In 1946 the Col-Tex stations in Oklahoma were rebranded Anderson-Prichard while the remaining territory continued to use the Col-Tex brand. In 1952 Anderson-Prichard purchased the extensive Kanotex marketing and introduced the APCO brand to replace Kanotex and Anderson-Prichard. The Col-Tex operation was sold off to Cosden, who continued to operate Col-Tex as a secondary brand until their merger with Fina in 1963. Control of APCO passed to Ashland Oil in the late 1950s and in 1978 Total Petroleum purchased the Apco operation. Total continues to brand dealer operated APCO stations today.

IDENTIFICATION SIGNS

COL-TEX W/ DEER 30"RD
(4) 1930-1935 $850
Red porcelain sign with white geartooth outline. Tan outlined green circle in center with detailed jumping deer on circle. White "COL-TEX" arched around top, "GASOLINE" around bottom.

ANDERSON PRICHARD A/P 30"RD
(4) 1940-1946 $400
White outlined red porcelain sign with white outlined blue center circle. Large "AP" on circle with white band across sign face. Blue "MOTOR FUELS" on band. White "ANDERSON" arched around top, "PRICHARD" below.

ANDERSON PRICHARD CHALLENGE 60"RD
(3) 1946-1953 $200
White outlined red porcelain sign with white outlined blue center circle. White band across sign face with blue script "Challenge" on band. White "ANDERSON" arched around top, "PRICHARD" below.

APCO PETROLEUM PRODUCTS 60x78
(2) 1953-1970 $75
Red trapezoidal porcelain sign with white band across upper sign face. Red "APCO" in white band with white "PETROLEUM/PRODUCTS" on lower red area.

CURB SIGNS

HIGHEST QUALITY COL-TEX 30"RD
(5) 1935-1940 $300
Round porcelain sign with red upper band, white middle band and blue lower band. Blue outlined red "COL-TEX" across center white band. White "HIGHEST" arched above "QUALITY" in upper red area, "PETROLEUM" above arched "PRODUCTS" in lower blue area.

HIGHEST QUALITY COL-TEX 30"RD
(5) 1940-1946 $300
Red outlined white porcelain sign with red "COL-TEX" across center. Blue "HIGHEST/QUALITY" above, "PRODUCTS" below.

COL-TEX ETHYL 30"RD
(4) 1930-1935 $350
Red porcelain sign with white geartooth outline. Black outlined white circle in center with Ethyl (EGC) logo with blue rays. White "COL-TEX" arched around top, "GASOLINE" around bottom.

KANOTEX - ARKANSAS CITY, KS

Founded as Superior Refining in Longton,KS in 1905, the Kanotex trademark was first used in 1909. The company moved to Arkansas City,KS in 1917 and later merged with Arkansas City based Lesh Refining. Innovative marketers, the KANOTEX trademark appeared in more that 20 midwestern states. In 1952 Anderson Prichard purchased Kanotex and rebranded the Kanotex operation, as well as their own Anderson-Prichard stations to APCO.

IDENTIFICATION SIGNS

KANOTEX GAS 30"RD
(5) 1925-1930 $850
White porcelain sign with red "KANOTEX" arched through sunflower and star logo at top. Red "GASOLINE" arched below.

CURB SIGNS

KANOTEX ETHYL 30"RD
(5) 1926-1935 $850
White porcelain sign with red "KANOTEX" arched through sunflower and star logo at top. Ethyl (EGC) logo with yellow rays below.

NATIONAL REFINING CO. - ARKANSAS CITY, KS

Early Kansas refiner that merged with Kanotex in 1928.

LUBRICATION SIGNS

LESH GOLDEN GLOW AUTO OILS 18x24
(5) 1920's $150
Gold outlined blue tin litho sign with gold sunburst in center. Gold "GOLDEN GLOW" arched around top, "AUTO OILS" across bottom with blue "Lesh" script logotype in center of sunburst.

WINSTUN - KANSAS CITY, MO: Winstun was a Kansas City area Kanotex jobber who branded several stations under the Winstun name in the late 1930's.

IDENTIFICATION SIGNS

WINSTUN GASOLINE MOTOR OIL 48"RD
(5) 1936-1940 $200
Steel framed porcelain sign with white band across center. "WINSTUN" in band, with outlined arch around top with white "GASOLINE," arch & "MOTOR OILS" around bottom.

VICKERS PETROLEUM CORPORATION - WICHITA, KS

Founded in 1917, Vickers was a mid-cotinenr refiner/marketer, oprating stations in the central United States. The company remained in the Vickers family until 1968 when it was purchased by Esmark, the conglomerate that had it's roots in Swift and Company, the Chicago meat packers. Under Esmark, the brand expanded into about a dozen midwestern states. Total purchased Vickers in 1980 and continues to operate stations under the Vickers brand today.

IDENTIFICATION SIGNS

VICKERS PETROLEUM PRODUCTS 42"RD
(4) 1936-1950 $150
White tin litho sign with red center circle. White band across lower area of red circle, superimposed over large black outlined white "V" that covers most of circle. Red "VICKERS" on band with black "PETROLEUM PRODUCTS" arched around bottom.

VICKERS PETROLEUM PRODUCTS 30"RD
(4) 1936-1950 $200

VICKERS PETROLEUM PRODUCTS 42"RD
(4) 1936-1950 $150
White porcelain sign with red center circle. White band across lower area of red circle, superimposed over large black outlined white "V" that covers most of circle. Red "VICKERS" on band with black "PETROLEUM PRODUCTS" arched around bottom.

VICKERS 72x72
(4) 1950-1968 $75
Black outlined white porcelain sign with red band across bottom. Two-tone black "V" in upper white area with red crown in open area of "V." White "VICKERS" on red band below.

KNIGHT OIL CO. - SPRINGFIELD, MO

Missouri marketer purchased by Vickers about 1959. Rebranded Vickers in the early 1960s.

IDENTIFICATION SIGNS

KNIGHT OIL COMPANY 48x48
(5) 1946-1960 $300
Red outlined white porcelain sign with blue horse and rider in center. Red "KNIGHT" across top, "PRODUCTS" across bottom.

PUMP SIGNS

KNIGHT REGULAR 11x12
(5) 1946-1960 $325
Blue outlined white die-cut porcelain shield shaped sign. Red/white banded red circle on upper sign face with blue/white horse and rider on circle. Blue ribbon across center of sign, with white "REGULAR" on ribbon. Red "GASOLINE" below ribbon with blue "KNIGHT OIL CO." arched around bottom.

KNIGHT PREMIUM 11x12
(5) 1946-1960 $325
Blue outlined white die-cut porcelain shield shaped sign. Red/white banded red circle on upper sign face with blue/white horse and rider on circle. Blue ribbon across center of sign, with white "PREMIUM" on ribbon. Red "GASOLINE" below ribbon with blue "KNIGHT OIL CO." arched around bottom.

BELL OIL AND GAS - TULSA, OK

Independent refiner/marketer, founded in 1918, that was purchased by Vickers in the late 1960s. Bell was primarily a wholesale marketter with only a few branded stations.

CURB SIGNS

BELL ETHYL 36"RD
(5) 1930'S $250
Blue outlined orange tin-litho sign with blue "BELL" with white dropshading. Blue bell below, with Ethyl (EGC) logo with yellow rays on bell.

BEN FRANKLIN REFINING CO. - ARDMORE, OK

Oklahoma refiner that operated a few Ben Franklin branded stations in the 1930s. Merged with Bell Oil prior to WW2. No collectible size Ben Franklin signs known to exist.

JAYHAWK OIL COMPANY - LAWRENCE, KS:

Kansas marketer purchased by Vickers about 1960.

JAYHAWK OILS 24x32
(4) 1940s $350
Blue outlined white tin litho sign with blue band across bottom. White "JAYHAWK OILS" on band with large red/yellow/blue "Jayhawk" bird in center.

UNION OIL CO. OF CALIFORNIA

Organized in 1890 in Santa Paula, California from the holdings of founders Lyman Stewart and Wallace Hardison and others, Union began selling California made gasoline to California motorists in 1913. Throughout it's history a dominant west coast marketer, Union was operating stations in 12 states at the time of the 1965 merger with Chicago based Pure Oil. By 1970 the Union 76 brand had replaced Pure and the company was marketing coast-to-coast. The corporate name became Unocal Corp in 1983 and the brand became Unocal 76 in 1985. In 1993 Unocal began consolidating it's marketing, eliminating those areas not profitably served. Much of the southeastern jobber market was allowed to participate in a co-operative and license the Pure brandname from Unocal. Pure stations exist in 10 states today. In the Great Lakes Area, Unocal marketing is in joint partnership with the Venezuelan Oil consortium under the heading Uno-Ven. They continue to operate stations under the Union 76 brand. The 76 brand has also been licensed to a co-operative of truck stop operators nationwide.

IDENTIFICATION SIGNS

UNION GASOLINE SPEED & POWER 26x30
(4) 1925-1932 $350
Blue outlined white die-cut shield shaped sign with red/white/blue shield at top of sign. White "UNION/OIL/COMPANY/OF CALIFORNIA" in blue area of shield. Blue outlined red UNION/GASOLINE" below with blue "SPEED & POWER" across bottom.

76 SQUARE 30x40
(4) 1932-1946 $250
Orange porcelain sign with two parallel blue lines across top and bottom. White outlined blue "76" in center.

76 72"RD
(2) 1970-1993 $100
Orange porcelain sign with white outlined blue "76" logotype in center. East coast territory useage.

76 42"RD
(4) 1970-1993 $50

76 72"RD
(2) 1970-1993 $25
Orange fiberglass sign with white outlined blue "76" logotype in center. East coast territory useage, installed over existing Pure signs.

BUILDING SIGNS AND LETTERS

UNION OIL DEALER STRIP 12x120

(3) 1950-1970 $125
UNION OIL DEALER STRIP 12x240
(3) 1950-1970 $125
Orange porcelain strip with white outlined blue "UNION OIL DEALER" across strip. Various sizes available.
MINUTE MAN SERVICE 12x240
(3) 1950-1970 $125
Orange porcelain strip with white outlined blue "MINUTE MAN SERVICE" across strip. Various size available.
UNION 76 STRIP 12x72
(2) 1970-1993 $50
Fiberglass strip with white area to left and orange area to right. Blue "UNI N" on white with "O" formed by orange circle with blue "76" logotype.

PUMP SIGNS

UNION GASOLINE SPEED & POWER 10x12
(3) 1926-1932 $225
Blue outlined white die-cut shield shaped sign with red/white/blue shield at top of sign. White "UNION/OIL/COMPANY/OF CALIFORNIA" in blue area of shield. Red "UNION/GASOLINE" below with blue "SPEED & POWER" across bottom.
UNION ETHYL GASOLINE 10x12
(3) 1926-1932 $225
Blue outlined white die-cut shield shaped sign with blue "UNION/OIL/COMPANY/OF CALIFORNIA" at top of sign. Red outlined white shield in center with blue "UNION" at top, "ETHYL" at bottom and Ethyl
(EGC) logo in center. Large red "GASOLINE" across bottom.
UNION GASOLINE TOPPER 4x12
(5) 1926-1932 $200
White porcelain strip with "UNION/GASOLINE" across center. Mounted on pump tops instead of illuminated globes.
UNION ETHYL GASOLINE TOPPER 4x12
(5) 1926-1932 $200
White porcelain strip with "UNION ETHYL" in outlined area over "GASOLINE" across center. Mounted on pump tops instead of illuminated globes.

UNION 76 GASOLINE 10"RD
(3) 1946-1955 $150
UNION 76 GASOLINE 11.5"RD
(1) 1946-1955 $150
Orange porcelain sign with blue "76" logotype in center. White "UNION" across top, "GASOLINE" across bottom.
UNION 7600 GASOLINE 10"RD
(3) 1946-1967 $150
UNION 7600 GASOLINE 11.5"RD
(1) 1946-1967 $150
Orange porcelain sign with blue "7600" logotype in center. White "UNION" across top, "GASOLINE" across bottom.
ROYAL 76 GASOLINE 10"RD
(3) 1955-1967 $150
ROYAL 76 GASOLINE 11.5"RD
(1) 1955-1967 $150
Orange outlined white porcelain sign with orange "76" logotype in center. Blue "ROYAL" across top, "GASOLINE" across bottom.

UNION 76 PLUS GASOLINE 11.5"RD
(3) 1967-EXP $200
Orange porcelain sign with white outlined blue "76" logotype beside white outlined blue underlined "plus" across center. White "UNION" across top, "GASOLINE" across bottom.
UNION DIESOL DIESEL FUEL 10"RD $175
UNION DIESOL DIESEL FUEL 11.5"RD $175
Blue porcelain sign with white outlined orange "DIESOL" across center. White "UNION" across top, "DIESEL FUEL" across bottom.
UNION G.M.DIESOL DIESEL FUEL 11.5"RD $175
Blue porcelain sign with white outlined orange "G.M./DIESOL" across center. White "UNION" across top, "DIESEL FUEL" across bottom.
76 UNIFUEL DIESEL FUEL 11.5"RD $175
Orange porcelain sign with blue "UNIFUEL" across center. Blue "76" logotype at top, small white "DIESEL FUEL" across bottom.
76 UNIFUEL DIESEL FUEL 11.5"RD $150
White porcelain sign with blue "UNIFUEL" across center. Blue "76" logotype at top, small blue "DIESEL FUEL" across bottom.
UNION 7600 MARINE GASOLINE 11.5"RD
(3) 1946-1955 $225
Orange porcelain sign with blue "7600" logotype in center. White "UNION" across top, "MARINE/GASOLINE" across bottom.
ROYAL 76 MARINE GASOLINE 11.5"RD
(3) 1955-1967 $225
Orange outlined white porcelain sign with orange "76" logotype in center. Blue "ROYAL" across top, "MARINE/GASOLINE" across bottom.
7600 OUTBOARD FUEL GASOLINE 11.5"RD
$225
Orange porcelain sign with white outlined blue "76" over with outlined blue "OUTBOARD/FUEL." White "GASOLINE" across bottom.

REGULAR 76 UNION GASOLINE 10x12
(2) 1967-1970 $150
Orange outlined white porcelain sign wth orange circle in center. White outlined blue "76/UNION" logotype on circle. Orange "REGULAR" across top, "GASOLINE" across bottom.
ROYAL 76 UNION GASOLINE 10x12
(2) 1967-1970 $150
Blue outlined white porcelain sign wth orange circle in center. White outlined blue "76/UNION" logotype on circle. Blue "ROYAL" across top, "GASOLINE" across bottom.
76 UNION UNIFUEL DIESEL FUEL 10x12
(3) 1967-1970 $175
White porcelain sign with small orange circle at top. White outlined blue "76" logotype on circle. Blue "UNIFUEL" over smaller "DIESEL FUEL" below logo.
REGULAR 76 UNION GASOLINE 13x18
(2) 1967-1970 $200
Orange outlined white embossed porcelain sign wth orange circle in center. White outlined blue "76/UNION" logotype on circle. Orange "REGULAR" across top, "GASOLINE" across bottom.

ROYAL 76 UNION GASOLINE 13x18
(2) 1967-1970 $200
Blue outlined white embossed porcelain sign wth orange circle in center. White outlined blue "76/UNION" logotype on circle. Blue "ROYAL" across top, "GASOLINE" across bottom.

NEON SIGNS

76 ON RECTANGLE 30x40
(4) 1932-1946 $650
Orange porcelain can type sign with two parallel blue lines across top and bottom. White outlined blue "76" in center. Neon "76" and one line above and below.

CURB SIGNS

UNION ETHYL GASOLINE 30"RD
(5) 1926-1930 $425
White porcelain sign with red outlined shield in center. Blue "UNION" arched around top, "ETHYL GASOLINE" around bottom with Ethyl
(EGC) logo in center.

LUBRICATION/MOTOR OIL SIGNS

ARISTO MOTOR OIL SHIELD 19x20
(3) 1925-1930 $450
Red/white outlined green porcelain shield with white band diagonal from lower left to upper right and red stripe vertical down center of sign face. Green outlined red "ARISTO" on diagonal band, with white "BEST/ALL WAYS" above band and "MOTOR/OIL" below band. White "UNION OIL COMPANY/OF CALIFORNIA" across bottom of sign face.
MOTORITE A UNION MOTOR OIL 19x20
(4) 1925-1930 $400
Red/white outlined blue porcelain die-cut shield with white band diagonal across from upper left to lower right. Red band vertical down center, with red "MOTORITE" on diagonal band. White script "200/extra/miles" above band, "A UNION/MOTOR/OIL" below band. White "UNION OIL COMPANY" across bottom.
TRITON
(CAN)-EXTRA MARGIN OF SAFETY 14"RD
(3) 1946-1950 $225
White outlined orange porcelain sign with detailed white/gold oil can on upper sign face. Blue "TRITON" arched across top of can, with orange/blue "76 Union" logo in center. Orange lettering on lower can. Blue outlined white "EXTRA MARGIN/OF SAFETY" below can.
TRITON
(CAN)-EXTRA MARGIN OF SAFETY 30"RD
(3) 1946-1950 $200
White outlined orange porcelain sign with detailed white/gold oil can on upper sign face. Blue "TRITON" arched across top of can, with orange/blue "76 Union" logo in center. Orange lettering on lower can. Blue outlined white "EXTRA MARGIN/OF SAFETY" below can.
ROYAL TRITON THE FINEST MOTOR OILS 14"RD
(4) 1955-1962 $225
White outlined purple porcelain sign with white "ROYAL/TRITON" arched around top and "MOTOR OILS" arched around bottom. Orange script "The Finest" across center of sign face.
ROYAL TRITON BELL SHAPED 28x36
(3) 1955-1962 $250
Purple tin litho bell shaped die-cut sign with white area across top. White "ROYAL/TRITON/MOTOR OIL" on lower sign face with purple lettering at top.
ROYAL TRITON 30x30
(4) 1960s $100
White tin litho sign with blue band across bottom. Purple "ROYAL/TRITON" on upper white area with white "THE AMAZING/PURPLE MOTOR OIL" on lower purple band.
ROYLA TRITON 10-30 CAN 18x24
(5) 1954-1960 $100
Die cut tin litho can shaped sign. Upper area is purple with white script "Royal" above white "TRITON/10-30" and Union logo. Purple "UNION/OIL COMPANY OF CALIFORNIA" across bottom.

SECONDARY SIGNS

AUTOMATIC TRANSMISSION SERVICE 11.5"RD
$250
Orange porcelain sign with white outlined blue "AUTOMATIC/TRANSMISSION/SERVICE" across center. Blue outlined white "AUTHORIZED" across center, with smaller white text across lower area.
STOP YOUR MOTOR ARISTO SHIELD 6x36
(5) $175
White porcelain sign with orange "STOP YOUR MOTOR" to left, "NO SMOKING" to right and red/white/green Aristo shield logo in center.
STOP YOUR MOTOR UNION SHIELD 6x36
(4) $175
White porcelain sign with orange "STOP YOUR MOTOR" to left, "NO SMOKING" to right and red and blue Union shield logo in center. White "UNION" in upper blue area, white "PRODUCTS" in lower red area.
STOP YOUR MOTOR UNION 76 6x36
(3) $150
White porcelain sign with orange "STOP YOUR MOTOR" to left, "NO SMOKING" to right and orange circle logo in center. White outlined blue "76" over "UNION" on logo.
PLEASE FOR YOUR SAFETY/NO SMOKING 12x15
(4) $150
White porcelain sign with orange "PLEASE" and blue "FOR YOUR SAFETY" across top above orange "NO SMOKING/STOP YOUR MOTOR." Orange and blue Minuteman character at upper left.
76 UNIFUEL DIESEL FUEL 24"RD
$200
White porcelain sign with blue "UNIFUEL" across center. Blue "76" logotype at top, small blue "DIESEL FUEL" across bottom.
76 OUTBOARD THE FINEST PRE-MIX 24"RD
(5) 1955-1965 $200
White porcelain sign with orange and blue "76/UNION" logo at top. Blue "OUTBOARD FUEL" in center above orange script "The Finest" and blue "PRE-MIX."
CERTIFIED CAR CONDITION SERVICE 22"RD
(3) 1958-1967 $150
White geartooth border around light orange porcelain sign wiht darker orange circle in center. White outlined blue "76/UNION" logotype on center circle with white/blue outlined dark orange "CERTIFIED" arched aorund top, "CAR CONDITION SERVICE" around bottom.

MINUTEMAN SERVICE 24"RD
(4) 1947-1960 $300
Orange porcelain signwith white outlined script "Minute Man" across top, "SERVICE" across bottom and blue/ white/tan characature of service station attendant (Union's "Minute Man") in center.

CLEAN REST ROOMS 24"RD
(3) 1950s $100
Blue outlined white porcelain sign with blue "CLEAN/REST/ ROOMS" across center.

WOMEN'S ROOM $75
Blue porcelain sign with blue circle at top center above white "WOMEN'S ROOM" below. White sillouette of womans head and hand with powder puff on blue circle. Blue die-cut arrows attaches to sign across bottom to point either direction, of for use without arrow on rest room door.

WE REDEEM UNION COUPONS 10x16
(5) 1925-1932 $225
White porcelain sign with blue outlined red/white/blue "UNION/OIL/COMPANY" shield in upper left corner. Blue "WE REDEEM" across top above blue outlined red "UNION/COUPONS." Small blue "SOLD BY/UNION OIL COMPANY/OF CALIFORNIA" below.

UNION CREDIT CARDS HONORED HERE 10x10
(5) 1930-1935 $225
Blue/white outlined die-cut porcelain flange shield with blue upper half and orange lower half. White "UNION/OIL COMPANY" in upper blue area with blue "CREDIT CARDS/Honored Here" on lower orange area.

NOTE: The following signs are unique in the petroleum industry in that Union Oil displayed the trademarks of those oil companies with which they had credit card exchange arrangements. On the collectors market they are commonly assumed to be from the company whose logo is pictured. They were used at Union stations only.

WE HONOR GULF CREDIT CARDS 24"RD
(3) 1955-1960 $150
Blue outlined white porcelian sign with orange/blue Gulf logo in center.Blue "WE/HONOR" above logo, "CREDIT/CARDS" below.

WE HONOR CONOCO CREDIT CARDS 24"RD
(3) 1955-1960 $150
Red outlined white porcelian sign with red/white Conoco logo in center. Green "WE/HONOR" above logo, "CREDIT/CARDS" below.

WE HONOR BA CREDIT CARDS 24"RD
(3) 1955-1960 $150
Red outlined white porcelian sign with red/white/green BA logo in center. Green "WE/HONOR" above logo, "CREDIT/CARDS" below.

WE HONOR ROYALITE CREDIT CARDS 24"RD
(3) 1955-1960 $150
Orange outlined white porcelain sign with orange/blue Royalite logo in center. Light blue "WE/HONOR" above logo, "CREDIT CARDS" below.

TRUCK SIGNS

UNION OIL COMPANY SHIELD 8"
(5) 1925-1930 $350
Blue/white outlined die-cut porcelain shield with blue upper half and red and white vertical stripes in lower half. White "UNION/OIL/COMPANY" in upper blue area.

UNION PRODUCTS 6"
(5) 1930-1935 $150
Blue/white outlined die-cut porcelain flange shield with blue upper half and orange lower half. White "UNION" in upper blue area with blue "PRODUCTS" on lower orange area.

HARBOR OIL COMPANY - LOS ANGELES, CA:
Originally a California refiner marketer, Union purchased Harbor about 1960 and operated it as a secondary brand until the mid 1970s.

IDENTIFICATION SIGNS

HARBOR PETROLEUM PRODUCTS 42x42
(4) 1938-1950 $2000
Irregular shaped "oval and shield" die-cut porcelain sign. Black/white/red outline around lower light blue shield area and around upper oval. Blue and white scene with seaplane on water in upper oval. Black outlined white and red "HARBOR" across center with yellow outlined black "PETROLEUM" above black outlined yellow "PRODUCTS" below.

UNION OIL COMPANY OF CANADA - VANCOUVER, BC: Canadian marketing affiliate of Union Oil Company of California.

IDENTIFICAITION SIGNS

UNION GASOLINE 60"RD
(5) 1930s $400
White porcelain sign with red/white outlined green shield in center. White "UNION/GASOLINE" on upper shield above large white "U" superimposed over orange vertical stripe.

LUBRICATION SIGNS

ARISTO UNION GASOLINE SPEED & POWER 26x30
(5) 1925-1930 $450
Blue outlined white die-cut shield shaped sign with red/ white/green "ARISTO" (UNION OIL CO OF CANADA) shield at top of sign. Blue outlined red UNION/GASOLINE" below with blue "SPEED & POWER" across bottom.

Independent Oil Companies

INDEPENDENT GASOLINE MARKETERS
The signs in the following section, titled "Independents," are from independent jobber brand or private brand operations. No ties to any major oil companies, except as noted, are known, although no doubt many early companies from which no history is known may have indeed sold out to or been rebranded by a major. If it is known that a company is still in business, that fact is noted, too. Listings are arranged

alphabetically by SIGN NAME (not company name) & a cross reference is provided when a sign is listed, out of aplhabetical order, with others from the same brand.

ARMOULD OIL COMPANY: Company history unknown.

LUBRICANTS SIGNS

ARMOULD OILS ARE BETTER 18x24
(5) 1926-1930 $225
Black tin litho sign with white "ARMOULD/OILS" across top, "ARE/BETTER" across bottom. White circle with opposing red quadrants at left in center, IOMA triangle & eagle logo at right in center.

ARVIDA OIL COMPANY - MIAMI, FL: A highly promoted marketing attempt from the early 1960s, Arvida set out to combine the besto features of both dealer operated and discount service stations. The concept did not work well and the few Arvida stations that had actually opened were sold to Gulf by 1965.

IDENTIFICATION SIGNS

ARVIDA 60x84
(5) 1959-1965 $100
White porcelain five point chevron-shield shaped sign with band across upper sign face. White "ARVIDA" on upper band, with large "V" in shield area below.

ULTRAMAR, INC - LONG BEACH,CA: A subsidary of London based Ultramar PLC, Ultramar purchased the Golden Eagle brand to enter US and Canadian markets and later, about 1980, purchased Beacon Oil Company, formerly The Caminol Company. They operate stations under both the Ultramar and Beacon brands today.

THE CAMINOL COMPANY-LONG BEACH, CA: Founded in California in 1931, Caminol marketed under the Beacon brand. They becam eBeacon Oil Company in the 1960's and were purchased by Britian's Ultramar PLC about 1980. Stations carry Beacon and Ultramar brands today.

CURB SIGNS

BEACON ETHYL 48"
(5) 1930-1935 $4000
Elaborate die-cut porcelain sign in shape of lighthouse with beacon rays, on base with waves washing against it. Yellow and blue lighthouse with yellow, red & white rays from beacon. Red outlined dark blue base area with yellow "ETHYL" over white "GASOLINE" across base. Red "A CAMINOL PRODUCT" across bottom. Red "BEACON" vertical down lighthouse with white circle at base. Ethyl
(EGC) logo in circle.

BEACON SECURITY 48"
(5) 1930-1935 $4000
Elaborate die-cut porcelain sign in shape of lighthouse with beacon rays, on base with waves washing against it. Yellow and blue lighthouse with yellow, red & white rays from beacon. Red outlined dark blue base area with yellow "SECURITY" over white "GASOLINE" across base. Red "A CAMINOL PRODUCT" across bottom. Red "BEACON" vertical down lighthouse.

SECONDARY SIGNS

BEACON RADIANT KEROSENE 12x18
(5) 1930s $350
Red outlined yellow porcelain sign with large red "KEROSENE" across center. Smaller red "BEACON/Radiant" above "KEROSENE," line and "A CAMINOL PRODUCT" below.

GOLDEN EAGLE: The Golden Eagle brand was originally introduced by Wasatch Refining Company for their stations in California. When Phillips purchased Wasatch in 1947, the California operation was sold to New York based Panama Refining and Petrochemical Company. They were one of the California based postwar self-service originators, operating pumpers throughout southern California. Golden Eagle was purchased by the British firm Ultramar in 1960, who eventually combined the Ultramar name with the Golden Eagle logo for use first in Canada, then later in California. About 1980 Ultramar purchased Caminol, by then renamed Beacon Oil Company, and they continue to operate stations in Canada under the Ultramar name and in California under both the Ultramar and Beacon brands.

PUMP SIGNS

GOLDEN EAGLE 14x14
(4) 1955-1970 $550
White outlined blue porcelain sign with orange band across bottom. Black outlined yellow eagle on upper blue area, with black "GOLDEN EAGLE" on orange band below.

BLUE AND WHITE - INDIANAPOLIS, IN: Local Indianapolis jobber brand founded in 1932 and still in operation today.

NEON SIGNS

BLUE AND WHITE SERVICE 72x94
(5) 1946-1990 $500
Round porcelain can type sign, punched for neon, with separate blue "banner" section across bottom. Round section is blue upper half and white lower half with white "BLUE" at top and blue "WHITE" below. Small "AND" in blue and white along color split in center. White "SERVICE" on blue banner.

BOLENE REFINING COMPANY, ENID, OK: Company history unknown.

CURB SIGNS

BOLENE BLUE SEAL GASOLINE 30"RD
(5) 1930s NO LISTING
Blue/white outlined blue porcelain sign with white center circle. Detailed blue & white ice flo scene with seal in center with blue script "Blue" above and "Seal" below. White "BOLENE" arched around top, "GASOLINE" around bottom.

BRUIN PETROLEUM, BUTLER, PA: Company history unknown.

FLANGE SIGNS

BRUINOIL/BRUIN GASOLINE 12x17
(5) 1926-1930 $2500
Red outlined yellow die cut tin litho flange sign. Black outlined red "BRUINOIL" arched around top, "BRUIN GASOLINE" around bottom with large brown bear in center. Black "The BEAR/OF/THEM ALL" to left of bear, PGCOA seal to right.

DORCHESTER GAS PRODUCING CO - AMARILLO, TX: History unknown.

PUMP SIGNS

CARGRAY GOLD 10"RD
(2) 1950s $125
Yellow outlined black porcelain sgin with yellow and black

winged tire at left on sign face. Black "CARGRAY" on tire, with yellow "GOLD" to right of tire.

CERTIFIED OIL CO-COLUMBUS, OH:

Currently operates over 100 stations in Ohio.

BUILDING SIGNS

CERTIFIED 32x72
(3) 1960-PRESENT $100
Blue outlined white porcelain crest-type shield sign with red "CERTIFIED" across center.

CLIPPER OIL COMPANY - SEATTLE, WA:

Company history unknown.

IDENTIFICATION SIGNS

CLIPPER GASOLINE 30x60
(5) 1946-1960 $2500
Green/white outlined yellow porcelain vertical oval sign with green/yellow/red airplane at top over green and white mountain scene. Red "CLIPPER" with speed lines across mountain scene below plane with yellow "GASOLINE" below.

COLONIAL OIL COMPANY-JACKSONVILLE, FL:

Founded in Jacksonville in 1932, Colonial was primarily a wholesaler prior to WW2. After the war, they established a chain of service stations in Florida & Georgia under the "Colonial Minuteman" brandname. Faced with supply shortages in the 1970's, Colonial closed most of their stations before selling out the balance of the operation to U-Save Automatic, operators of the "USA Gasoline" stations, in 1977.

IDENTIFICATION SIGNS

COLONIAL MINUTEMAN 42x84
(4) 1946-1977 $175
Red outlined white porcelain oval sign with blue "COLONIAL" arched across top. Small red "MINUTEMAN" just below. Series of red lines across bottom, interrupted by white area in center. Blue sillouette of minuteman head in white area.

PUMP SIGNS

COLONIAL MINUTEMAN REGULAR GASOLINE 14x16
(3) 1946-1977 $75
White/red outlined white porcelain sign with red lines across center. Blue sillouette of minuteman head in white area in center of red lines. Old english "Colonial" across top, above lines & sillouette, with red "MINUTEMAN" just below. Blue "REGULAR/GASOLINE" across lower sign face.

COLONIAL MINUTEMAN ETHYL GASOLINE 14x16
(3) 1946-1970 $75
White/red outlined white porcelain sign with red lines across center. Blue sillouette of minuteman head in white area in center of red lines. Old english "Colonial" across top, above lines & sillouette, with red "MINUTEMAN" just below. Blue "ETHYL/GASOLINE" across lower sign face.

CONTINENTAL REFINING CO-OIL CITY, PA:

Early lubricants manufacturer, founded in 1885, who ventured into gasoline marketing in the 1920s. By WW2 they had concentrated their gasoline marketing in eastern Canada only, but continued to sell lubricants in the United States until they closed in 1952.

CURB SIGNS

CORECO 24"RD
(5) 1926-1930 $475
Yellow/black outlined yellow porcelain sign with black outlined red "CORECO" arched around top, "MOTOR OILS" around bottom. Black PGCOA seal in center with "PERMIT No 35" below.

CORECO GASOLINE 24"RD
(5) 1926-1930 $475
Yellow/black outlined yellow porcelain sign with black outlined red "CORECO" arched around top, "GASOLINE" around bottom. Black keystone with smaller lettering in center.

SECONDARY SIGNS

USE CORECO GASOLINE PEP & POWER-NO CARBON 10x30
(5) 1925-1930 $450
Red outlined yellow self-framed tin litho sign. Large black keystone at left side with yellow "CORECO" arched across above "ESTABLISHED/1885." Black "USE" beside black outlined red "CORECO/GASOLINE" above black "PEP AND POWER - NO CARBON/CONTINENTAL REFINING CO., OIL CITY, PA" beside keystone.

CORECO MOTOR OILS & GREASES 10x30
(5) 1925-1930 $450
Red outlined yellow self-framed tin litho sign. Large black keystone at left side with yellow "CORECO" arched across above "ESTABLISHED/1885." Black outlined red "CORECO" above "MOTOR OILS/GREASES" above black "CONTINENTAL REFINING CO., OIL CITY, PA" beside keystone.

PENNSYLECT CORECO MOTOR OILS 20x30
(5) 1930s $400
Red porcelain signwith black outlined yellow map of Pennsylvania in center. Black band across center of Pennsylvania with yellow "CORECO" on band. Black outlined red "PENNSYLECT" above band, "MOTOR OILS" below, with PGCOA seal in lower right corner. Yellow "GOOD and TOUGH" across top of sign.

COUGAR OIL COMPANY - SELMA, AL:

Local marketer still in operation. History unknown.

IDENTIFICATION SIGNS

COUGAR 72"RD
(4) 1950s-1980s $600
Yellow porcelain signw ith black band across bottom. Detailed brown cougars head on upper sign face awith yellow "COUGAR" below.

SITE OIL CO. - CLAYTON, MO

Operated under the Cresyl brand in the 1930s and 1940s and evolved into the chain discounter Site after WW2. Gasoline grades continued to carry the "Cresyl" brandnames until the introduction of unleaded fuels in the 1970's. Site continues to operate stations throughout the midwest today.

PUMP SIGNS

CRESYL REGULAR 11x12
(3) 1946-1960 $150
Blue porcelain triangular porcelain sign, point down, with rounded points. White outlined yellow "CRESYL" above white "REGULAR" across upper area of sign. Small yellow circle at bottom with yellow "KNOCK/FREE" and lines in circle.

CRESYL 98 11x12
(3) 1946-1960 $150
Yellow outlined orange porcelain triangular porcelain sign, point down, with rounded points. Black outlined yellow "CRESYL" above black "98" across upper sign face. Small yellow circle at bottom with Ethyl
(EC) logo in circle.

CRESYL 110 11x12
(3) 1946-1960 $150
Black outlined yellow porcelain triangular porcelain sign, point down, with rounded points. Black outlined orange

"CRESYL" above black "110" across upper sign face. Small black outlined yellow circle at bottom with Ethyl
(EC) logo in circle.

CRYSTAL OIL COMPANY: A Texas based company, history unknown.

IDENTIFICATION SIGNS

CRYSTAL GASOLINE HERE 30"RD
(5) 1920s $300
Orange porcelain sign with orange outlined white crescent area at bottom of sign. White outlined green "CRYSTAL" arched across upper area of sign with green "GASOLINE" over orange "HERE" on lower white area.

CURB SIGNS

CRYSTAL ETHYL 30"RD
(5) 1930s $350
Black outlined white porcelain sign with black outlined white center circle. Green "CRYSTAL" arched around top, "GASOLINE" around bottom with Ethyl
(EGC) logo with yellow rays in center circle.

CRYSTAL FLASH PETROLEUM - INDIANAPOLIS, IN: Discount petroleum marketer founded in 1930 by the Feshenfeld family. Dual headquarters in both Indianapolis & in Grand Rapids,MI. Continues to market through around 20 stations in Indiana today.

IDENTIFICATION SIGNS

CRYSTAL FLASH JUST A LITTLE BETTER 47x99
(5) 1930s $1800
Elaborate die-cut white porcelain sign in the shape of and picturing a railroad tank car. Red outlined white tank with black details and black undercarriage. Black "The" on tank manway, with black "Crystal Flash" across tank car. Small black "LINE" below. Red script "Just a Little Better" in white area below tankcar.

CRYSTAL FLASH 48x96
(5) POSTWAR $300
White blue outlined white porcelain can-type sign with red lightning bolt through center. Red bolt-on extensions of lightning bolt above and below sign face. Blue "Crystal-Flash" across center of sign.

CURTS OIL COMPANY - MUSKOGEE, OK: Oklahoma regional brand dating from the 1960s.

PUMP SIGNS

CURTS POWER PACKED REGULAR 11x14
(5) 1965-1980 $150
Self framed aluminum litho sign with white face. Green outlined red "CURTS/OIL CO." across top with silver & green world globe in center at left. Small rocket diagonal above globe, therough "CURTS." Green script "Power/Packed" to right of globe, with red "REGULAR" across bottom.

CURTS POWER PACKED ETHYL 11x14
(5) 1965-1980 $150
Self framed aluminum litho sign with red face. Green outlined gold "CURTS/OIL CO." across top with silver & red world globe in center at left. Small rocket diagonal above globe, therough "CURTS." White script "Power/Packed" to right of globe, with gold "ETHYL" across bottom.

DARSEY OIL COMPANY: Company history unknown.

IDENTIFICATION SIGNS

DARSEY DOC GASOLINE 30x60
(5) 1940s $200
White porcelain sign with green "DARSEY" across top and black "GASOLINE across bottom. Large red "DOC" over black arrow in center

PENNSYLVANIA OIL PRODUCTS REFINING COMPANY - WARREN, PA:
Company history unknown.

LUBRICANTS SIGNAGE

ELDRED OIL FLANGE 10x14
(5) 1924-1930 $350
Black outlined orange porcelain sign with black outlined grey triangle in center. Black "ELDRED" across upper triangle with Pennsylvania in center with black "MOTOR," "OIL" below. PGCOA seal to left, other seal to right of triangle.

ESSARCO PETROLEUM CO., ST LOUIS, MO:
Company history unknown.

IDENTIFICATION SIGNS

ESSARCO 72 GASOLINE 24"RD
(5) 1930s $350
Orange outlined white porcelain sign with large orange "72" in center. Blue "ESSARCO" arched around top, "GASOLINE" around bottom. Small orange "ESSARCO PETROLEUM CO." arched around top outline, "ROOSEVELT BUILDING/ST LOUIS, MO." around bottom.

ETHYL CORPORTATION - RICHMOND, VA: In 1922 the General Motors research labs in Dayton, OH discovered that tetraethyl lead could be added to gasoline to increase its anti-knock properties. Higher compression engine development had been hindered by the lack of an avaialble motor fuel. Refiners Oil Company, a Dayton independent, was chosen to first market this blended gasoline. Ethylizers, mixing chambers that dispensed tetraethyl lead into gasoline, were added to pumps at several Refiners locations. The marketing experiment was deemed a success and in September 1923, Standard Oil of Indiana signed an agreement with General Motors to distribute the new gasoline throughout its territory. In early 1924, both GM and Standard looked at a partnership to produce and market the new product, but GM rejected Standard's offer and instead teamed up with Standard Oil of New Jersey (Esso). Esso had also been experimenting with tetraethyl lead, and their operations complemented GM's better than anything Standard had to offer.
In August 1924, the Ehtyl Gasoline Corporation was formed from GM and Esso's joint operation. In May 1925, Ethyl sales were temporarily suspended due to some lead poisoning concerns after several people involved in the manufacture of Ethyl died.The US sergeon general, after more than a year's research, allowed Ethyl to resume operations with the understanding that all blending would be done at the refinery and that Ethyl mixers added to gasoline pumps would be removed. Sales resumed in June 1926, and gasoline marketers everywhere applied for an Ethyl franchise. Strict guidelines for refining gasoline to be blended with Ethyl were set up, and it was not until 1933 that tetraethyl lead could be addd to a regular-grade gasoline.In April 1942, the Ethyl Gasoline Corporation became simply the Ethyl Corporation, and their logo was redesigned at that time. The company continues to market gasoline additives today.

PUMP SIGNS

ETHYL 8"
(1) 1926-1942 $75
Black oultined white porcelain sign with large black and yellow Ethyl
(EGC) logo in center. Yellow rays extending from Ethyl logo.

Black outlined yellow box above logo with black "WITH" in box.

ETHYL CURVED 8"
(1) 1926-1942 $75
Curved black oultined white porcelain sign with large black and yellow Ethyl
(EGC) logo in center. Yellow rays extending from Ethyl logo. Black outlined yellow box above logo with black "WITH" in box.

ETHYL 3"
(1) 1942-1964 $35

ETHYL 8"
(2) 1942-1964 $50
Black oultined white porcelain sign with large black and yellow Ethyl
(EC) logo in center. Yellow rays extending from Ethyl logo. Black outlined yellow box above logo with black "WITH" in box.

ETHYL CURVED 8"
(1) 1942-1964 $75
Curved black oultined white porcelain sign with large black and yellow Ethyl
(EC) logo in center. Yellow rays extending from Ethyl logo. Black outlined yellow box above logo with black "WITH" in box.

EZ SERVE STATIONS - HOUSTON, TX: Founded in the late 1960's self-serve boom, EZ Serve was an early operator of franchise convenience store and unattended fueling site operations. They expanded into convenience store operations in the 1980's. They purchased Atlanta based Magic Market in 1992, including the former TOC retail operation. Stations today operate under the EZ Serve, Magic Market, Direct & Red Diamond brandnames.

PUMP SIGNS

EZ SERVE 10x15
(3) 1968-1980 $90
White porcelain sign with irregular red shape at top. White area on red with blue "E-Z" on white area. White "SERVE" below. Blue outlined gold ribbon at bottom with blue "GASOLINE" on gold area.

AUTOTRONICS,INC - HOUSTON, TX: Founded in Houston in 1968, Autotronics originally offered gasoline installations, branded "Fill-Em Fast," at existing convenience stores. Tanks and pumps would be added at the corner of a convenience store lot, with the operator receiving a commission on gasoline sold. During the gasoline shortages of the 1970's, Autotronic became associated with Sigmor Corporation, a large regional Shamrock jobber organization. With assured supply, they began operating canopied pumper locations under the Fill Em Fast brand from Texas to Virginia. After Shamrock purchased Sigmor in the late 1980s, Fill Em Fast marketing territory was greatly consolidated and most locations were closed.

IDENTIFICATION SIGNS

FILL EM FAST 32x48
(2) 1968-1989 $100
Yellow outlined red porcelain sign with black outlined yellow rounded shield in center. Black "Fill-em'/fast" over black outlined red "SAVE" beside smaller red "SELF/SERVICE" on yellow shield.

PUMP SIGNS

FILL EM FAST GASOLINE 9.5x16
(2) 1968-1989 $75
Yellow outlined red porcelain sign with black outlined yellow rounded shield in center. Black "Fill-em'/fast" over black outlined red "GASOLINE" on yellow shield.

FILL EM FAST DIESEL 9.5x16
(2) 1968-1989 $75
Yellow outlined red porcelain sign with black outlined yellow rounded shield in center. Black "Fill-em'/fast" over black outlined red "DIESEL" on yellow shield.

FORD MOTOR COMPANY - DEARBORN, MI: While General Motors was involved in developing the gasoline additive tetraethyl lead, Ford was promoting a benzol based motor fuel as an alternative to the low octane gasoline of the day. Genuine Ford Benzol was sold through select Detroit area Ford dealers. By 1940 Ford had began promoting a benzol product to independent Michigan marketers under a franchise type program. Their "BENZOL GAS" was sold alongside other oil company branded products by a special licensing arrangement.

CURB SIGNS

BENZOL GAS 24"RD
(4) 1940-1950 $325
Blue porcelain sign with series of blue and white diagonal stripes on lower half. White "BENZOL" arched around top over "GAS," with small white and red star to either side.

DEYO OIL COMPANY - GALVESTON, TX: Company history unknown.

IDENTIFICATION SIGNS

GALTEX THE BETTER GASOLINE 42"RD
(5) 1930s $350
Red outline and white band around red porcelain sign; White "GALTEX" across center with "THE BETTER/GASOLINE" above and "MOTOR OIL" below. Red "DEYO OIL COMPANY, INC" arched around top, "PRODUCERS REFINERS MARKETERS" around bottom in white band.

GATE PETROLEUM-JACKSONVILLE, FL

Founded by former Billups executives in 1963 when Hess purchased Billups, Gate operates a chain of discount service stations and convenience stores troughout the southeast.

PUMP SIGNS

GATE REGULAR 12x12
(4) 1965-1990 $100
White outlined red porcelain sign with red detailed white fence gate on upper sign face over white band. Red "GATE" across gate, with red "REGULAR" in band.

GATE PREMIUM 12x12
(4) 1965-1985 $100
White outlined blue porcelain sign with blue detailed white fence gate on upper sign face over white band. Blue "GATE" across gate, with blue "PREMIUM" in band.

GATE UNLEADED 12x12
(4) 1975-1990 $100
Black outlined yellow porcelain sign with black detailed white fence gate on upper sign face over white band. Black "GATE" across gate, with black "UNLEADED" in band.

GATE PREMIUM UNLEADED 12x12
(4) 1985-1990 $100
White outlined blue porcelain sign with blue detailed white fence gate on upper sign face over white band. Blue "GATE" across gate, with blue "PREMIUM/UNLEADED" in band.

GATE KEROSENE 12x12
(4) 1965-1990 $100
White outlined green porcelain sign with green detailed

Cosden and Trillings signs. *Bruce Bednar.*

Auntie Chatter - scarce embossed tin.

Musgo porcelain dealer sign.

Rebel Oil dealer sign. *Al Alford.*

Clipper Gasoline porcelain sign.

Cities Service tin sign for oil rack.

Golden Eagle porcelain pump plate. *Leo Mathieu.*

Cities Service Koolmotor tin oil rack sign.

Miscellaneous porcelain pump plates. *Bruce Bednar.*

Houston Motor Oil porcelain sign. *Al Alford.*

Marine Gasoline porcelain building sign. *Dunbar's Gallery.*

Wolf Oil Co. embossed tin.

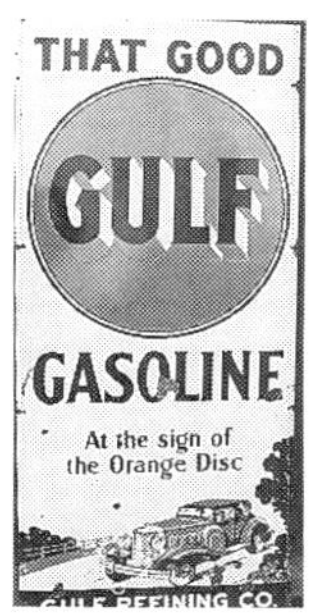

That Good Gulf porcelain sign. *Leo Mathieu.*

Gulf Refining porcelain sign. *Al Alford.*

Gulf Refining porcelain sign. *Bruce Bednar.*

white fence gate on upper sign face over white band. Green "GATE" across gate, with green "KEROSENE" in band.

GATE REGULAR/PREMIUM 10x20
(5) 1965-1985 $125
One piece twin pump sign. Left half is white outlined red porcelain sign with red detailed white fence gate on upper sign face over white band. Red "GATE" across gate, with red "REGULAR" in band. Right half is white outlined blue porcelain sign with blue detailed white fence gate on upper sign face over white band. Blue "GATE" across gate, with blue "PREMIUM" in band.

GATE PREMIUM/REGULAR 10x20
(5) 1965-1985 $125
Reverse of above, for opposite side of pump. NOTE: This series of Gate signs for two-product pumps is unique in that the two grades are identified on one sign, making it necessary to have a front sign and a back sign, reversed left and right, for opposite sides of a pump.

GATE REGULAR/UNLEADED 10x20
(5) 1985-1990 $125
One piece twin pump sign. Left half is white outlined red porcelain sign with red detailed white fence gate on upper sign face over white band. Red "GATE" across gate, with red "REGULAR" in band. Right half is black outlined yellow porcelain sign with black detailed white fence gate on upper sign face over white band. Black "GATE" across gate, with black "UNLEADED" in band.

GATE UNLEADED/REGULAR 10x20
(5) 1985-1990 $125
Reverse of above, for opposite side of pump.

GATE REGULAR/PREMIUM UNLEADED 10x20
(5) 1985-1990 $125
One piece twin pump sign. Left half is white outlined red porcelain sign with red detailed white fence gate on upper sign face over white band. Red "GATE" across gate, with red "REGULAR" in band. Right half is white outlined blue porcelain sign with blue detailed white fence gate on upper sign face over white band. Blue "GATE" across gate, with blue "PREMIUM/UNLEADED" in band.

GATE PREMIUM UNLEADED/REGULAR 10x20
(5) 1985-1990 $125
Reverse of above, for opposite side of pump.

GATE UNLEADED/PREMIUM UNLEADED 10x20
(5) 1985-1990 $125
One piece twin pump sign. Left half is black outlined yellow porcelain sign with black detailed white fence gate on upper sign face over white band. Black "GATE" across gate, with black "UNLEADED" in band. Right half is white outlined blue porcelain sign with blue detailed white fence gate on upper sign face over white band. Blue "GATE" across gate, with blue "PREMIUM/UNLEADED" in band.

GATE PREMIUM UNLEADED/UNLEADED 10x20
(5) 1985-1990 $125
Reverse of above, for opposite side of pump.

NORTHWEST REFINING COMPANY - CUT BANK, MT:

NORTHWEST REFINING COMPANY - CUT BANK, MT: When Standard Oil of New Jersey, through their Carter Oil division, purchased Northwest Refining's "Grizzly" marketing network in 1942, Northwest concentrated on unbranded marketing, primarily for wartime needs, until 1946. In 1946 they re-entered branded gasoline marketing under the Glacier name.

PUMP SIGNS

GLACIER 10"RD
(5) 1946-1950 $450
White porcelain sign with blue outline of Montana across sign face. Blue and white mountain scene inside outline, with red script "Glacier" across scene. Black "GASOLINE" across below Montana outline.

GLOBE OIL CO. - MCPHERSON, KS

Mid-continent refiner marketer operating under the Globe brand throughout the central US.

IDENTIFICAITON SIGNS

GLOBE 42"RD
(4) $1200
Blue outlined yellow porcelain sign with white center circle. Blue cars and planes circling center in outer yellow area. White outlined blue band arched across sign with white "Globe" on band. Blue North & South America sillouette on center white area with blue cars and planes circling center in outer yellow area.

GLOBE THE UNIVERSAL GASOLINE 30"RD
(4) $850

GLOBE THE UNIVERSAL GASOLINE 42"RD
(4) $750
Blue outlined yellow porcelain sign with white center circle. Blue cars and planes circling center in outer yellow area. White outlined blue band arched across sign with white "Globe" on band. Blue North & South America sillouette on center white area with Blue "The/UNIVERSAL" below blue band with blue "GASOLINE" arched around bottom.

GLOBE GASOLINE 30"RD
(4) $850

GLOBE GASOLINE 42"RD
(5) $750
Blue outlined red porcelain sign with white center circle. White outlined blue band arched across sign with white "Globe" on band. Blue North & South America sillouette and longitude/latude lines on center white area with Blue "GASOLINE" below blue band.

CURB SIGNS

GLOBE 24"RD
(5) $850
Blue outlined red porcelain sign with white center circle. White outlined blue band arched across sign with white "Globe" on band. Blue North & South America sillouette and longitude/latude lines on center white area with Blue "GASOLINE" below blue band.

GLOBE ETHYL 30"RD
(5) $850
Blue outlined yellow porcelain sign with white center circle. Blue cars and planes circling center in outer yellow area. White outlined blue band arched across sign with white "Globe" on band. Blue North & South America sillouette on center white area with blue cars and planes circling center in outer yellow area. Large Ethyl
(EGC) logo with yellow rays on lower sign face.

GOLDEN BEAR OIL COMPANY

SECONDARY SIGNS

GOLDEN BEAR OIL COMPANY 15"RD
(5) 1930s $400
White tin litho sign with blue center circle. Yellow bear with black details in center circle. Blue "GOLDEN BEAR OIL COMPANY" arched around top, "PETROLEUM PRODUCTS" arched around bottom.

POWER OIL COMPANY: Company history unknown.
IDENTIFICATION SIGNS
GOLDEN FLASH GASOLINE 30"RD
(5) 1935-1940 $225
Yellow outlined white porcelain sign with red "GOLDEN/FLASH/GASOLINE" across center of sign face. Black "POWER OIL" arched around top, "COMPANY" around bottom.

GOLDEN WEST OIL CO - SAN ANTONIO, TX
History unknown.
PUMP SIGNS
GOLDEN WEST
$900
White/black outlined round porcelain sing with yellow sun in center. Yellow and white rays up from sun with reflection of sun on blue mountains and water scene below. Black outlined white banner across top with black "GOLDEN WEST" across banner. White outlined dark blue "OIL/COMPANY" on sun

GOLDENWEST: Company history unknown.
IDENTIFICATION SIGNS
GOLDENWEST 30"RD
(5) 1930s no listing
Black/white outlined orange porcelain sign with black outlined white band across center. Orange outlined black "GOLDENWEST" on band with white outlined black "GASOLINE" below and white outlined black triangles and half-circle design above. Orange used on this sign is a deep red-orange.

GOOD LUCK OIL COMPANY, DALLAS, TX
IDENTIFICATION SIGNS
GOOD LUCK SERVICE 19x31
(5) 1946-1960 $350
Green scalloped border around white porcelain oval sign. Large green horseshoe in center with yellow diamonds to either side. Red "GOOD" arched to left of horseshoe, "LUCK" to right with black "SERVICE" across bottom.
PUMP SIGNS
USE GLOCO HI'R OCTANE 8x12
(2) 1946-1960 $125
Green scalloped border around white porcelain oval sign. Red "GLOCO" with red shading across center with black "USE" above and black "HI'R OCTANE" below.
SUPER GLOCO ETHYL 8x12
(2) 1946-1960 $125
Green scalloped border around white porcelain oval sign. Red "GLOCO" with red shading across center with black "SUPER" above and black "ETHYL" below.
SUPER GLOCO ETHYL EGC 12"RD
(4) 1935-1946 $125
Green scalloped border around white porcelain sign. Large red "GLOCO" with red shading across center of sign with black "SUPER" above and black outlined white circle below. Ethyl
(EGC) logo with yellow rays in circle.
GLO-LUBE ETHYL HI-TEST 12x14
(3) 1960s $75
Rectangular sign with red band across top, white band across center and blue band across bottom. White "GLOLUBE: on red band, blue "ETHYL" on white band and white "HI-TEST" on blue band.
NOTE: At first glance it would seem this sign is from another company with a similar brandname. However, the "GLOLUBE" lettering matches the "GLOCO" logotype used on the other signs.

RUSSIAN OIL COMPANY - CHICAGO, IL: Company history unknown.
CURB SIGNS
USE GUYLER OILS 26x29
(5) 1920's $750
Description not available.

GRAYBURG OIL COMPANY - SAN ANTONIO, TX: Company history unknown.
FLANGE SIGNS
GRAYBURG OIL 12x14
(5) 1920s $450
Red outlined gray porcelain flange sign. Red triangle in gray center with red "GASOLINE/LUBRICANTS/KEROSENE" around triangle and "PETROLEUM/PRODUCTS" in triangle. White "GRAYBURG OIL COMPANY" arched around top, "SAN ANTONIO, TEXAS" around bottom.

HAMMAN: Company and history unknown.
PUMP SIGNS
HAMMAN GASOLINE 10x10
(5) 1940's $225
Blue/white outlined red octagon shaped sign with white outlined blue band across center. White "HAAMAN" on blue band with white script "Extra Good" in upper red area, "Gasoline" below.

HICKS OIL & GAS - ROBERTS, IL: Founded in 1948, they marekted gasoline under the "Hicks" brand until 1950 when the Hicksatomic brand was introduced, playing on the postwar "atomic" age. They operated stations throughout Illinois until the environmental regulations of the late 1980s forced them to greatly cut back their operations. They concentrate on Propane & LP Gas sales today, with only two retail outlets, branded "Hicks Gas" in operation.
IDENTIFICATION SIGNS
HICKSATOMIC 42x84
(4) 1950-1970 $250
Green outlined white porcelain oval sign with large red sunburst in center. Green "Hicksatomic" in progressively larger lettering across center of sign face.

HI PLUS: Company and history unknown.
PUMP SIGNS
HI PLUS PREMIUM 8"RD
(5) 1950s $200
White poorcelain sign with vertical red rectangle in center. Black band across center with white "HI-PLUS/PREMIUM" on black band.

RAYMER OIL COMPANY - STATESVILLE, NC:
Raymer Oil is a branded Phillips 66 jobber with stations in the central piedmont area of North Carolina. In the 1960's they had a few stations operating under their own "Home Gas" brand.
PUMP SIGNS
HOME OIL COMPANY REGULAR 10x10
(5) 1965-1970 $150
Red porcelain sign with white outline and white band across lower sign face. White "HOME/OIL" above band with red "REGULAR" across band. Small white "RAYMER OIL CO." interrupts outline at bottom.
HOME OIL COMPANY PREMIUM 10x10
(5) 1965-1970 $150
Blue porcelain sign with white outline and white band across

lower sign face. White "HOME/OIL" above band with blue "PREMIUM" across band. Small white "RAYMER OIL CO." interrupts outline at bottom.

HOUSTON OIL COMPANY: Company history unknown.

IDENTIFICATION SIGNS

HOUSTON GASOLINE 48"RD
(5) 1930s no listing
Black outlined white porcelain sign with black outlined red "HOUSTON" arched around top, "GASOLINE" around bottom and green with black details of Sam Houston on a horse in the center.

HOUSTON MOTOR OIL 42"RD
(5) 1930s no listing
Black outlined white porcelain sign with black outlined red "HOUSTON" arched around top, "MOTOR OIL" around bottom and green with black details of Sam Houston on a horse in the center.

L.M.HUGHEY & COMPANY - TAMPA, FL: Founded in Tampa in 1967, L.M.Hughey operated a chain of discount stations from Virginia south to Florida. The Virginia & North Carolina stations were based in Waynesboro, VA and operated under the Imperial brand. Tennessee stations carried the Cargo name, while Florida locations were branded Charge or Kash n' Karry. In 1985 they began converting most of the locations to "JOY" food stores, selling the brand gasoline appropriate for that region. In 1988 the Derby brandname replaced the regional brands on gas pump identification, and by 1989 most locations had branded Coastal (along with all other Derby outlets nationwide). Environmental problems at some stations caused them to become ineligible for the Coastal brand and in 1993 these few stations rebranded "Joy," selling Imperial gasoline. Florida locations were not involved in the Coastal branding and they remain as Joy stores, selling Joy gasoline today.

PUMP SIGNS

IMPERIAL REGULAR 12x14
(5) 1967-1985 $50
Metallic gold anodized aluminum litho sign with blue over red oval on upper sign face. White band across center of oval with blue "IMPERIAL" on band. Blue "HI-TEST/REGULAR" below oval.

IMPERIAL PREMIUM 12x14
(5) 1967-1985 $50
Metallic gold anodized aluminum litho sign with blue over red oval on upper sign face. White band across center of oval with blue "IMPERIAL" on band. Blue "PREMIUM" below oval.

KASH N KARRY REGULAR 12x14
(5) 1967-1985 $50
Metallic gold anodized aluminum litho sign with blue over red oval on upper sign face. White band across center of oval with blue "KASH N' KARRY" on band. Blue "HI-TEST/REGULAR" below oval.

KASH N KARRY PREMIUM 12x14
(5) 1967-1985 $50
Metallic gold anodized aluminum litho sign with blue over red oval on upper sign face. White band across center of oval with blue "KASH N' KARRY" on band. Blue "PREMIUM" below oval.

CHARGE REGULAR 12x14
(5) 1967-1985 $50
Metallic gold anodized aluminum litho sign with blue over red oval on upper sign face. White band across center of oval with blue "CHARGE" on band. Blue "HI-TEST/REGULAR" below oval.

CHARGE PREMIUM 12x14
(5) 1967-1985 $50
Metallic gold anodized aluminum litho sign with blue over red oval on upper sign face. White band across center of oval with blue "CHARGE" on band. Blue "PREMIUM" below oval.

CARGO REGULAR 12x14
(5) 1967-1985 $50
Metallic gold anodized aluminum litho sign with blue over red oval on upper sign face. White band across center of oval with blue "CARGO" on band. Blue "HI-TEST/REGULAR" below oval.

CARGO PREMIUM 12x14
(5) 1967-1985 $50
Metallic gold anodized aluminum litho sign with blue over red oval on upper sign face. White band across center of oval with blue "CARGO" on band. Blue "PREMIUM" below oval.

J&L STATIONS - VERNON HILLS, IL: Founded in Chicago in 1952, J&L grew rapidly after 1960, increasing from 14 locations to some 70 outlets, including some former Enco units scattered into the southeast, by the time of the 1973 gas shortage. Since that time shortages and environmental requirements has reduced the station count by 30% or so, but the company is still going strong, with over 50 outlets today.

J&L 5x6
(4) 1970-1985 $100
Yellow porcelain sign with two white trapezoids, points together, top and bottom. Green "J&L" in upper white area, red "GAS" below. Small sign used on pump dial faces.

JOHNSON HIGH TEST COMPANY - FREEPORT, IL: Chicago area marketer that may have been purchased by or as a jobber converted to Mobil.

IDENTIFICATION SIGNS

JOHNSUNITE HIGH TEST GASOLINE 36x72
(5) 1930s $225
White porcelain sign with red "JOHNSUNITE/TRUE" over blue "GASOLINE" across sign face.

HI DUTY MOTOR OIL 36x72
(5) 1930s $225
Tandem sign with above Johnsunite Gasoline sign. White porcelain sign with blue outlined red circle in center. White band diagonal across center with red "HI-DUTY" in band. White "NEW IMPROVED" arched around top, "TOUGH FILM" around bottom.

CURB SIGNS

HI DUTY 24"RD
(5) 1930s $250
White porcelain sign with blue outlined red circle in center. White band diagonal across center with red "HI-DUTY" in band. White "NEW IMPROVED" arched around top, "TOUGH FILM" around bottom.

LUBRICANTS SIGNAGE

HI DUTY 12x72
(5) 1930s $100
Blue porcelain vertical panel with white "HI-DUTY" vertical on sign.

KOCOLENE, INC - SEYMOUR, IN: Founded in Seymour in 1938, Kocolene was purchased by the Myers family in 1943. After WW2 the chain expanded greatly, operating several hundred stations in Indiana, Ohio, Pennsylvania, Kentucky & Tennessee. They continue to operate convenience stores throughout their marketing territory today.

PUMP SIGNS

KOCOLENE HI-TEST REGULAR 12"RD
(5) 1950s $250
White porcelain sign with blue ripple border. Red "Kocolene" logotype over blue "HI-TEST" over red "REGULAR" in center of sign.

KOCOLENE GENUINE ETHYL 12"RD
(5) 1950s $250
Red porcelain sign with yellow wreath design around outer edge. Yellow "Kocolene" logotype across top above black "GENUINE." Yellow "ETHYL above small yellow circle at bottom. Ethyl
(EC) logo on circle.

KUNZ OIL COMPANY - MINNEAPOLIS, MN: Founded in 1888 as a lubricants manufacturer, Kunz got involved in gasoline marketing in the mid 1910s. They continue to operate a small number of branded Kunz stations today.

CURB SIGNS

KUNZ GASOLINE 30"RD
(5) 1920s no listing
Round porcelain sign with "KUNZ/GASOLINE" across top over red/green car and trees scene. No other details known.

GOLD MEDAL OILS 24"RD
(5) 1920s no listing
Green/red outlined dark green sign with red and green automobile and tree scene in center. Red outlined yellow "GOLD/MEDAL" across top, "OILS" across bottom, with red "SINCE 1888" arched around bottom.

LAURELEAF OIL COMPANY: Wyoming company, history unknown.

IDENTIFICATION SIGNS

GO WITH LAURELEAF GASOLINE 30x60
(5) 1935-1940 $650
Red porcelain sign with large white "LAURELEAF" across center above "GASOLINE." Smaller white "GO" at top, with white sedan above small "with" beside "GO."

PUMP SIGNS

LAURELEAF 75 W/ POLY 12"RD
(5) 1935-1940 $350
Red outlined white porcelain sign with large green outlined red "75" in center. Green "LAURELEAF" above "75," "with Poly" below.

LIBERTY GASOLINE: Details unknown. This is a West Coast marketer and there is no known connection between them and the Illinois based discounter of the same name.

IDENTIFICATION SIGNS

LIBERTY 24x48
(5) 1930's $2200
Black/white outlined orange porcelain vertical rectangle sign with die-cut corners with black band across bottom. Detailed black & white statue of liberty in orange area with white "LIBERTY/GASOLINE" on lower black area.

BARNETT OIL COMKPANY: Company history unknown.

PUMP SIGNS

LIFE 10x12
(4) 1935-1940 $1100
Green outlined white porcelain sign with red arched band at top of yellow cone in center of sign face and red band across bottom. Red/white/green/black elf in center of yellow cone. White "LIFE" in upper band, "GASOLINE" in lower with green "BARNETT OIL COMPANY" below lower band.

LIFE ETHYL 10x12
(4) 1935-1940 $850
Green outlined white porcelain sign with red arched band at top of white cone in center of sign face and red band across bottom. Green "ETHYL" and Ethyl
(EGC) logo in center of white cone. White "LIFE" in upper band, "GASOLINE" in lower with green "BARNETT OIL COMPANY" below lower band.

LILY WHITE GASOLINE - LIMA, OH: Company history unknown.

IDENTIFICATION SIGNS

USE LILY WHITE GASOLINE 18x24
(5) 1920s $450
Red outlined yellow porcelain sign with black "Lily White" arched across top. Small black "USE" above, with black "GASOLINE/ FROM/PENNSYLVANIA/CRUDE" below.

LUBRI-GAS INTERNATIONAL, INC - CHICAGO, IL : Company history is sketchy, but as best as can be determined this company supplied an additive to other petroleum marketers, similar to "Ethyl" or "Solvenized" gasolines. One known markter of "Lubri-Gas" was Mid-West Refining Co of Muskegon,MI, which later franchised the White Rose brand from National Refining.

SECONDARY SIGNS

LUBRI-GAS CAMEL 19x28
(4) 1935-1945 $400
White tin litho sign with black outlined yellow camel on black area across bottom. Red "LUBRI-GAS" on bottom black area with black ."..CLEANS AND LUBRICATES../AS IT POWERS YOUR MOTOR" across top above red script "Protects/The Motor." Other lettering on camel saddle.

GEAUGA OIL COMPANY, CLEVELAND, OH: Company history unknown.

IDENTIFICATION SIGNS

USE MAGIC GASOLINE 18x24
(5) 1920s $250
Self framed cream colored tin sign with red metal band pump globe picetured in center. Black script "Magic" on globe face with "GASOLINE" arched around bottom and red lightning bolts in center. Black "USE" across top of sign above globe likeness.

MARINE GASOLINE: This is a California brand, not a product for boats, but details about the company are unknown.

IDENTIFICATION SIGNS

MARINE GASOLINE 36x60
(5) 1940's $950
White outlined blue porcelain die cut vertical rectangle sign with cut corners, with large white outlined light blue water scene at top. Blue and white detailed boat on water. White "MARINE/GASOLINE" below picture.

MARINE GASOLINE POSITIVE OCTANE 36x60
(5) 1940's $950
White outlined blue porcelain die cut vertical rectangle sign with cut corners, with large white outlined light blue water scene at top. Blue and white detailed boat on water. White "MARINE/GASOLINE" below picture. Small white "Positive Octane" below.

MARINE OIL CO. - ST. LOUIS, MO

History unknown. Jobber operated stations were branded Marine while the 40 or so company operated discount stations were branded Mars. Mars stations were found from Chattanooga to California with Marine stations within 300 miles of St. Louis.

PUMP SIGNS

MARS REGULAR 14x16
(3) 1955-1970 $75
White porcelain sign with large red circle at top. Blue band diagonal across circle with white "MARS" on band. Blue "REGULAR" across bottom.

MARS PREMIUM 14x16
(3) 1955-1970 $75
White porcelain sign with large red circle at top. Blue band diagonal across circle with white "MARS" on band. Blue "PREMIUM" across bottom.

SECONDARY SIGNS

MARS STAMPS 30x42
(3) 1960-1970 $75
Red outlined white tin litho sign with large red circle in center. Blue band diagonal across circle with white "MARS" on band. Blue "SAVE" across top "STAMPS" across bottom.

MARINE STAMPS 30x42
(3) 1960-1970 $75
Red outlined white tin litho sign with blue & white oval "MARINE" logo with seahorse in center. Red "SAVE" across top "STAMPS" across bottom.

MELROSE GASOLINE: Company history unknown.

PUMP SIGNS

MELROSE GASOLINE 12"RD
(5) 1950-1960 $100
White porcelain sign with large blue "M" in center. White "MELROSE" with black dropshading across "M" above yellow "GASOLINE"

MERCURY OIL COMPANY - SHREVEPORT, LA:

Company. History unknown.

PUMP SIGNS

MERCURY SUPER REGULAR 105 GASOLINE 8x12
(5) 1950-1960 $125
Red outlined white porcelain sign with blue outlined red oval at top. White band through center of red oval. Blue "USE" above oval with blue "SUPER/REGULAR" below. Red "105" above blue "GASOLINE" on lower sign face.

MEADVILLE CORP. - HAVERFORD, PA

Meadville operated large multi-pump discount stations in mid-Atlantic metropolitan areas under the Save-Way and Safe-Way brands. In the 1950s, they acquired Reading, Pennsylvania based Merit and they continue to operate stations under the Merit brand today.

PUMP SIGNS

MERIT ADVANCED REGULAR 10x14
(5) 1959-1968 $200
White porcelain sign with black and white characature of driver in car in center. Red "MERIT" across top of sign with red "Advanced REGULAR" below.

MERIT 100 PLUS 10x14
(5) 1959-1968 $200
White porcelain sign with black and white characature of driver in car in center. Blue "MERIT" across top of sign with blue "100 PLUS" below.

MIDWAY: Company and history unknown.

PUMP SIGNS

MIDWAY SUPERIOR GASOLINE
$200
Rectangular white porcelain sign with series of red/white/blue outlines. Blue double lines across separate sign into thirds. Red "MIDWAY" in upper area with blue "SUPERIOR" in center and red "NO EQUAL/AT IT'S PRICE" across bottom.

C.E.MILLS OIL COMPANY, SYRACUSE, NY

SECONDARY SIGNS

MILLS EXCELSIOR 14x24
(5) 1920-1925 $250
White outlined blue porcelain sign with white "WE SELL/MILLS EXCELSIOR/MOTOR OILS/AND GASOLINE/ C.E.MILLS CO SYRACUSE, N.Y." on sign face.

DUCKY DRAKE 20x24
(4) 1940s $400
Dark blue outlined red porcelain sign with white line drawing of duck's body with man's head covering most of sign. White "DUCKY/DRAKE" in lower right corner.Supposedly a characature of the company owner.

DUCKY DRAKE 20x24
(4) 1940s $400
Yellow outlined green porcelain sign with yellow line drawing of duck's body with man's head covering most of sign. Yellow "DUCKY/DRAKE" in lower right corner.

MOORE'S OIL COMPANY - ALABASTER, AL:

Company history unknown.

PUMP SIGNS

MOORE'S FIRST GRADE REGULAR SUPREME 12x16
(3) 1950s $100
White outlined green porcelain sign with white band in center. White script "Moore's" in upper green area, "Supreme" in lower green area with large yellow circle in white area with green lightning bolt. Red band across center of circle with white "REGULAR" on band, green script "First" above and "Grade" below.

MOORE'S ETHYL
(EC) SUPREME 12x16
(3) 1950s $100
White outlined red porcelain sign with white band in center. White script "Moore's" in upper red area, "Supreme" in lower red area with large black outlined white circle in center. Ethyl
(EC) logo with yellow rays in circle.

MUSKEGON OIL CO. - MUSKEGON, MI: Michigan marketer in operation from 1927 until 1929. Details unknown.

IDENTIFICATION SIGNS

MUSGO 36"RD
(5) 1927-1929 $4000
Black outlined orane porcelain sign with black outlined white band across center and white center circle. Multicolored, detailed Indian in center circle with black outlined red "GASO" to left on band and "LINE" to right. Black outlined green "MUSGO" arched around top with black outlined green "MICHIGAN'S/MILE MAKER" acrhed aorund bottom.

FLETCHER OIL CO, MONROE, NC: Founded about 1960 in North Carolina, Fletcher Oil branded numerous independent stations throughout North and South Carolina

Mobil set up. *Jim Hollabaugh.*

Wall of Mobil.

Mobilgas hanging porcelain dealer sign. *Dunbar's Gallery.*

Mobiloil tin sign. *Dunbar's Gallery.*

Miscellaneous signs adorn a garage wall. *Dennis McClellan.*

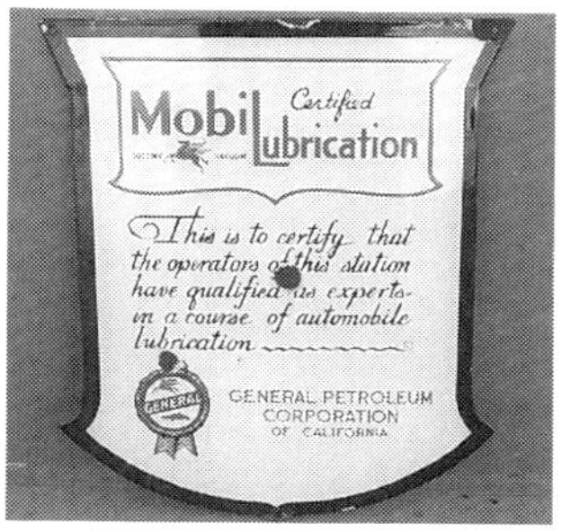

Rare Mobil/General Lubrication porcelain sign. *Leelan Glanzer.*

Gilmore porcelain sign, rare. *Allan Darr.*

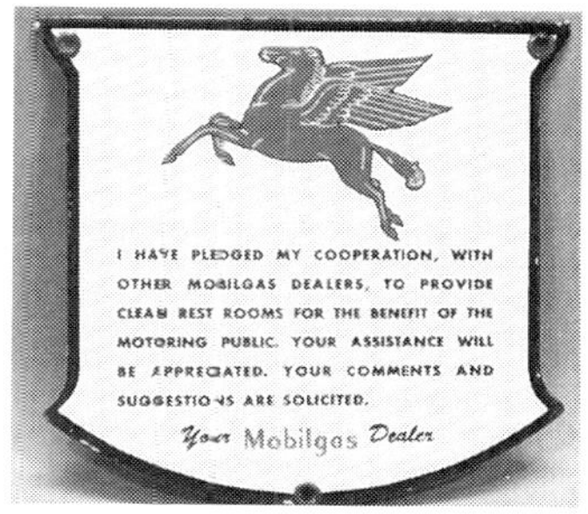

Mobil Rest Room Certification sign. *Dunbar's Gallery.*

Magnolia/Socony 30" porcelain sign. *Bruce Bednar.*

Pair of Socony Lubester signs - still mounted. *Don Christensen.*

The Grizzly Group - rare signs.

Husky miscellaneous porcelain pump plates.

Mobiloil embossed tin sign. *Dunbar's Gallery.*

Mobiloil Lubester porcelain sign *Leelan Glanzer.*

and Georgia until the early 1980's. They currently have a few company operated stations in central NC.

IDENTIFICATION SIGNS

MUSTANG 72"RD
(3) 1960-1985 $850
White outlined aqua porcelain sign with white outlined orange center circle. Detailed brown and black rearing horse in center circle. White "MUSTANG" arched around top, "OIL" around bottom.

NATIONAL INDEPENDENT OIL COMPANY - YORK, PA: History unknown.

LUBRICANTS SIGNAGE

NICOLENE 16x24
(4) 1926-1930 $600
Mint green porcelain sign with red outlined white "NICOLENE" arched across upper sign face. Red outlined white script "We Sell" above, with arrow in script pointing to Red and white "NATIONAL INDEPENDENT OIL COMPANY" logo with green keystone at upper right. Red outlined white "MOTOR OIL/GUARANTEED 100%/PURE PENNSYLVANIA" on lower sign face with PGCOA seal at lower left.

NORTHLAND GASOLINE - WATERLOO, IA
Primarily a lubricants manufacturer, Northland was involved with a small number of gasoline stations in the 1920s and 1930s. Manufactures specialty lubricants today.

LUBRICANTS SIGNAGE

NORTHLAND MOTOR OIL PURE PENNSYLVANIA W/IOMA & PGCOA 22x30 $500
Black/white/red outlined black porcelain sign with white outlined red script "Northland" diagonal across sign. IOMA logo in upper left corner, PGCOA seal in lower left and red and white Northland shield logo in lower right. White "There IS a difference" across top of sign face, "MOTOR OIL/Pure Pennsylvania" across bottom.

O'NEIL OIL & PAINT COMPANY - MILWAUKEE, WI: Wisconsin brand affiliated with IOMA. Founded in 1893, they distributed gasoline throuhout southeastern Wisconsin until the late 1920s.

LUBRICANTS SIGNAGE

O'NEILS VELVET MOTOR OILS 36"RD
(5) 1925-1930 $450
Red outlined orange porcelain sign with black script "Velvet" logotype diagonal across center. Black "O'NEIL'S/Every drop 1tested" at top, with black "MOTOR/OILS" below.

OMAR REFINING CO. - TULSA, OK
History unknown.

CURB SIGNS

OMAR ETHYL 30"RD
(4) $350
White porcelain sign with black outlined red "Omar" logotype across top. Ethyl
(EGC) logo with yellow rays in center, with black outlined red "GASOLINE" arched around bottom.

H.L. HUNT REFINING CO. - TUSCALOOSA, AL
Gasoline marketing division of the H.L.Hunt family interests. A few stations in Alabama, operated by jobber Ramoco, remain branded Parade today.

PUMP SIGNS

HLH PARADE PRODUCTS 7x15
(3) 1960-1970 $50
White porcelain sign with blue outline ring. Red "PARADE" with speed lines across center with blue "HLH" at top and "PRODUCTS" below

PLATOLENE - CARMI, IL: Founded in Terre Haute,IN before WW2 as R.J. Oil Refining, they operated a refinery at Princeton, IN and stations throughout southern Indiana & Illinois. Stations were originally branded simply "500." The Platolene name was added in the late 1940's. The refining operation was abandoned in the 1960's and operation was shifted to Carmi,IL, where they continue to operate Platolene 500 stations today, some with dual-branding by the majors.

PUMP SIGNS

PLATOLENE 500 12"RD
(4) 1946-1966 $275
White/black outlined red porcelain sign with white outlined black banner across lower sign face. White "PLATOLENE" on black band. Large white "500" with black dropshade blocking and black & white checkered flag above banner.

SECONDARY SIGNS

PLATOLENE 500 24x36
(3) 1970s $100
White tin litho sign with red outlined blue oval in center. White "PLATOLENE" on blue area with red "500" in white area above blue.

PRIDE OIL COMPANY, KNOXVILLE, TN: Founded in the 1930's, Pride is a Knoxville based discount operation that had stations under their own "Pride" brand as well as dual branding "Pride-Bay" stations from the 1960s through the 1980s. They continue today as branded Phillips 66 jobber and operating Pride convenience stores.

IDENTIFICATION SIGNS

PRIDE 60"RD
(5) 1950-1980 $100
Black outlined white porcelain sgin with series of red concentric circles on sign face. White band across center with black "PRIDE" on band.

PRIDE OIL COMPANY OF OREGON: Oregon marketer, history unknown.

PUMP SIGNS

PRIDE OF OREGON REGULAR 10x18
(4) 1950s $125
White/red outlined white porcelain sign with green "PRIDE" over black "OF OREGON" across top. Red "REGULAR" across bottom, with yellow statue on black & white pedestal in center.

PRIDE OF OREGON ETHYL 10x18
(5) 1950s $125
White/red outlined white porcelain sign with green "PRIDE" over black "OF OREGON" across top. Red "ETHYL" across bottom, with yellow statue on black & white pedestal in center.

POWER: Montana marketer, history unknown.

PUMP SIGNS

POWER 12"RD
(5) 1940s $150
Black outlined white porcelain sign with elaborate red and white shield in center. Red/white outlined shape of Montana in center of shield with white "POWER" with black dropshading across center. Red and white banner across bottom of shield.

ATLAS OIL CORP. - NEW ORLEANS, LA

Deep south discounter that rebranded "TCR" (for Texas City Refining) in the 1960s.

IDENTIFICATION SIGNS

REBEL 72"RD
(5) 1960-1970 $450
White porcelain sign with large red/white/blue rebel flag in center. Blue "REBEL" across above flag, "OIL" below.

MAPCO-WEST MEMPHIS, AR

Mapco is an independent refiner that purchased Nashville, TN based Red Ace Petroleum and Memphis, TN based Delta Oil in the 1970s. By 1985 most locations had been rebranded Delta and in the late 1980s the Delta stations were converted to convenience stores and branded Mapco Express or Delta Express.

BUILDING SIGNS

RED ACE 48x96
(5) 1946-1980 $400
Large red porcelain die-cut "x" with white "RED/ACE" in center.

ROYALINE OIL CO. - ROYAL OAK, MI

History unknown. Still in operation.

IDENTIFICATION SIGNS

ROYALINE HI TEST GASOLINE MOTOR OIL 30"RD
(5) 1920s $1800
Blue outlined white porcelain sign with blue center circle. Blue "ROYALINE" arched around top, "GASOLINE" around bottom with "HI-TEST" in white area above. White "PENNSYLVANIA/MOTOR OILS" arched around top of blue area with black "ROC" logo in white center circle.

REFINERS PRIDE - CHICAGO, IL: Chicago area discounter active into the 1980's.

PUMP SIGNS

REFINERS PRIDE ETHYL 8"RD
(5) 1946-1960 $225
Black outlined white porcelain sign with red outlined white center circle. Black outlined red "REFINERS PRIDE" arched around top, "GASOLINE" around bottom with Ethyl (EC) logo in cneter circle.

RITTERS SERVICE STATIONS, AUSTIN, TX:

Company history unknown.

PUMP SIGNS

RITTERS GASOLINE 9x14
(4) 1955-1970 $150
White porcelain sign with green oval extending from "G" in script green "Gasoline" cross bottom of sign. Red "RITTERS" in center of oval.

SPEEDWING CORP. - ALTON, IL

Midwestern independent operating under the Speedwing, Royal 99 and Pana brandnames.

PUMP SIGNS

ROYAL 99 8x10
(4) 1950s $250
Blue/white outlined blue porcelain die-cut three-point shield shaped sign. White band diagonal from upper left to lower right with red "ROYAL" diagonal in band. White flower above band, "99" below.

SAV'N SAMS STATIONS: Company history unknown.

PUMP SIGNS

SAV'N SAMS REGULAR GASOLINE 10"RD
(4) 1950s $350
Red/white outlined red porcelain sign with white upper half. Red, white & black cowboy design with dollar bills in upper white area with white "Sav'n SAMS/REGULAR/GASOLINE" on lower red area.

SAV'N SAMS ETHYL GASOLINE 10"RD
(4) 1950s $350
Red/white outlined red porcelain sign with white upper half. Red, white & black cowboy design with dollar bills in upper white area with white "Sav'n SAMS/ETHYL/GASOLINE" on lower red area.

SAV'N SAMS SUPER ETHYL GASOLINE 10"RD
(4) 1950s $350
Red/white outlined red porcelain sign with white upper half. Red, white & black cowboy design with dollar bills in upper white area with white "Sav'n SAMS/SUPER ETHYL/GASOLINE" on lower red area.

SEARS, ROEBUCK AND COMPANY, CHICAGO, IL: Sears merchandised automotive products under the Allstate and Cross Country brandnames through their catalogs from the turn of the century. Sears opened their first retail store in 1926 and by the early 1930's free-standing Sears stores in major cities had automotive service centers in adjacent buildings. Sears sold gasoline under the Allstate brand at these centers from the mid 1930's until the late 1980s when environmental regulations eliminated the practice. Competitors Montgomery Ward and J.C.Penney also offered automotive service and sold gasoline (Wards under the "Riverside" brand) from the late 1950's until about 1990.

PUMP SIGNS

ALLSTATE REGULAR 10"RD
(5) 1946-1960 $150
Blue porcelain sign with white outlined orange band across center, superimposed over light blue map of the United States. White "ALLSTATE" oband with white "REGULAR" in blue area below map.

THE SECURITY OIL COMPANY, WICHITA, KS:

Marketed under the Security brand in the early 1930's, they introduced the Wings brand in 1934 and operated staitons in Kansas under that brand well into the 1970's.

IDENTIFICATION SIGNS

WINGS 30x48
(5) 1946-1960 $225
Black outlined white chevron shaped tin litho sign with red logotype "wings" across upper sign face over smaller red "MORE/FOR YOUR/MONEY." Black geese and cloud scene below.

PUMP SIGNS

SECURITY OIL 15"RD
(5) 1930-1934 $400
White/green outlined white porcelain sign with green outlined white center circle. Red "SECURITY" arched around top, "GASOLINE" around bottom with red anchor on green waves in center.

CURB SIGNS

WINGS MOTOR OIL 30"RD
(5) 1940s $250
Red/white outlined red sign with white "wings" logotype across center above white "MOTOR/OIL."

Socony porcelain flange sign. *Pete Clark.*

Socony Motor Oil porcelain sign. *Bruce Bednar.*

Socony Gasoline porcelain sign -rare.

White Star Refining Co.- rare 30" porcelain sign. *C. Hook.*

Socony Crank Case Service 32"x36" porcelain sign. *Bruce Bednar.*

White Star Hudson Service porcelain signs. *Neil Stockwell.*

White Star Staroleum porcelain dealer sign.

Miscellaneous signs. *Bruce Bednar.*

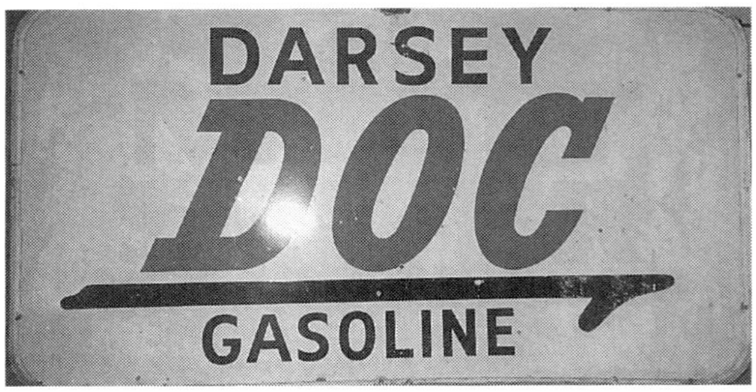

Darsey Doc tin sign. *James Kelley.*

Veltex Gasoline porcelain pump sign.

WNAX - scarce embossed tin fence sign.

Pair of Colonial pump plates. *Bruce Bednar.*

SERVICE DISTRIBUTING COMPANY - ALBEMARLE, NC

North Carolina's largest independent, marketing through over 100 stations across the state. Founded with a single station in Albemarle in 1933, they introduced the Service-Plus brand at the end of WW2. Currently operates under both Service Plus and Servco.

BUILDING SIGNS

SERVICE DISTRIBUTING CO.,INC. 10"
(3) 1946-PRESENT $100
Art-deco style cast aluminum lettering as shown used on building facade.

THE COLUMBUS OIL CO. - COLUMBUS, OH

History unknown.

IDENTIFICATION SIGNS

SILVER FLASH 16x24
(5) 1930-1935 $250
Die-cut dark green porcelain colonial shield shaped sign with light green diagonal band across center. Red logotype "Silver/Flash" on diagonal band.

SKYLARK: Comapny and history unknown.

SKYLARK GASOLINE 12x30
(5) 1930s $650
Porcelain sign with script "Skylark" diagonal across center. White "GASOLINE" below, with clouds across top of sign and small airplane at right in center.

SMITHOLENE: Company and history unknown.

PUMP SIGNS

SMITHOLENE AVIATION BRAND GASOLINE 10"RD
(3) 1946-1950 $2000
Dark blue outlined white porcelain sign with red outlined grey center circle. Dark blue "SMITH-O-LENE" arched around top, "GASOLINE" around bottom. Detailed red and blue airplane in center with blue outlined red "AVIATION" above plane, "BRAND" below.

SOUTHLAND OIL CO. - YAZOO CITY, MS:

Founded in 1941 as Paluxy Asphalt Company, the company became The Southland Company (later Southland Oil Company) in 1950 and introduced the Southland gasoline brand to Mississippi motorists that year. They continue in asphalt refining today and still operate a small number of Southland gas stations.

IDENTIFICATION SIGNS

SOUTHLAND
(5) 1950s NO LISTING
Southland used a porcelain sign with a die-cut cat in the 1950's but exact details are not available.

PUMP SIGNS

SOUTHLAND REGULAR 12x12
(3) 1950-1970 $125
White porcelain sign with two green concentric circles forming outer ring and two green concentric circles forming inner ring. Red "SOUTHLAND" superimposed across center with small italic "Regular" below.

SOUTHLAND ETHYL 12x12
(3) 1950-1970 $125
White porcelain sign with two green concentric circles forming outer ring and two green concentric circles forming inner ring. Red "SOUTHLAND" superimposed across center with small Ethyl (EGC) logo with yellow rays below "SOUTHLAND." Sign does not pre-date 1950 even though earlier Ethyl logo is shown.

A MISSISSIPPI PRODUCT 6x12
(4) 1950-1970 $75
White/green outlined white porcelain sgin with red "A/ MISSISSIPPI/PRODUCT" across sign.

NOTE: The following signs were used in states where grade identification, not just a trademark brandname, was required to be displayed on the pumps.

REGULAR 2x9
(4) 1950-1970 $50
White/green outlined white porcelain sgin with red "REGULAR" across sign.

PREMIUM 2x9
(4) 1950-1970 $50
White/green outlined white porcelain sgin with red "PREMIUM" across sign.

UNITED DIVIDEND OIL CO-CANON CITY, CO

Founded in Canon City, Colorado in 1930, they operated stations under the Speedway brand until the 1970s.

SECONDARY SIGNS

MORE PEP MORE POWER SPEEDWAY THE BETTER GAS
STAMPED TIN $175
Red outlined white embossed tin litho "nose" sign with red "MORE PEP MORE POWER" across the top, "SPEEDWAY" across center and "THE BETTER GAS" across the bottom. Charicature with nose at right side.

NEON SIGNS

SPEEDWAY THE BETTER GAS NEON 60x96
(5) 1935-1960 $1100
Porcelain irregular shape can-type neon sign with white "SPEEDWAY" arched across top. Small "The Better" in the center with large white "GAS" below,

SUNLAND REFINING CO

History unknown.

LUBRICANTS SIGNAGE

SUNLAND MOTOR OIL 11x16
(4) 1950s $225
Die-cut oil can shaped porcelain sign. White sigwith red rim around can "top" and black line/red band around bottom. Black "SUNLAND" arched across top above script "Friction-eezed." White "MOTOR OIL" on lower red band.

F.B.RAE OIL CO. - ROCHESTER, NY: Company history unknown.

IDENTIFICATION SIGNS

SUN RAE No Listing
Black outlined red porcelain sign with intricate red/yellow/blue sunrise and rolling hills scene in center. Black outlined white "GASOLINE" arched around top, "MOTOR OIL" around bottom with black outlined yellow "SUN-RAE" across blue area of center scene. Smaller yellow "F.B.RAE OIL CO.INC./ROCHESTER, N.Y." across red area at bottom of scene.

SWIFTY OIL COMPANY, SEYMOUR, IN: Founded after WW2 by other members of the Myers family who operate Kocolene, Swifty operates stations in Ohio, Indiana and Kentucky today.

PUMP SIGNS

SWIFTY HI-TEST REGULAR 12x14
(4) 1960S-1989 $75

Yellow/red outlined yellow porcelain sign with red "SWIFTY" across top over black "HI-TEST" over red "REGULAR"
SWIFTY GENUINE ETHYL 12x14
(3) 1960s-1989 $75
Red/yellow outlined red porcealin sign with yellow "SWIFTY" across top and "ETHYL" across bottom. Black "GENUINE" across center beside small yellow circle with Ethyl (EC) logo in center

TANKAR STATIONS - NORFOLK, VA

Founded in Sheffield, Alabama in 1933, Tankar operated a chain of discount service stations in Virginia, West Virginia and Alabama, with station buildings built out of actual railroad tank cars. Company shifted headquarters to Norfolk in 1939. Tankar began operating automatic car washes in 1966 and gradually replaced the gas stations with car washes. Last station closed in Portsmouth, Virginia in 1979, although they continue to operates car washes today.

BUILDING SIGNS

TANKAR STATIONS 12"
(5) 1941-1979 $200
White porcelain letters, "TANKAR STATIONS," on black porcelain rails.

UNION TEXAS PETROLEUM - HOUSTON, TX

Texgas was a unit of Allied Chemical, operating stations in the south central states.

PUMP SIGNS

TEXAS/GAS CORPORATION 7x12
(4) 1955-1960 $125
Black outlined white flat diamond shaped porcelain sign with red script "Texas" above black "GAS CORPORATION" across center of diamond.
TEXGAS GASOLINE 7x12
(4) 1960-1970 $125
Black outlined white flat diamond shaped porcelain sign with red script "Texgas" above black "GASOLINES" across center of diamond.
PLATINUM REGULAR 4x12
(4) 1955-1970 $100
White outlined blue porcelain sign with white band across bottom. White script "Premium" on upper blue area with red "REGULAR" on lower white band.
TEXGAS PLATINUM PLUS SUPER PREMIUM 12x12
(3) 1955-1970 $75
Blue outlined white porcelain sign with blue banner across top above blue outlined white diamond logo. White script "Platinum" on banner with large red script "Plus" across diamond. Red "SUPER/PREMIUM" below diamond logo.

THONI OIL/MAGIC BENZOL GAS STAS INC. - MIAMI, FL

Founded in the 1930s in Nashville, TN as Magic Benzol Gas Stations, They were operating about 50 stations under the Thoni's name throughout the south by the 1950s. Continued to operate a widely scattered chain of stations until founder Richard Thoni died in 1990 and the stations were forced to close to settle matters of the estate.

OIL RACK SIGNS

THONIS MOTOR OIL 14x24
(5) 1950-1960 $200
Green porcelain sign with white oval and stars Thoni logo with green lettering at upper left. White script "High Grade" to right of logo, all over larger white "MOTOR/OIL 25c QT." across bottom.

TIDEWATER OIL SERVICE - MATTHEWS, VA

Rural Virginia jobber that branded several dozen stations under the TIWOSER brand from 1932 until 1952. Branded Pure in 1952, they market Unocal products today.

PUMP SIGNS

TIDEWATER REGULAR 10x12
(4) 1935-1952 $100
White porcelain sign with red lower half. Red "TIDEWATER/REGULAR" across upper sign area with white "TIDEWATER/OIL SERVICE/OWNED & OPERATED BY: /GEORGE A PHILPOTTS" on lower red area.
TIDEWATER PREMIUM 10x12
(4) 1935-1952 $100
White porcelain sign with blue lower half. Blue "TIDEWATER/PREMIUM" across upper sign area with white "TIDEWATER/OIL SERVICE/OWNED & OPERATED BY: /GEORGE A PHILPOTTS" on lower blue area.

TIME OIL COMPANY, LOS ANGELES, CA:

History unknown. Operates hundreds of stations along the Pacific Coast.

IDENTIFICATION SIGNS

TIME GASOLINE 42x84
(3) 1950-1970 $650
Red outlined white porcelain diecut "upside down tombstone" sign with blue square on upper sign face. White clock (with blue outline where it extends into white area) on upper blue area with red center circle, red numbers and blue hands. White "TIME" arched across top with red dropshading. Red and white pendulum on lower white area with blue "GASOLINE" across just below clock dial face.

PUMP SIGNS

TIME PREMIUM GASOLINE 9.25x14 $400
Red outlined white porcelain diecut "upside down tombstone" sign with blue square on upper sign face. Red outlined white clock, on upper blue area with red center circle, red number markers and blue hands. White "TIME" arched across top with red "PREMIUM" below clock over blue "GASOLINE."
TIME SUPER GASOLINE $400
Red outlined white porcelain diecut "upside down tombstone" sign with blue square on upper sign face. Red outlined white clock, on upper blue area with red center circle, red number markers and blue hands. White "TIME" arched across top with red "SUPER" below clock over blue "GASOLINE."
TIME PREMIUM GASOLINE CONDENSED $400
Red outlined white porcelain diecut "upside down tombstone" sign with blue square on upper sign face. Red outlined white clock, on upper blue area with red center circle, red number markers and blue hands. White "TIME" arched across top with condensed red "PREMIUM" below clock over blue "GASOLINE."
TIME SUPER GASOLINE CONDENSED $400
Red outlined white porcelain diecut "upside down tombstone" sign with blue square on upper sign face. Red outlined white clock, on upper blue area with red center circle, red number markers and blue hands. White "TIME" arched across top with condensed red "SUPER" below clock over blue "GASOLINE."

TAYLOR OIL CO. - WINSTON SALEM, NC

Roby Taylor, Richfield jobber in Winston-Salem NC founded the Travelers discount chain in 1946, eventually operating nearly 50 stations throughout central and eastern North Carolina. In 1960 their best locations were rebranded Etna & were switched from dealers to company direct operation, as they continue to be today. In 1968 Taylor sold the balance of the Travelers operation to Mid-State Oil Company in Lexington, NC, a Sunoco "jobber" operation that was actually a Sun Oil subsidary. Most Travelers stations eventually rebranded to Sunoco. In 1990 Sunoco decided to withdraw from marketing in North Carolina and their Mid-State subsidary was sold off as independent. Mid-State became a branded BP jobber, but numerous dealer stations supplied by Mid-State were branded Travelers in 1992.

IDENTIFICATION SIGNS

TRAVELERS GASOLINE 60x96
(5) 1946-1968 $750
Steel framed oval porcelain sign with yellow upper half and blue lower half. Blue "TRAVELERS" arched around top, with blue airplane flying above blue car in upper yellow area. Yellow "GASOLINE" arched below in blue area.

BUILDING SIGNS

TRAVELERS LETTERS 16"
(5) 1946-1968 $100
Blue porcelain letters "TRAVELERS" on white porcelain rails.

PUMP SIGNS

TRAVELERS GASOLINE 9x12
(5) 1946-1968 $225
Oval porcelain sign with yellow upper half and blue lower half. Blue "TRAVELERS" arched around top, with blue airplane flying above blue car in upper yellow area. Yellow "GASOLINE" across below in blue area.

TRAVELERS ETHYL 9x12
(5) 1946-1968 $225
Oval porcelain sign with yellow upper half and blue lower half. Blue "TRAVELERS" arched around top, with blue airplane flying above blue car in upper yellow area. Yellow "ETHYL" across below in blue area.

TRAVELERS KEROSENE 9x12
(5) 1946-1968 $225
Oval porcelain sign with yellow upper half and blue lower half. Blue "TRAVELERS" arched around top, with blue airplane flying above blue car in upper yellow area. Yellow "KEROSENE" across below in blue area.

TRILLINGS: Company and history unknown.

PUMP SIGN

TRILLINGS HIGH TEST 16"RD
(5) 1930s $150
White outlined blue porcelain sign with white half-circle on upper sign face. White "TRILLINGS" arched around top, with white "HIGH TEST" across center below half-circle.

TROJAN: Company and history unknown.

TROJAN GASOLINE 14x28
(5) 1920s $300
White porcelain sign with black outlined "T" in center. Black outlined red "TROJAN" arched aorund top, "GASOLINE" around bottom.

U FILL'EM STATIONS: Company history unknown.

PUMP SIGNS

U FILL'EM 10"RD
(5) 1960s $225
Blue outlined white porcelain die-cut chevron shaped sign, point down, with large red script "U" in center. Red "FILL'EM" across "U."

UNION GAS SYSTEM: Company history unknown.

IDENTIFICATION SIGNS

UNION GAS SYSTEM 30"RD
(5) 1920s $350
Round porcelain sign with red outer band around yellow center circle. Yellow "UNION" arched around top, "SYSTEM" around bottom with red "GAS" in elongated lettering in center.

F.O.FLETCHER OIL CO. - BOISE, ID: Operated stations in the Pacific Northwest and Rocky Mountain area. Still in operation, with several stations remaining in Idaho.

IDENTIFICATION SIGNS

VELTEX 30"RD
(4) 1930s $500
Orange outlined blue porcelain sign with large orange center circle. White "FLETCHER OIL" arched around top of blue area, "COMPANY" around bottom with large blue "V" in centerr. White outlined orange band across "V" with blue "VELTEX" on band.

VELTEX 30"RD
(4) 1940s $500
Orange porcelain sign with blue/white outline around orange center area. Large white outlined blue "V" in center with white/blue outlined orange band across center. Blue "VELTEX" on center band.

PUMP SIGNS

VELTEX GASOLINE 12.5x15
(4) 1930s $450
White porcelain sign with orange circle on upper sign face. Blue "V" on circle with white outlined orange band across center. White outlined blue "VELTEX" on band. Orange extension below circle with white outlined blue "GASOLINE" in bottom band of extension.

VELTEX SUPER ETHYL 12.5x15
(4) 1930s $450
White porcelain sign with orange circle on upper sign face. Blue "V" on circle with white outlined orange band across center. White outlined blue "VELTEX" on band. Orange extension below circle with white outlined blue "SUPER ETHYL" above blue "GASOLINE" in bottom band of extension.

CURB SIGNS

VELTEX 16"RD
(4) 1930s $550
Orange outlined blue porcelain sign with large orange center circle. White "FLETCHER OIL" arched around top of blue area, "COMPANY" around bottom with large blue "V" in centerr. White outlined orange band across "V" with blue "VELTEX" on band.

VENTURA REFINING CO - VENTURA, CA: Company history unknown.

PUMP SIGNS

VENTURA GASOLINE DIE CUT BELL 10x12
(5) 1930-1935 NO LISTING
Red porcelain die-cut bell shaped sign with yellow details top and bottom. White "VENTURA/GASOLINE" between yellow areas, with white "VENTURA REFINING CO" across bottom of sign.

FLANGE SIGNS

INSIST ON VENTURA OIL 10x16
(5) 1930-1935 $325
Blue outlined white porcelain sign with large red "V" in center. blue "Insist on" across top, "PARAFFIN BASE OIL" across

bottom with blue band and circle across "V." White "VENTURA" on band with white "MOTOR/OIL" on circle.

ANDERSON PETROLEUM - ANDERSON, SC: Dating from the early 1930's, Anderson Petroleum marketed under the Viking brand until about 1940. They became a Dixie Distributors affiliate before WW2 and remained branded Dixie well into the 1960's.

IDENTIFICAITON SIGNS

VIKING 30x72
(5) 1930-1941 $300
Black outlined white porcelain sign with large red "V" with black dropshading in center. Red "VIKING" with black dropshading across center.

THE HOUSE OF GURNEY-YANKTON, SD

Founded about 1930 by radio promoter D.B.Gurney as a cooperative of gasoline station owners in the northern plains states. Using his radio station as a promotional source, Gurney expanded the "fair-price" WNAX chain until WW2 forced restrictions on gasoline consumption and Gurney turned his interests to securing other "fair-price" items for plains farmers.

ADVANCE SIGNS

WNAX FAIR PRICE GAS & OIL 16x20
(5) 1930s $225
Black outlined white embossed tin litho sign with black arrow across top. Center of arrow interrupted by a blacnk space to paint in number and black "MI.TO." Black "HELLO FOLKS" above, with large "WNAX" above "FAIR PRICE/GAS AND OIL" below.

WAGGONER REFINING CO., INC.: Company history unknown.

IDENTIFICATION SIGNS

WAGGONER REFINING CO 3-D's PETROLEUM PRODUCTS
42" $2500
Red outlined white porcelain sign with large red bull in center. Three red upside down "D"s below bull, symbolic of a brand. Black "WAGGONER REFINING CO.,INC." arched around top, "PETROLEUM PRODUCTS" around bottom.

WEBSTER BASKET COMPANY - WEBSTER, NY: Company history unknown.

SECONDARY SIGNS

WEBACO CREDIT CARDS 14x24
(5) 1940s $225
Red outlined yellow tin litho sign with blue outlined yellow oval on upper sign face. Red "WEBACO" in oval with blue "CREDIT HONORED CARDS" across the bottom. "CREDIT" and "CARDS" at 45 degree angles to "HONORED"

WELCO - CHICAGO, IL: Company history unknown.

PUMP SIGNS

WELCO DELUXE REGULAR 10"RD
(5) 1930s $200
Orage porcelain sign with white band across center and small white outlined blue half circle below band. Blue outlined red "WELCO" on upper wign face with white "DELUXE REGULAR" on blue half circle and white "GASOLINE" arched in orange area below.

WESTLAND OIL CO. - MINOT, ND

Marketed the famous Buffalo gasoline in the northern plains states.

IDENTIFICATION SIGNS

WESTLAND SERVICE 72"RD
(4) 1950's-1970's $3000
Red outlined white porcelain sign with red outlined green center circle. Detailed red and black charging buffalon in center with black "WESTLAND" arched around top, "SERVICE" around bottom.

NEON SIGNS

WESTLAND SERVICE 72"RD
(4) 1950's-1970's $3000
Red outlined white porcelain sign with red outlined green center circle. Detailed red and black charging buffalon in center with black "WESTLAND" arched around top, "SERVICE" around bottom. Constructed as two piece porcelain faces over steel frame. Neon tubing reading "GAS" installed over center of sign.

SECONDARY SIGNS

THE SIGN OF THE BUFFALO 14x24
(4) 1950s $500
Red outlined white porcelain sign with red "THE SIGN/OF THE/BUFFALO" across center. Used as secondary identification on poles supporting primary station signs.

WHITLOCK OIL COMPAPNY: Company history unknown.

PUMP SIGNS

REGULAR WHITLOCK LEADED GASOLINE 10"RD
(4) 1935-1940 $225
White porcelain sign with black outlined red outer rring and rings around upper and lower half circles, formain white band across center. Black outlined red "WHITLOCK" in center band, with black outlined yellow "REGULAR" in upper half circle and black "LEADED" over black outlined yellow "GASOLINE" in lower half circle.

WHITLOCK SUPER GASOLINE ETHYL
(EGC) 10"RD
(5) 1935-1940 $225
White/black/red outlined white porcelain sign with black outlined red "WHITLOCK" across upper sign face. Black script "Super Gasoline" across center with Ethyl
(EGC) logo on lower sign face.

WILCOX OIL COMPANY - TULSA, OK: Company history unknown.

CURB SIGNS

WILCOX ETHYL 30"
(5) 1928-1935 $375
Blue outlined white porcelain sign with red "WILCOX" arched around top, "GASOLINE" around bottom. Ethyl
(EGC) logo with yellow rays in center.

WILLIAMS OIL COMPANY: Company history unknown.

IDENTIFICATION SIGNS

WILLIAMS 42"RD
(5) 1930s $300
Round porcelain sign with white center circle. Large "W" in center circle with band across horizontal center. White "WILLIAMS" arched around top, "GASOLINE" around bottom and white "WILLIAMS OILS" across center band.

OFFICIAL SIGNS: The following porcelain signs are listed as official signs as they are grade certification markings from the ststates listed or are the "CONTAINS LEAD" signs required as a health warning on pumps dispensing leaded gasoline.

NC REGULAR GRADE 8"RD
(2) 1934-1946 $75
White porcelain sign with red outline and two red concentric circles, with red solid center circle. Red outlined white band across center. Red "NORTH CAROLINA" arched around top, "GRADE" around bottom, with "REGULAR" in center band.
NC PREMIUM GRADE 8"RD
(2) 1134-1946 $100
White porcelain sign with blue outline and two blue concentric circles, with blue solid center circle. Blue outlined white band across center. Blue "NORTH CAROLINA" arched around top, "GRADE" around bottom, with "PREMIUM" in center band.
NC MOTOR GRADE 8"RD
(2) 1934-1946 $100
Yellow porcelain sign with black outline and two black concentric circles, with black solid center circle. Black outlined white band across center. Black "NORTH CAROLINA" arched around top, "GRADE" around bottom, with "MOTOR" in center band.
NC REGULAR GRADE CURVED 8"RD
(2) 1934-1946 $100
Curved white porcelain sign with red outline and two red concentric circles, with red solid center circle. Red outlined white band across center. Red "NORTH CAROLINA" arched around top, "GRADE" around bottom, with "REGULAR" in center band.
NC PREMIUM GRADE CURVED 8"RD
(2) 1934-1946 $125
Curved white porcelain sign with blue outline and two blue concentric circles, with blue solid center circle. Blue outlined white band across center. Blue "NORTH CAROLINA" arched around top, "GRADE" around bottom, with "PREMIUM" in center band.
NC MOTOR GRADE CURVED 8"RD
(2) 1934-1946 $125
Curved yellow porcelain sign with black outline and two black concentric circles, with black solid center circle. Black outlined white band across center. Black "NORTH CAROLINA" arched around top, "GRADE" around bottom, with "MOTOR" in center band.
LOUISIANA APPROVED 10x12
(4) 1930s $125
Black outlined white porcelain sign with black "LOUISIANA/APPROVED/GASOLINE" on sign face.
LOUISIANA ARTIFICIALLY COLORED 10x12
(4) 1930s $150
Black outlined white porcelain sign with black "ARTIFICIALLY COLORED/LOUISIANA/APPROVED/PREMIUM/GASOLINE" on sign face.
CONTAINS LEAD AVOID SPILLING 7x8.5
(2) 1926-1946 $30
Black outlined white porcelain sign with black "CONTAINS/LEAD" across top. Black "(TETRAETHYL) AND/IS TO BE USED AS/MOTOR FUEL ONLY/NOT FOR CLEANING/OR ANY OTHER USE/AVOID SPILLING" across sign.
CONTAINS LEAD TETRAETHYL 6x7
(1) 1932-PRESENT $12
Black outlined white porcelain sign with black "CONTAINS/LEAD" across center. Black "FOR USE AS A/MOTOR FUEL ONLY" across top "(TETRAETHYL)" across bottom
CONTAINS LEAD ANTIKNOCK COMPOUNDS 6x7
(1) 1962-PRESENT $10
Black outlined white porcelain sign with black "CONTAINS/LEAD" across center. Black "FOR USE AS A/MOTOR FUEL ONLY" across top "(ANTIKNOCK COMPOUNDS)" across bottom

Independent Lubrication Companies

INDEPENDENT LUBRICANTS MANUFACTURERS:
The following signs are those used by independnet lubricants manufacturers. As far as is known, none of the following companies made any significant inroads into the operation of service stations or any gasoline marketing. They simply refined, packaged, and marketed lubricants, motor oils, and greases. As such, the sign classifications listed differ somewhat from those associated with other petroleum marketers. Specifically, there are no "identification signs" as such, and most signs come under the headings "Secondary Signs" or "Curb Signs."

ALEMITE CORPORATION - CHICAGO, IL: Pioneers in lubrication equipment and processes, Alemite began marketing motor oils and greases in the 1910s. They became a part of Stewart-Warner Corp about 1970 and discontinued lubricants marketing shortly thereafter.
CURB SIGNS
ALEMITE MOTOR OIL 18x30
(5) 1930s $250
Red/blue outlined yellow porcelain sign with blue outlined red "ALEMITE" over blue "100% PENNSYLVANIA" above blue outlined red "MOTOR OIL." Small PGCOA seal below, with blue "ALEMITE" to left and "PROCESS" to right.

BRADFORD OIL REFINING CO - BRADFORD, PA: Company history unknown.
CURB SIGNS
ALL-PENN MOTOR OIL W/PGCOA SEAL 22x30
(5) 1930's $400
Orange porcelain tombstone shaped sign with intricated blue and white border. White outlined blue "ALL PEN" arched across top, "MOTOR OIL" straight across bottom with PGCOA seal in center. Small blue "PENNSYLVANIA'S BEST/MEETS EVERY TEST" at top, white "BRADFORD OIL REFINING CO/BRADFORD,PA" in blue area of bottom borber.
SECONDARY SIGNS
BRADFORD'S ALL-PEN 24x30
(5) 1930s $350
Orange outlined black porcelain sign with large yellow and

red wing design in center. Black "ALLPEN" on wing with yellow "BRADFORDS" across top, "MOTOR OIL" below. Other small red and yellow lettering.

LEWIS OIL COMPANY - ST LOUIS, MO: History unknown.

LUBRICANTS SIGNS

AUNTIE CHATTER MOTOR OIL 16x16

(5) 1920-1926 $650

White embossed tin litho sign with black circle covering most of sign face. White/black outlined yellow center circle with black & white drawing of elderly woman in center. White outlined black "AUNTIE" across top, "Chatter" below. Black "LEWIS" to left of woman, "ORIGINAL" to right. White "MOTOR OIL" arched around upper black area, "FOR FORDS" below.

W.C.ROBINSON & SON - BALTIMORE, MD: Founded as a lubricants manufacturer in 1832, Robinson introduced the Autoline Oils brand in the early years of this century.

CURB SIGNS

AUTOLINE OIL 24x32

(5) 1930s $350

Red porcelain sign with black outlined white "AUTOLINE/ OIL" across center. Black "for your motor's sake" arched across bottom.

BELL OIL COMPANY, ST LOUIS, MO: Company history unknown.

SECONDARY SIGNS

BELL OIL CO LUBRICANTS 6x24 No Listing

Red porcelain sgin wtih white bell at left end. Large white "BELL OIL Co's LUBRICANTS" across strip above small white "ST.LOUIS,MO"

CRESCENT OIL COMPANY - ST LOUIS, MO: Company history unknown.

CURB SIGNS

CRESCENT MOTOR OIL 21x28

(5) 1930s $2000

Blue outlined white porcelain tombstone sign with yellow/blue/white crescent moon and stars logo. Yellow "CRESCENT" arched across top of logo with other smaller writing on logo. Red and black PGCOA seal at top.

D-A LUBRICANTS COMPANY - INDIANAPOLIS, IN: Manufacturer of heavy duty lubricants for railroad and industrial applications, DA continues to distribute lubricants nationwide today,

SECONDARY SIGNS

D-A AUTHORIZED DEALER 10x24

(5) 1950s $300

Black outlined red porcelain oval with white gears meshing design at top. Black outlined yellow "D-A/LUBRICANTS" across center with white "AUTHORIZED DEALER" and other lettering below.

UNITED OIL CO - BALTIMORE, MD: Company history unknown.

CURB SIGNS

DURALENE D+ 19.5x27.5

(5) 1950s $150

Yellow tin litho sign with black band across top and bottom. Yellow "DURALENE" on upper band, smaller writing on lower band. Large "V" in center, with other smaller writing on sign.

FRANKLIN MOTOR OIL: Company history unknown.

CURB SIGNS

FRANKLIN PENNSYLVANIA MOTOR OIL 22x28

(5) 1920s NO LISTING

Red outlined yellow tombstone shaped sign with logo at top. Red "FRANKLIN" across below loqo, with black "PENNSYLVANIA" above red "MOTOR OIL" below. Small PGCOA seals to either side at bottom.

THE GLOBE REFINING COMPANY, CLEVELAND, OH: Company history unknown.

LUBRICANTS SIGNAGE

GLOBE-SEAL MOTOR OIL 9.5x14

(5) 1920s NO LISTING

Black outlined yellow tin litho sign with black outlined white geartooth circle at top. Yellow and black scene in center of circle with black "GLOBE SEAL" arched around top, "MOTOR OIL" below. Black "SEALS THE PISTONS" across below circle.

STAR OIL COMPANY: Company history unknown.

LUBRICANTS SIGNAGE

GOLD STAR PURE PENNSYLVANIA 30"RD

(5) 1930s NO LISTING

White porcelain sign with blue center circle. Blue "GOLD STAR" arched around top, "PURE/PENNSYLVANIA" around bottom. PGCOA seal to either side at midpoint. Yellow star in center with white oil derrick on star. Small "STAR OIL CO" arched at base of star.

A.W.HARRIS OIL COMPANY - PROVIDENCE, RI: Company history unknown.

HARRIS OILS AMERICA'S LEADING LUBRICANTS 16x28

(4) 1930s $1800

Black/yellow outlined black porcelain sign with brown keg across center. Black outlined white "OILS" on keg, with yellow "HARRIS" acorss top, "AMERICA'S/LEADING/LUBRICANTS" across bottom.

HARRIS OIL FLANGE 18x19

(5) 1920s $1500

Black porcelain flange sign with brown keg across center. Black outlined white "OILS" on keg, with yellow "HARRIS" acorss top, "AMERICA'S/LEADING/LUBRICANTS" across bottom.

HARRIS OILS PURE PENNSYLVANIA 18x24

(5) 1930s $2200

Steel half-frame around black outlined yellow porcelain die cut sign. Black outlined red "HARRIS OILS" across top above black "PURE PENNSYLVANIA" across center. Yellow, brown and black "HARRIS OIL" oil barrel logo on die cut area extending from bottom.

HY VIS OIL COMPANY - WARREN, PA: Founded as Conewango Refining Company, little is known of the early history.

LUBRICANTS SIGNAGE

SUPER REFINED HYVIS MOTOR OIL 16x42

(3) 1930s $200

Black porcelain sign with red line across bottom. Yellow script "Super refined" above yellow and black PGCOA seal at leftm with white "HYVIS/MOTOR OIL" and red "O" charachature at right.

HY VIS MOTOR OIL SUPER REFINED 16x24

(3) 1930s $175
White outlined black porcelain sign with red line across center. White "HYVIS/MOTOR OIL" and red "O" charachature above red line. Yellow script "Super refined" across bottom below line with yellow and black PGCOA seal at right.
HY VIS MOTOR OIL SUPER REFINED 16x24
(3) 1930s $175
White outlined black tin litho sign with red line across center. White "HYVIS/MOTOR OIL" and red "O" charachature above red line. Yellow script "Super refined" across bottom below line with yellow and black PGCOA seal at right.
HYVIS MOTOR OIL PURE PENNSYLVANIA 16x24
(3) 1940s $175
White outlined black porcelain sign with red line across center. White "HYVIS/MOTOR OIL" and red "O" charachature above red line. Yellow script "Super refined Pure Pennsylvania" across bottom below line with yellow and black PGCOA seal at right.
HY VIS MOTOR OIL SUPER REFINED PENN 16x24
(3) 1950s $175
White outlined black porcelain sign with red line across center. White "HYVIS/MOTOR OIL" and red "O" charachature above red line. Yellow "SUPER-REFINED PENNSYLVANIA" across bottom below line with yellow and black PGCOA seal at right.
HY VIS MOTOR OIL 12x72
(4) 1930s $100
Black/white outlined black tin litho sign with white "HI-VIS" vertical down sign. Red "O" charachature in center covered by "V" and "I." White "MOTOR/OIL" across bottom above large PGCOA seal/

CHARLES F.KELLOM & Co - PHILADELPHIA, PA: Company history unknown.

LUBRICANTS SIGNAGE

USE INVADER MOTOR OIL 30"SQ
(5) 1930s $450
Yellow porcelain sign with blue/white knight on horse on red area at bottom. PGCOA seal to left, with blue "USE/INVADER/MOTOR OIL" across top of sign.
INVADER LUBESTER SIGN 8"RD
(5) 1930S $200
Yellow porcelain sign with black "INVADER" across center and "MOTOR OILS" arched around bottom. PGCOA seal at top of sign.
INVADER PERFECTED LUBRICATION 12x24
(5) 1940-1950 $150
Yellow outlined black tin litho embossed sign with yellow "INVADER/MOTOR OIL" across black area. Yellow/red/black knight on horse at right end. Black "PERFECTED LUBRICATION" across bottom.
INVADER MOTOR OIL 28x30
(5) 1940s $150
Yellow porcelain sign with red outlined black half circle at top. Red/yellow/black knight on horse on black area with black "INVADER/MOTOR OIL" across bottom of sign.

MacMILLAN PETROLEUM - NEW YORK, NY

Lubricants manufacturer that operated a few gas stations from the 1930s until the 1980s.
MACMILLAN "M" 30"
(3) 1940s-1960s $225
White outlined red die-cut "M" with white "MACMILLAN/RING-FREE/OILS" across center.

DRYER, CLARK AND DRYER - OKLAHOMA, CITY, OK: Founded in 1879, other company history unknown.

LUBRICANTS SIGNAGE

MOTHER PENN LUBESTER SIGN 6x12
(5) 1930s $600
Red outlined white porcelain sign with blue arched areas across top and bottom and red outlined white circle extending from top center with blue woman's face in center. Blue script "Mother Penn" above narrow blue band and red "MOTOR OIL" across center. White "ALL PENNSYLVANIA" on center blue band with white "1879" to either side of center on upper blue band, "DRYER, CLARK & DRYER OIL CO." on lower blue band.
MOTHER PENN MOTOR OIL 13x20
(4) 1930s $600
Red outlined white porcelain sign with blue arched areas across top and bottom and red outlined white circle extending from top center with blue woman's face in center. Blue script "Mother Penn" above narrow blue band and red "MOTOR OIL" across center. White "ALL PENNSYLVANIA" on center blue band with white "1879" to either side of center on upper blue band, "DRYER, CLARK & DRYER OIL CO." on lower blue band.
MOTHER PENN PANEL 24x48
(5) 1940s $450
Blue porcelain sign with red outlined white area across center with blue arched areas across top and bottom. Blue script "Mother Penn" above narrow blue band and red "MOTOR OIL" across center. White "ALL PENNSYLVANIA" on center blue band with white "1879" in center on upper blue band, "DRYER, CLARK & DRYER OIL CO." on lower blue band.

NEON SIGNS

MOTHER PENN 36x60
(5) 1930s $2000
Red outlined white porcelain sign with blue arched areas across top and bottom and red outlined white circle extending from top center with blue woman's face in center. Blue script "Mother Penn" above narrow blue band and red "MOTOR OIL" across center. White "ALL PENNSYLVANIA" on center blue band with white "1879" to either side of center on upper blue band, "DRYER, CLARK & DRYER OIL CO." on lower blue band.

SWAN-FINCH OIL CORPORATION - NEW YORK, NY: A unit of the Standard Oil Trust, Swan-Finch was involved in lubricants manufacturing in the 1930s.
MOTUL MOTOR OIL 30"RD
(5) 1930s $350
Red flange sign with black band across center. White "MOTUL" across center band, with white winged gear logo at top. "SWAN" in left wing, "FINCH" in right, with script "Established 1853" below wings. White "MOTOR OILS" below band, with white "SWAN FINCH OIL CORPORATION/NEW YORK NEWARK, NJ CHICAGO BUFFALO/SAN FRANCISCO" arched around bottom.

NOURSE OIL COMPANY - KANSAS CITY, MO:
Company history unknown.
NOURSE 30"RD
(4) 1930s $1800
Black outer band around white porcelain sign. Orange and black Nourseman in center with shield. Black "Guaranteed/MOTOR OIL" above man, with white "NOURSE" arched around top, "BUSINESS IS GOOD" around bottom. Smal

IOMA logo below man.
NOURSE OILS GREASES BUSINESS IS GOOD 14x48
(5) 1930s $600
Black/white outlined black tinlitho sign with white "NOURSE" vertical on sign above white "Oils/Greases." Small white norseman and white "BUSINESS/IS GOOD" at bottom of sign.

FRONTIER MANUFACTURING COMPANY - NORTH TONAWANDA, NY: Company history unknown.
OAK MOTOR OIL 15x26
(4) 1930s $850
White outlined red porcelain sign with brown/green oak tree in center. Black outlined white "OAK" with green inlines across top above black outlined white "MOTOR OIL." White band across bottom with black "SUMMER AND WINTER ONE GRADE FOR ALL CARS" across band.
OAK MOTOR OIL 15x26
(4) 1930s $850
White outlined red porcelain sign. Black outlined white "OAK" with green inlines across top above black outlined white "MOTOR OIL." White "30c per qt" across bottom.
OAK MOTOR OIL 16"RD
(4) 1930s $850
White outlined red porcelain sign with brown/green oak tree in center. Black outlined white "OAK" with green inlines across top above black outlined white "MOTOR OIL."

WHITE & BAGLEY, WORCESTER, MA: Founded in Worcester, MA in 1888 as a lubricants manufacturer, they introduced one of the first nationally marketed motor oils about 1906 and continued to market them until after a 1988 merger with Baltimore based Dryden Oil Company. Several "Oilzum Man" logos have been used over the years, and they are described below for reference in establishing which sign is being described. All dates are approximate!

NOTE: Despite rumors to the contrary, NO Oilzum gas stations can be confirmed at this time. It appears that they were a lubricants manufacturer only.

OILZUM KID LOGO 1905-1909: Black outlined white child's face with black details. Black outlined orange hat with black "Oilzum" arched down across center over "for MOTOR CARS/and/MOTOR BOATS." Black "THE WHITE & BAGLEY CO/WORCESTER MASS.U.S.A." across top of hat. Small black "TRADE MARK/REG.U.S.PAT.OFF" just above "OILZUM."
EARLY LOGO 1909-1920s: Logo superimposed over a large black outlined orange "O." Black outlined white face with black details. Black outlined orange hat with black "Oilzum" arched down across center over "MOTOR OILS/LUBRICANTS." Black "THE WHITE & BAGLEY CO/WORCESTER MASS.U.S.A." across top of hat. Small black "TRADE MARK/REG.U.S.PAT.OFF" just above "OILZUM."
SECOND LOGO 1920s-1935: Black outlined white face with black details. Black outlined orange hat with black "Oilzum" arched down across center over "MOTOR OILS/LUBRICANTS." Black "THE WHITE & BAGLEY CO/WORCESTER MASS.U.S.A." across top of hat. Small black "TRADE MARK/REG.U.S.PAT.OFF" just above "OILZUM."
THIRD LOGO 1935-1946: Black outlined white face with black details. Black outlined orange hat with black "Oilzum" arched down across center.
FOURTH LOGO: 1946-1951Black outlined white face with black details. Black outlined orange hat with black outlined white "Oilzum" arched down across center.
FIFTH LOGO 1951-1988: Orange rounded corner triangle, point down, with black outlined white "OILZUM" arched above black outlined and detailed drivers face and hat with nothing across hat.

IDENTIFICATION SIGNS
OILZUM AMERICA'S FINEST OIL 30x42
(4) 1910s $1500
Black self framed tin litho sign with black/orange outlined white panel across top where dealer's name would be painted in in black. Black/orange outlined white "Oilzum" across bottom with "early logo" superimposed over "O." Orange and black can at right with white "America's Finest Oil" above left of can.
OILZUM DEALER 22x32
(4) 1920s $1000
Sheet brass self framed sign with logo #2 above dealers imprinted name in black.

CURB SIGNS
OILZUM 24"RD
(3) 1920s $4500
Black outlined orange porcelain sign with black center circle. Large logo #2 in center, with black "If Motors Could Speak we wouldn't need to Advertise" arched around top, "The Cream of Pure Pennsylvania Oil" arched around bottom.
OILZUM 24"RD
(3) 1920s $4500
Black outlined orange porcelain sign with black outline around white center circle. Large logo #2 in center, with black "If Motors Could Speak we wouldn't need to Advertise" arched around top, "The Cream of Pure Pennsylvania Oil" arched around bottom.
OILZUM AMERICAS FINEST OIL 22x28
(4) 1930s $3500
Orange outlined black porcelain sign with logo #2 at top over white "America's/Finest Oil"
OILZUM MOTOR OIL 30x30
(4) 1950s $250
Triangular shaped orange porcelain sign with white "OILZUM" with black dropshade across top above black "MOTOR OIL." Black logo #5 below.

LUBRICANTS SIGNAGE
OILZUM AMERICA'S FINEST OIL 12x36
(4) 1910-1920 $300
Black self framed tin litho sign with black/orange outlined white "Oilzum" across bottom with "early logo" superimposed over "O." Orange and black can at right with white "America's Finest Oil" above left of can.
OILZUM MOTOR OIL 12x36
(4) 1935-1946 $300
Black outlined orange self framed tin litho sign with white outlined black "Oilzum" across center over black "MOTOR OIL," with logo #3 superimposed over "O."
OILZUM CAN 42"
(5) 1950s $450
White tin litho sign in the shape and design of a quart oil can. Large orange triangle in center with black outlined white "OILZUM" arched across top above logo #5. Blue band across bottom with white "MOTOR OIL" across band. Blue script "Choice of Champions" across lower sign area.
OILZUM MOTOR OIL CHOICE OF CHAMPIONS 14x28
(2) 1951-1960 $250
White self framed tin litho sign with orange logo #5 at right. Light blue band across bottom with white "Choice of Cham-

pions" across band. Black "MOTOR/OIL" to right of logo
OILZUM MOTORCYCLE OIL 14x28
(2) 1951-1960 $250
White self framed tin litho sign with orange logo #5 at right. Light blue band across bottom with white "Choice of Champions" across band. Black "MOTORCYCLE/OIL" to right of logo

FLANGE SIGNS

OILZUM MOTOR OIL 13x17
(3) 1951-1960 $350
Orange triangular tin litho flange sign with black flange area. Black outlined white "OILZUM" across top above "MOTOR OIL" Oilzum logo numnber 5 at bottom.

PENCOIL: PENNSYLVANIA PETROLEUM CO - KANSAS CITY, MO: Company history unknown.

PENCOIL OF THE HIGHEST QUALITY 16x24
(5) 1926-1930 $325
White porcelain sign with black "Pencoil" above large red keystone in center of sign. Black and white unusual version of PGCOA seal on keystone. Red "OF THE HIGHEST QUALITY" across bottom of sign.

PENCOIL 8"RD
(5) 1930-1935 $250
Red outlined white porcelain sign with black "Pencoil" across top above red and black PGCOA seal. Black "PENNSYLVANIA PETROLEUM CO. KANSAS CITY,USA" arched around bottom.

PENZENE: Company and history unknown.

PENZENE 20x32
(5) 1920s No Listing
Yellow outlined white porcelain sign with large red liberty bell and yoke on sign. White outlined black "PENZENE" across yoke, with white "AT/THE SIGN/OF/THE BELL" on bell. Small PGCOA seal on bell clapper at bottom.

PENN-O-TEX - MINNEAPOLIS, MN: Company history unknown.

RAJAH MOTOR OIL 20x28
(5) 1930s NO LISTING
Yellow die cut tin litho sign with blue lower section. Blue detailed mans face and urban with yellow script "Double Mileage" above yellow "RAJAH/MOTOR/OIL." Yellow "A PENN-O-TEX PRODUCT" arched aorund bottom.

REAL PENN: Company and history unknown.

REAL PENN MOTOR OILS 24"RD
(5) 1930s $450
Red outlined green porcelain sign with black outlined red "REAL PENN" acrhed across top, "MOTOR OIL" around bottom. Black/white outlined red diamond in center with black and white detailed quaker head in center. Red and black PGCOA seal at left, Bradford seal at right.

REAL PENN 12x60
(5) 1930s $300
Black outlined green self framed tin litho sign with white outlined red diamond in center with white William Penn head in center. White outlined red "REAL PENN" vertical down sigh with logo at top and PGCOA seal at bottom.

ST LOUIS OIL COMPANY, ST LOUIS, MO: Company history unknown.

CURB SIGNS

SENTINEL MOTOR OIL 28x28
(4) 1930s No Listing
Yellow/blue outlined white porcelain sign with blue line drawing of soldier at left. Blue outlined yellow "SENTINEL/MOTOR/OIL" diagonal across center. Blue and yellow shield in center with yellow "PENN" over white "100" on shield. PGCOA seal to right of shield. Blue "IN/NON-REFILLABLE/TANKS" across bottom. Blue "You're SURE its PURE" across top.

E.F.HOUGHTON - PHILADELPHIA, PA: Lubricants marketer selling products under the Vital brandname

CURB SIGNS

VITAL MOTOR OILS 24"RD
(5) 1925-1930 $650
Orange porcelain sign with white outlined black band across center. White "VITAL/MOTOR OILS" on band with black "BEST MOTOR OILS ARE VITAL" arched over "HOUGHTONS" in upper orange area. "100% Best Grade/ Pennsylvania Crude/E.F.HOUGHTON & CO/ALL OVER THE WORLD" in lower orange area.

WARR-PENN REFINING CO, WARREN, PA: Company history unknown.

CURB SIGNS

WARR-PENN MOTOR OIL 24"RD
(5) 1930s $450
White porcelain sign with large blue keystone in center. White PGCOA seal at top above white "WARR-PENN/MOTOR OIL." Smaller white text below. Blue "VISCOSITY" across top, "LONGETIVITY" on left and "LUBRICITY" on right of keystone.

ILLINOIS OIL PRODUCTS, INC - ROCK ISLAND, IL: Founded in 1908, no other details known.

WELCH GUARANTEED MOTOR OIL 10x24
(4) 1950s $225
Yellow tin litho sign with black/yellow/green can at right end beside green outlined black circle. Green "PER/QUART/31c" on circle. Black outlined green "WELCH" across top, "MOTOR OIL" across bottomw ith black "Guaranteed" between. Black "ILLINOIS OIL PRODUCTS INC. SINCE 1908 ROCK ISLAND,ILL." across bottom.

WELCH PENN MOTOR OIL 10x24
(4) 1950s $225
Black tin litho sign with white "WELCH-PENN/MOTOR OIL" with yellow dropshade across center of sign. Black and yellow PGCOA seal below to either side, with white "FOR HI-COMPRESSION LUBRICATION" between seals. White "ILLINOIS OIL PRODUCTS INC. ROCK ISLAND,ILL." across bottom.

WELCH GUARANTEED 12x20
(5) 1960s $225
Black outlined white pentagon shaped sign with red car, red and white tractor tailer on road behind green tractor in field across bottom of sign. Black/white/green/red can pouring oil at left, with black outlined green "WELCH" above black script "Guaranteed" across top.

BRANDNAME REFERENCES

AETNA ....SEE ASHLAND
ALADDIN ....SEE CO-OP/FS
AMERICAN LIBERTY ....SEE FINA
AMOCO ....SEE AMOCO
ANDERSON-PRICHARD.. ....SEE TOTAL
APCO ....SEE TOTAL
ARRO ....SEE STANDARD
ASHLAND ....SEE ASHLAND
ASSOCIATED ....SEE FLYING A
ATLANTIC ....SEE ATLANTIC
BARECO ....SEE DX/Barnsdall
BARNSDALL ....SEE DX
BAY ....SEE TENNECO
BEELINE ....SEE HUSKY/Frontier
BELL ....SEE TOTAL/Vickers
BEN FRANKLIN ....SEE TOTAL/Vickers
BENZO-GAS ....SEE PHILLIPS
BETHOLINE ....SEE RICHFIELD(NY)
BILLUPS ....SEE SIGNAL
BONDED ....SEE EXXON/Gaseteria
BONDED ....SEE MARATHON
BULK ....SEE GULF
CALDWELL & TAYLOR ....SEE SOHIO
CALSO ....SEE CHEVRON
CANFIELD ....SEE SOHIO
CARTER ....SEE EXXON
CAVALIER ....SEE TOTAL/Leonard
CENEX ....SEE CENEX
CHAMPLIN ....SEE CHAMPLIN
CHEKER ....SEE MARATHON
CHEVRON ....SEE CHEVRON
CHIEF ....SEE PHILLIPS
CITIES SERVICE ....SEE CITIES SERVICE
CITIZENS 77 ....SEE TENNECO/Bay
CLARK ....SEE CLARK
CLIFF BRICE ....SEE SHAMROCK
CLOVERLEAF ....SEE KERR MCGEE
COMET ....SEE ASHLAND/Tresler
COLONIAL(GA) ....SEE COLONIAL
COLONIAL/PROGRES ....SEE MARATHON/Rock Is.
COLONIAL BEACON ....SEE EXXON
COL-TEX(APCO) ....SEE TOTAL
COL-TEX(COSDEN) ....SEE FINA
CONOCO ....SEE CONOCO
CONSOLIDATED ....SEE MARATHON
CO-OP(DBL CIRCLE) ....SEE CO-OP'S
CORYELL ....SEE GULF
COSDEN ....SEE FINA
COVEY ....SEE SINCLAIR
CREW-LEVICK ....SEE CITIES SERVICE
CROWN CENTRAL ....SEE CROWN
DANCE ....SEE ASHLAND/Payless

HUSKY ....SEE HUSKY
ILLINOIS FS ....SEE CO-OP's
IMPERIAL REF ....SEE MARATHON/Cheker
IMPERIAL(MI) ....SEE MARATHON/Cheker
INDEPENDENT ....SEE MOBIL
INDEPENDENT ....SEE PHILLIPS
INDEPENDENT OILS ....SEE IOMA
INDIAN ....SEE TEXACO
INGRAM ....SEE SPUR
IOMA ....SEE IOMA
IOWA GAS ....SEE DX
IQS ....SEE ASHLAND
JENNEY ....SEE CITIES SERVICE
JOHNSON/BRIL BRONZ ....SEE EXXON/Gaseteria
KANOTEX ....SEE TOTAL
KAYO ....SEE CONOCO
KENDALL ....SEE KENDALL
KERMAC ....SEE KERR MCGEE
KERR MCGEE ....SEE KERR MCGEE
KEYSTONE ....SEE PENNZOIL
KNIGHT ....SEE TOTAL
KRIEGER ....SEE CONOCO/Douglas
KYSO ....SEE STANDARD OIL
LAMSON ....SEE RICHFIELD
LARCO ....SEE SINCLAIR
LEONARD ....SEE TOTAL
LINCO ....SEE MARATHON
LION ....SEE LION
LINCOLN OILS ....SEE CITIES SERVICE
LITENING ....SEE EXXON
LORECO ....SEE CITIES SERVICE
MFA ....SEE CO-OP'S
MAGNOLIA ....SEE MOBIL
MAJOR'S ....SEE HUDSON
MALCO ....SEE CONOCO
MANHATTAN ....SEE PHILLIPS
MARATHON ....SEE MARATHON
MARLAND ....SEE CONOCO
MARTIN ....SEE MARTIN
McCLANNAHAN ....SEE TOTAL/Leonard
METEEOR ....SEE PURE
METRO ....SEE MOBIL
MIDLAND ....SEE CO-OP'S
MID-WEST ....SEE TOTAL/Leonard
MILEAGE ....SEE CONOCO
MILEAGE ....SEE CS/Warner-Quinlan
MILEAGE MART ....SEE KERR MCGEE
MISSOURI FARM SERV ....SEE CO-OP'S
MOBILGAS(VACUUM) ....SEE MOBIL
MOBILGAS(SV) ....SEE MOBIL
MOBIL ....SEE MOBIL
MONAMOTOR ....SEE DX/Barnsdall
MUTUAL ....SEE KERR McGEE
NORTH STAR ....SEE ASHLAND

Index